T5-ARJ-874

NEW VENTURE INVESTMENT: CHOICES AND CONSEQUENCES

NEW VENTURE INVESTMENT: CHOICES AND CONSEQUENCES

Edited by

Ari Ginsberg
Stern School of Business, New York University, NY 10012-4211, USA

and

Iftekhar Hasan
Lally School of Management, Rensselaer Polytechnic Institute, NY 12180-3590, USA

2003

ELSEVIER

Amsterdam — Boston — Heidelberg — London — New York — Oxford — Paris
San Diego — San Francisco — Singapore — Sydney — Tokyo

ELSEVIER B.V.
Sara Burgerhartstraat 25
P.O. Box 211,
1000 AE Amsterdam
The Netherlands

ELSEVIER Inc.
525 B Street, Suite 1900
San Diego,
CA 92101-4495
USA

ELSEVIER Ltd
The Boulevard, Langford
Lane Kidlington,
Oxford OX5 1GB
UK

ELSEVIER Ltd
84 Theobalds Road
London
WC1X 8RR
UK

First edition 2003

Library of Congress Cataloging in Publication Data
A catalog record is available from the Library of Congress.

British Library Cataloguing in Publication Data
A catalogue record is available from the British Library.

ISBN: 0-444-51239-X

♾ The paper used in this publication meets the requirements of ANSI/NISO Z39.48-1992 (Permanence of Paper). Printed in The Netherlands.

List of contributors

Ilgaz Arikan
School of Management, Boston University

Michele Bagella
Facoltá di Economia, Universitá Tor Vergata, Roma

Leonardo Becchetti
Facoltá di Economia, Universitá Tor Vergata, Roma

Fabio Bertoni
Dipartimento di Ingegneria Gestionale, Politecnico di Milano

Tom Berglund
Department of Finance, Swedish School of Economics

Douglas J. Cumming
School of Business, University of Alberta

Bill Francis
College of Business Administration, University of South Florida

Ari Ginsberg
Stern School of Business, New York University

Giancarlo Giudici
Facoltá di Economia, Universitá Tor Vergata, Roma

Iftekhar Hasan
Lally School of Management, Rensselaer Polytechnic Institute

Chengru Hu
School of Management, Rutgers University

Edvard Johansson
The Research Institute of the Finnish Economy, Helsinki, Finland

Pieter Knauff
School of Management, University of Amsterdam

Yarom Landskroner
Stern School of Business, New York University

Dimi Leshchinskii
Finance Department, Groupe HEC

Jeffrey G. MacIntosh
Faculty of Law, University of Toronto

Barbara Martini
Facoltá di Economia, Universitá Tor Vergata, Roma

Jacob Paroush
Department of Economics, Bar-Ilan University

Peter Roosenboom
Department of Finance, Erasmus University

Annika Sandström
Department of Finance, Swedish School of Economics

Zur Shapira
Stern School of Business, New York University

Sridhar Seshadri
Stern School of Business, New York University

Richard Sylla
Stern School of Business, New York University

Christopher L. Tucci
Swiss Federal Institute of Technology, EPFL-CDH-ILEMT-CSI

Tjalling van der Goot
Department of Finance, University of Amsterdam

Joakim Westerholm
School of Business, University of Sydney

Contents

New Venture Investment: Choices and Consequences

New venture founders and their sponsors seek to create economic value by finding and commercializing new and better ways of doing things. Their common goal, which also defines the purpose of the entrepreneurial process itself, requires a solid grasp of the choices involved in attempting to create economic value under highly uncertain conditions and of the key decision making elements that influence these choices. It also requires an understanding of the consequences of new venture investment and of the various contextual factors that influence investment decisions and outcomes.

In seeking to detect generalities and to make abstracted sense of decisions faced by entrepreneurs and new venture investors, academics commonly classify the problem as a special case of some theory or model that originates in the specific discipline- or field-based knowledge they possess (Davidsson, 2002). The explanations that academic researchers provide and the predictions they make are therefore likely to be framed in terms of the types of variables, theoretical perspectives, levels of analysis, and research methodologies with which they are familiar.

To explore the intellectual underpinnings of new venture investment, we have gathered and organized a set of twelve papers that provide scholarly analysis of the choices involved in new venture investment as well the various contextual factors that influence investment outcomes. To insure a more robust and hopefully interesting scholarly treatment of such problems, we have included a broad range of theoretical frameworks and methodological approaches. To provide an international perspective we have also included studies of new venture investment in other countries besides the United States. As illustrated in Exhibits 1 and 2, the first six chapters examine the ways in which key decision making elements influence the investment-related choices made by venture capitalists and entrepreneurs. The second six chapters examine the ways in which various contextual factors influence the outcomes of new venture investments.

In their chapter "*Asymmetry of Information and of Beliefs in Venture Capital*," Yoram Landskroner and Jacob Paroush analyze the interaction of a venture capitalist and the entrepreneur during the early stages of fund raising for potential start-ups. In focusing on the central role played by venture capital funds in the financing of new entrepreneurial start-ups, the authors highlight the fact that a high degree of uncertainty and asymmetric information exists between the entrepreneurs and venture capitalist. Building on this supposition, Landskroner and Paroush develop a model for analyzing the behavior of an entrepreneur and the response of a venture capitalist in the context of a first round financing environment. Their model shows that the entrepreneur has superior information concerning the outcome of the venture and uses an appropriate incentive scheme to attract investment from the venture capitalist. This scheme makes capitalists' more optimistic about the venture and influences their decision to supply equity and debt financing. As an extension of their model, the authors attempt to determine the optimal amount of external financing raised and the optimal capital structure of the firm.

In contrast to the rational modeling approach taken by economics and finance scholars Landskroner and Paroush, operations management and decision theorists Sridhar Seshadri, Zur Shapira and Chrisopher Tucci argue in their chapter "*Venture capital investing and the Calcutta Auction*" that rational models of firm valuation are grossly overrated. Their analysis sheds new light on previously held conceptions regarding the rationality of VC investment decisions. After pointing out that such factors as market myopia, technology races, herding behavior, and cash flows are largely responsible for over-valuation, they introduce an alternative approach that incorporates the seemingly non-rational investment behavior into a "Calcutta auction" model. This model suggests that in considering investment under conditions of high uncertainty, the behavior of investors can be explained through the examination of specific behavioral components without resorting to traditional irrationality arguments. The authors argue that the Calcutta auction appears to be a good mechanism to explain markets that are characterized by uncertainty, information asymmetry and by small number of participants as well as to explain actual bids put up by VCs. Their analysis highlights three key assumptions: The first is that the externalities of market development efforts affect firm valuation in electronic markets. The second is that both irrationality, with small investments - and rationality, with large investments – may exist in newly discovered markets. Finally, analogous to the Calcutta auction, the total VC investment in an industry seems to depend on two factors, the initial "value" of the undeveloped market as well as the fraction of individual VC investment that works to develop the market for other entrants.

Having examined the ways in which VC investment choices may be influenced by relevant decision making models, we turn next to an examination of how the cognitive mindsets of entrepreneurs influence the choices they make in dealing with investment opportunities during different stages of their firm's life cycle.

In their chapter, "*The Entrepreneur's Initial Contact with a Venture Capitalist -Why Good Projects May Choose to Wait*," Tom Berglund and Edvard Johansson seek to explain the absence of services by venture capitalists to client firms at early stages of the firm's life cycle. They observe that the complementarity of financing and advice provide the venture capitalist with an advantage compared to firms specialized in either of the two. They also point out that the entrepreneur is not interested in the total value of an investment project, but rather in the part that he can secure for himself. By being aware that a venture capitalist may find it more profitable to reject the project and transfer useful parts of the idea to other clients than to join forces with the entrepreneur, the profit-maximizing entrepreneur may find it in her best interests to wait and develop the project herself before involving the venture capitalist. This allows the entrepreneur to improve his bargaining position and to secure a larger share of the expected profits of the business idea once the venture capitalist becomes involved. The optimal length of the delaying venture financing will be determined by the point where the marginal increase in contractibility exactly balances the marginal deterioration in total value of the project due to competitive pressure. Using comparative statics the authors show that the delay in involving a venture capitalist will be longer the less rapidly the NPV deteriorates, the more skilled is the entrepreneur, and the slower is the rate of increase in contractibility. In further pointing out that the market for a venture capitalist's services at the start-up stage is highly prone to failure, Berglund and Johansson also predict the likely performance outcome of the entrepreneur's decision not to delay: Since entrepreneurs with projects that have above average expected profitability will postpone their initial contact with the venture capitalist, entrepreneurs that do contact a venture capital firm at the start-up stage will tend to have a lower, or below average, expected profitability.

In contrast to the analysis provided by Berglund and Johansson, which focuses on the question of timing faced by an entrepreneur dealing with a VC investor, the next chapter "*How should entrepreneurs choose their investors*" by Dima Leshchinskii, addresses the question of what type of investor to deal with in the first place. Investors differ by their abilities to screen and monitor innovative projects, to bring aboard alternative managers, and to practice a portfolio approach. This chapter shows how these differences change the expected net present value of an entrepreneurial project and determine the choice of investor by the entrepreneur. Different factors can be critical for the project's payoff at each particular stage of its development. Therefore, for each stage the entrepreneur may attract investors with the highest level of expertise in the relevant area. For example, professional angels and venture capitalists obtain better information about a project's potential outcomes and emerge as providers of capital when this information is crucial. Venture capitalists and corporate investors, sometimes including the parent companies, are the industry's experts who have access to a pool of professional managers. Because they can furnish less costly managerial replacement, they should be the main source of capital at the stage when top-quality management is the decisive factor. In contrast, VC investment can lead to a higher probability of success, especially at the R&D phase.

The last two chapters in this section take us from the questions of when and from whom to seek new venture financing to theoretical and empirical analyses of exit strategy choices faced by entrepreneurs who have already established their new business.

In their chapter "*In quest of equity partners: the determinants of the going public-large blockholder choice*," Michele Bagella, Leonardo Becchetti, and Barbara Martini present a theoretical analysis of the determinants affecting the controlling shareholders choice between going public, remaining private, and looking for a blockholder when they need external financing. In the model they propose, these choices are strictly connected through a bargaining framework between controlling shareholders and the large blockholders in which the going public choice represents the controlling shareholders' outside option. Another important feature of the model is that controlling shareholders have to monitor managerial activity and choose their optimal effort in monitoring them. Under the going public choice, they remain residual claimants of firm profits after satisfying the reservation utility of small shareholders and the reservation wage of the manager. For this reason, optimal effort is proportional to the ex post property right share. In contrast, under the block holder choice, controlling shareholder bear the costs of monitoring managerial effort while they benefit from their effort only in proportion to their bargaining power. A key claim of this chapter is that the main advantage of going public for controlling shareholders is the stronger bargaining position they possess as they face dispersed small shareholders, and the main disadvantage is that small shareholders suffer more from informational asymmetries and are risk averse. This suggests that the compensation they require to participate in the venture may be too costly for controlling shareholders under high financial volatility, bullish stock market conditions, or when the firm is not well known to them. The authors further posit that not only is the blockholder choice socially preferred to the going public choice in general, but the desirability of the going public choice is further reduced in a weak institutional environment with illegal collusion under the block holder choice, which comes at the expense of small shareholders.

The last chapter in this section - "*How should an entrepreneurial firm be sold? Auctions versus negotiations*," by Ilgaz Arikan moves us from the question of whether and how to do an initial public offering to the choices faced by the entrepreneur on how to sell the new venture. Whereas economists Bagella, Becchetti, and Martini anchor their hypotheses in agency theory models that focus on monitoring costs and information asymmetries, management scholar Arikan uses the theory of auctions and negotiations to explain the optimal choice between market mechanisms in an entrepreneurial context. Two major markets exist for the sale of a new venture: initial public offering (IPO) versus mergers and acquisitions (M&A) markets. In his chapter, Arikan argues that choosing between these two market mechanisms rests on five factors: bargaining power, resource value, market thickness, risk propensity and search costs. Using a nested logit model he tests this general discrete choice using a sample of IPOs and M&As of privately held entrepreneurial firms between 1975–1999. He finds that, all else being equal, entrepreneurial firms with high bargaining power are more likely to choose negotiations (M&A) versus auctions (IPOs). Firms that represent high private values (e.g., in high-tech industries) are more likely to be sold through auctions versus negotiations. As market thickness increases, the likelihood of entrepreneurial firms being sold through M&A decreases. However, this finding is reversed for firms with higher private values. For firms with high debt ratios, the likelihood of M&A increases compared to IPOs. The likelihood of M&A also increases as venture capital activity in the focal industry increases.

The second section of this book examines how different types of contextual factors or characteristics affect the consequences of new venture investment. We begin with a historical perspective of the effects of a country's financial system on venture capital investment activity. In his chapter "*Venture Capital in Financial Systems: Historical and Modern Perspectives*," financial historian Richard Sylla demonstrates the influence of Schumpeter's insights on economic development on his historical analysis of important trends in entrepreneurship and new venture financing. In Schumpeter's (1934) celebrated theoretical analysis of economic development (a synonym for modern economic growth), the partner of the entrepreneur is the banker, and there is a strong implication that development will not occur unless both characters are present and working effectively together. In pursuing that inference in his own research, Sylla reaches the conclusion that the nature and effectiveness of venture financing at any time or place in history, including the present, depends on context. To be more specific, the nature and effectiveness of venture financing depends on the nature and effectiveness of the overall financial system of which venture financing is just a part, and often a minor part. In this essay he illustrates and amplifies his contention that venture financing is financial-system-context specific. Sylla's analytical framework corresponds to the time-series and cross-section approaches used by econometricians in testing hypotheses and estimating parameters of their models with economic data. His time series analysis consists of cases of new venture financing in several eras of U.S. history. His cross sectional analysis consists of cases of new venture financing in different parts of the world today. Each type of evidence tends to support the argument that the nature and effectiveness of new venture financing depends on context. There are two policy implications of this argument, one negative and the other more positive. The negative implication is that because new venture financing is context-specific, one cannot at all easily transfer venture-financing practices from one context—say, a country that is a paragon of entrepreneurship and economic growth—where those practices function well – to another context, namely, a context in which the overall financial system is different from, and possibly less developed than the financial system of the paragon country. A more positive policy implication is that, in seeking to improve financial systems throughout the world, the efforts of national and international development agencies, the latter including prominently the World Bank and the IMF, are, as the British say, spot on. If these agencies achieve the hoped-for successes, then new venture financing possibilities should proceed apace with financial development. Of course, countries themselves need to take the lead in this; there is only so much the external development agencies can do. Both, however, can learn much about designing better financial systems and avoiding worse ones from studying financial history.

Having examined how VC investment behavior may be influenced by the way in which a country's financial system is designed, we turn next to a specific case of government intervention to influence the performance of labor-sponsored venture capital corporations in Canada.

In their chapter "*Canadian Labor-Sponsored Venture Capital Corporations: Bane or Boon*?" Douglas J. Cumming and Jeffrey G. MacIntosh examine a governmental assistance program run by various Canadian governments, both provincial and federal, that takes the form of indirect subsidies to technology enterprises via tax subsidization of the investors in venture capital funds called "labour-sponsored venture capital corporations," or LSVCCs. The mechanism for inducing investment in LSVCCs has been the provision of generous tax subsidies to investors, consisting of a combination of tax credits and deductibility of the investment from income. Their study suggests that the tax expenditures that underlie the LSVCC programs do not represent a useful expenditure of public monies, and that the various government sponsors should seriously consider abandoning the LSVCC programs. They begin with a sketch of the Canadian venture capital industry, which is followed by a description of organizational structure of the LSVCCs and of the various statutory constraints that they operate under. Then they briefly compare this structure and these constraints with those applicable to the LSVCCs' private sector counterparts and suggest that the LSVCC structure is highly inefficient and likely to lead to a high level of agency costs vis-à-vis funds investors. Consistent with the relevant research on the topic the authors report a lower LSVCC performance. While the fixed and variable components of fund manager compensation (the management expense ratio, or MER, and the carried interest, respectively) are comparable to those of private funds, LSVCCs have performed very poorly compared to both Canadian and U.S. private funds, and various Canadian and U.S. market indices. Other research suggests that they have even underperformed short-term bank deposits and treasury bills. While this suggests that the tax expenditures that underlie the LSVCC programs have not been wisely spent, many have claimed that the LSVCC programs are justifiable even in the face of poor returns, on the basis that they have significantly augmented the pool of Canadian venture capital investments. However, the authors point to evidence in related research that strongly suggests that LSVCCs have crowded out other types of venture capital thereby leading to an overall reduction in the aggregate pool of Canadian venture capital. They conclude that the Canadian LSVCC program has been a costly failure for its government sponsors.

In the remaining chapters of this section, we shift from an examination of country level factors to an examination of the ways in which market and firm-level contextual factors influence the nature and duration of VC investment as well as IPO performance outcomes.

In their chapter "*'New' Stock Markets in Europe: a 'New' Exit for Venture Capital Investments*," Fabio Bertoni and Giancarlo Giudici explore the role of venture capitalists in new market (NM) companies before, during and after the listing and also attempt to identify differences between venture-backed IPOs versus other companies, and European VC-backed IPOs versus their US counterparts. Analyzing a sample of 575 firms listed on European 'New Markets' from 1996 to 2001, they find that VC-backed IPOs are significantly smaller than other IPOs. However, although they exhibit lower sales and earnings and raise less equity capital, they are not significantly younger and more underpriced. In comparing the European experience with the US experience during the 1996 to 2000 period, Bertoni and Giudici highlight several similarities, e.g., NM companies are not significantly younger than US companies and the initial mean underpricing is surprisingly similar; they also point out some interesting differences, e.g. NM companies are smaller and less profitable, insider ownership is concentrated in the hands of CEOs, and the presence of VCs is less frequent and significant. Their analysis suggests a number of interesting findings: First, VCs tended to retain shares in companies in which their marginal contribution to the creation of value also might be significant after the listing (that is, in young companies with a scarce capability to generate cash, characterized by further growth opportunities). Second, VCs tended to sell shares of the smallest companies in their portfolio. Last, VCs took advantage of the market euphoria towards technology stocks. These findings challenge the traditional view that going public coincides with private equity investors' exit from the firm and strengthens the hypothesis that the listing on European NMs is not considered a stage subsequent to private equity financing, but a further relevant 'public' source of funds alongside venture capital. The role of private VCs in NMs is perceived as strategic even after the IPO: their permanence provides a 'certification effect' strengthened by lock-up provisions.

In their chapter "*Post-Issue Performance of Hot IPOs*," Annika Sandström and Joakim Westerholm investigate the operating performance of Finnish initial public offerings during the years 1984–2000. Their study, which reports, an average excess return of 27% at the end of the first day trading in the Helsinki Stock Exchange, provides new evidence on the performance of Finnish IPOs during a period when the market experienced a relatively high number of companies going public. Analyzing the changes in operating performance over a five year period, Sandström and Westerholm found an initial high level of underpricing on the first day of offering followed by a significant and continuous decrease of performance as portrayed by a variety of accounting ratios over the next fine year period. The experience of high underpricing on the first day of offerings followed by a continuous decline of performance during the subsequent five years is consistent with most prior literature of IPO performance in other countries.

In their chapter "*Is Accounting Information Relevant to ValuingEur opean Internet IPOs*?" Pieter Knauff, Peter Roosenboom, and Tjalling van der Goot investigate the relevance of accounting information to valuing European Internet IPOs during the years 1998–2000 both before and after the Internet bubble burst in April 2000. Using market value as the proxy for firm value, they study the relationship between accounting variables and IPOs and find results that are contrary to the widely held belief that accounting information is of limited use when valuing the IPO shares of Internet companies. The authors provide a number of contributions to the literature by conducting empirical research on the relevance of accounting information for valuation of Internet firms in the European markets. The paper examines the value drivers underlying Internet IPOs from two perspectives: the offer price and the stock price at the end of the first trading day. The authors also extend their analysis to non-accounting information that is available at the time of the IPO. Consistent with previous findings in the U.S. markets, they find that free float (i.e., the percentage of shares being sold in the IPO) is negatively related to market value. However, they found no significant positive relationship between market value and the percentage of post-IPO ownership of the largest shareholder. Based on their analysis, they conclude that market value is negatively related to earnings in the Internet bubble period before April 2000 and that the free float is more value relevant in the bubble period.

In the last chapter in this section "*Deliberate Underpricing and Price Support: Venture Backed and Non-ventured backed IPOs*," Bill B. Francis, Iftekhar Hasan, and Chengru Hu examine the premarket underpricing phenomenon within a group of venture-backed and a group of non venture-backed initial public offerings

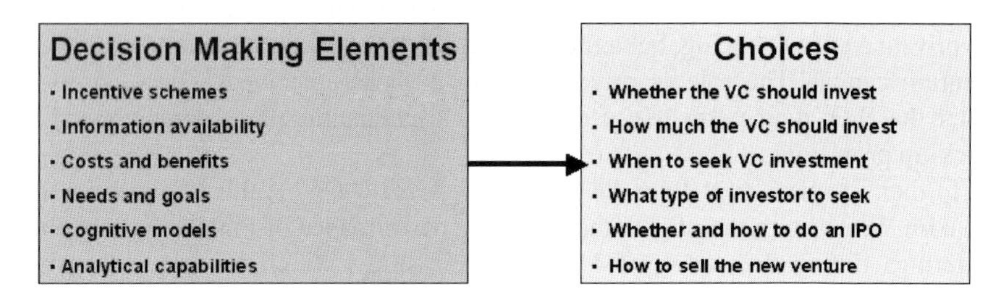

Exhibit 1. Theoretical relationships examined in chapters 1–6

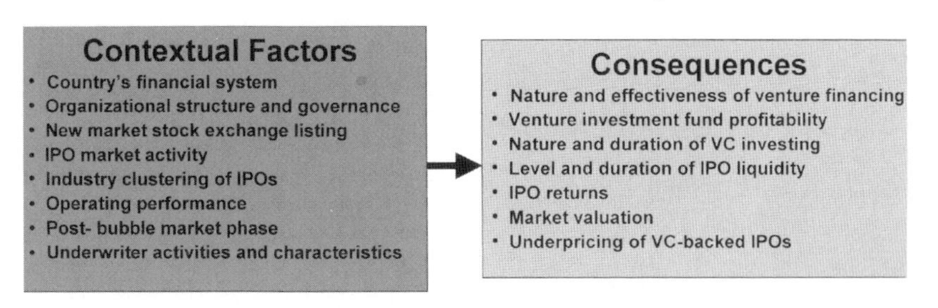

Exhibit 2. Theoretical relationships examined in chapters 7–12

(IPOs). Using an economic frontier analysis, their chapter attempts to address two issues. First, it investigates whether the difference in underpricing between VC-backed IPOs and non VC-backed IPOs is due to deliberate underpricing in the pre-market or to other factors such as underwriter price support in the after market. Second, we examine the role deliberate underpricing plays in the IPO process as it relates to initial day returns and price stabilization. The evidence indicates that in contrast with prior studies, VC-backed IPOs tend to be more underpriced than non VC-backed IPOs and the degree of pre-market underpricing is significantly higher for VC-backed IPOs relative to their non-VC counter part. The results suggest that the documented difference in underpricing in the literature between VC-backed and non VC-backed IPOs cannot be fully attributed to post-market phenomena. Rather, a significant portion may be due to deliberate underpricing as a means of compensating investors. In addition, the evidence indicate that for both VC- and non VC-backed IPOs deliberate underpricing in the pre-market leads to a higher probability of price stabilization in the after-market; and in that scenario, there is also a higher likelihood of price support for non VC-backed IPOs.

REFERENCES

Davidsson, Per (2002). What Entrepreneurship Research can do for Business and Policy Practice, *International Journal of Entrepreneurship Education*, *1*(1), 1–20.

Schumpeter, Joseph, (1934). *The Theory of Economic Development*, Cambridge: Harvard University Press.

Editors
Ari Ginsberg and Iftekhar Hasan

New Venture Investment: Choices and Consequences
A. Ginsberg and I. Hasan (editors)

Chapter 1

Asymmetry of Information and of Beliefs in Venture Capital[†]

YORAM LANDSKRONER[a,b,*] and JACOB PAROUSH[c]

[a] *Stern School of Business, New York University, New York, USA*
[b] *School of Business Administration, The Hebrew University of Jerusalem, Jerusalem, Israel*
[c] *Department of Economics, Bar-Ilan University, Ramat Gan, Israel*

ABSTRACT

Venture capital is a specialized source of financing for new firms/start-ups, which do not yet have access to traditional forms of external financing because of a high degree of uncertainty and problem of asymmetry of information. In this paper we analyze the interaction between a venture capitalist ("capitalist") and the initial owner of an entrepreneurial entity in which she/he invests ("entrepreneur"). We focus on the agency problems and derive an optimal incentive scheme. In our model the capitalist provides a combination of equity and debt financing while the entrepreneur provides equity financing which serves a dual purpose. This equity finances part of the project and is also a signal affecting the beliefs ("optimism") of the capitalist.

[†] A previous version of this paper entitled "Venture Capital: Structure and Incentives" was published in the *International Review of Economics and Finance* (1995), 4(4): 317–332.

* Corresponding author. Tel.: +1-212-998-0913(O).
E-mail address: ylandskr@stern.nyu.edu (Y. Landskroner).

1. INTRODUCTION

The venture capital industry, which has developed mostly in the US, is relatively new. It has experienced very high growth in the last two decades; growing from virtually zero in the mid 1970s to $17.2 billion annual flows into capital venture funds in 1998. The industry in the U.S. consists of about 500 funds with a total capital of about $50 billion. These funds have been a major source of financing of new ventures in start-ups (entrepreneurship) at their early stages of development (see Gompers and Lerner, 1999).[1]

Two important properties of the ventures are high degree of uncertainty and asymmetric information between the entrepreneurs and capitalist. The industries in which venture capitalists invest are mostly characterized by high degree of uncertainty such as biotechnology, communications and electronics. These industries typically also have substantial intangible assets, which are difficult to value and operate in markets with volatile conditions. These conditions create many opportunities for self-benefit behavior of the parties. These characteristics of high risk and high return firms have led to the development of the specialized form of venture capital financing.[2]

New business entities raise funds from different sources. Initial financing to a new project may be provided by the entrepreneur, usually a specialist in the field of the project, who contributes entrepreneurship (and other talents) as well as capital. Eventually a successful firm will meet the increasing financing needs by raising external funds from traditional sources such as bank loans and public offering of its stock and bonds (or by merging into another corporation). Venture capital funds are invested in the interim stage: after the funds of the entrepreneur are exhausted but before the firm can borrow from banks or 'goes public' because of problems of high degree of uncertainty and asymmetric information. In general, venture capital is provided in the form of equity financing but may also appear in the form of debt or a combination of the two (including preferred stock and convertible securities).[3] Venture capital organizations enter into two types of contracts: an internal contract with the investors that provide the funds to the venture capital organization and an external one with the firm (entrepreneur) in which it invests. Sahlman (1990)

---

[1] Venture capitalists also provide funds for firms at other stages such as financing of leveraged buyouts and acquisitions. In this paper however we focus on the early stage financing.

[2] Private investors have set up different organizational structure to provide venture capital. The three types of firms are: limited partnerships, corporate subsidiaries and small business investment corporations. The largest and fastest growing organizations are the limited partnerships. Under this arrangement private investors, including institutional investors, purchase limited partnership interests in a venture capital fund, which is managed by a venture capitalist (the general partner).

[3] The risks inherent in the early stages, and the lack of collateral made these businesses in many cases unacceptable customers for the traditional commercial lending institutions who provide exclusively debt financing.

describes and analyzes the structure of venture capital organizations and the relationship between the parties involved. There is a growing literature on the financing of new ventures through venture capital dealing mainly with the internal contract (Berglof, 1994; Chan, Siegel & Thakor, 1990; Gompers, 1995; Hellman, 1998).

In this paper we focus on the external contract between the venture capitalist and the entrepreneur in the "first round" of venture capital financing. We analyze the interaction between the venture capital fund (the 'capitalist'), and the initial owner ('entrepreneur'), in the entrepreneurial venture (firm) in which they invest. We address the agency problem and asymmetric information which exists between the capitalist and the entrepreneur. In the model we assume that the entrepreneur is better informed and signals the capitalist attempting to affect his/her optimism. We focus on the compensation system derived under conditions of uncertainty regarding the outcome of the venture.[4]

We consider a firm (project) in the early stage of its development where no prior knowledge (observations) about it exists, and thus it is without access to traditional capital sources. A single transaction between two parties, entrepreneur and capitalist is examined—this first round of financing may be followed in the future by additional rounds of financing. The entrepreneur (owner) provides his/her own capital and raises outside capital to be used by the firm. The capitalist receives compensation in the form of a fixed income (debt), and a share of the firm's income in return for his (equity) investment while the entrepreneur receives a share of the firm's income.

In the absence of competitive capital markets, it is assumed that the entrepreneur and the capitalist face increasing financing cost in raising capital. An agency model of "capital structure" is analyzed, the purpose of which is to determine the optimal amount of external financing raised by the firm, and the optimal capital structure of the firm (debt vs. equity). The inclusion of both debt and equity financing is one of the interesting features of this paper. Additionally, the optimal distribution of the enterprise's income in the "incentive scheme" is determined.[5]

The owner and the capitalist have different expectations, before entering into their relationship, as to the income their joint venture will generate. It is presumed here that the owner has relatively superior information about the distribution of returns ("inside information") but not perfect information since this is a new venture without prior observations. Over time and repeated play, the entrepreneur will gain information and correct his biases. At this stage however, when information is heterogeneous and unequally distributed, individuals can exploit superior knowledge.

A crucial point of the paper is that the owner can influence the beliefs of the capitalist (make him or her more/less "optimistic") through her observed actions

[4] We abstract from other features such as the management role that the capitalist many play.

[5] Stiglitz (1974) derived a model of risk sharing and incentives in a study of sharecropping. This model extended the understanding of the operations of the closely held firm.

(such as her investment). That is, the owner's capital serves a dual purpose: as an input of the firm in generating its income and also as a signaling device to attract external capital. In modeling the signaling aspect of the problem, the entrepreneur is assumed to be a Stackelberg leader in choosing his capital input while the capitalist follows in choosing the amount and form of financing. Thus, the signaling approach is incorporated into this analysis as it relates to the financing decision of the firm. Leland and Pyle (1977) in common with the model presented here, consider the owner's signal to outsiders to be his equity investment in the firm.[6] In the present model however, the firm does not have access to competitive capital markets and there is a presumption of an increasing financing cost to the investors.

In Section 2 we present the model and in Section 3, the characteristics of the optimal solutions. The case of multiple venture capitalists (outside investors) is presented and analyzed in Section 4. Conclusions are presented in Section 5.

2. THE MODEL

This section presents a model of a single transaction between two parties: an entrepreneur (the founder and initial owner) and a venture capitalist, who finance a joint venture and share its income. The entrepreneur supplies equity capital, y, and "entrepreneurship" input (not modeled explicitly). The capitalist supplies capital, x, in two forms, equity and debt, denoted x_1 and x_2, respectively.

The enterprise generates an uncertain income $r(y+x)$ where r is one plus a random rate of return. Total income is distributed between the capitalist (Z_p) and the entrepreneur (Z_e).

$$Z_p = \gamma r x_1 + \beta x_2 \tag{1}$$

$$Z_e = r(y+x) - \gamma r x_1 - \beta x_2 = ry + (1-\gamma) r x_1 + (r-\beta) x_2 \tag{2}$$

The parameters (β, γ) are contractually specified in advance. In fact, β, γ are determined by the entrepreneur to raise outside capital for the venture, they are thus an incentive (compensation) scheme for the capitalist. Given the incentives (parameters) of the contract, the capitalist responds by determining both his total capital contributed to the venture, x, and its decomposition into x_1 and x_2. Thus, β, γ are parameters in the capitalist's revenue function and are controllable factors in the entrepreneur's cost function. Typically, $\beta > 1$ because it has a dimension of one plus an interest rate while $0 < \gamma < 1$ because the entrepreneur, who contributes his

[6] The first application of signaling to finance theory has been done by Ross (1977). Downes and Heinkel (1982) empirically examined the relationship of firm value and two types of signals: the fraction of ownership retained by an entrepreneur and the divided policy of a firm.

"entrepreneurship" in addition to capital, is compensated by receiving more than his share of equity (and the capitalist less), i.e. the proportions of the entrepreneur and the capitalist in the return of equity are:

$$\frac{y+(1-\gamma)x_1}{y+x_1} > \frac{y}{y+x_1} \quad \text{and} \quad \frac{\gamma x_1}{y+x_1} < \frac{x_1}{y+x_1}$$

The capitalist's income, Z_p, consists of two parts: interest income βx_2, and uncertain income on his equity investment, $\gamma r x_1$.[7] The entrepreneur's income Z_e consists of three components: income from his equity investment ry, the entrepreneurship return which is deducted from the capitalist's equity return, $(1-\gamma)rx_1$ and the uncertain return on financial leverage $(r-\beta)x_2$. The unique feature here is the "entrepreneurship fee" $(1-\gamma)rx_1$ is added to his income, and is paid out of the capitalist equity income.

It is also assumed that each of the parties faces an increasing opportunity cost in raising his capital. The entrepreneur's financial opportunity cost is denoted by $c(y)$ and the capitalist's by $d(x)$. It is further assumed that marginal costs are strictly positive and increasing, that is, $c' > 0, d' > 0, c'' > 0, d'' > 0$. Note that the return on venture capital is assumed to have constant returns to scale and the cost of capital displays increasing marginal cost. We can easily and without loss of generality reverse the assumptions and assume diminishing returns to scale of the return of the capitalist and constant cost of capital. However we prefer the present formulation because of imperfections in the capital market. The total net income from the joint venture is $r(y+x) - \gamma r x_1 - \beta x_2 - c(y)$ for the entrepreneur and $\gamma r x_1 + \beta x_2 - d(x_1 + x_2)$ for the capitalist. It is assumed that the cost of financing (risk premium) increases with the amount raised, but is not affected by the form of financing: debt or equity.

The two parties differ in their preferences and beliefs. More precisely, they are assumed to have different attitudes towards risk and different perceptions or expectations of the risk and returns of the project. The utility functions of the entrepreneur and the capitalist, defined over the net return, are denoted by U and V, respectively where $U' > 0,\ U'' \leq 0,\ V' > 0,\ V'' \leq 0$. The subjective density (probability) functions of the rate of return are denoted by $f(r)$ and $g(r)$ for the entrepreneur and capitalist, respectively.

Assuming that both are Von-Neumann Morgenstern expected utility maximizers, the capitalist's problem is to maximize:

$$\int V(\gamma r x_1 + \beta x_2 - d(x))g(r)\mathrm{d}r \tag{3}$$

[7] In this framework we have assumed that the borrowing x_2, is done by the entrepreneur against his personal account. Thus we assume no default risk on the debt of the firm. This simplifies the analysis without affecting the optimization process.

with respect to x_1 and x_2 for every given contract (β, γ). Given the capitalist response functions $x_1(\beta, \gamma)$ and $x_2(\beta, \gamma)$, the entrepreneur's problem is to maximize:

$$\int U(r(y + x_1 + x_2) - \gamma r x_1 - \beta x_2 - c(y)) f(r) \mathrm{d}r \tag{4}$$

Equation (4) is maximized w.r.t. y, (β, γ). In other words facing an increasing cost of capital the entrepreneur offers incentives to attract financing of the project by the capitalist. The incentives have two parameters a proportion γ of the uncertain return on the project r, and a fixed interest rate on the loan $(\beta - 1)$. Given (β, γ) the capitalist two decisions are how much to invest in the project and in what form: equity or debt. Typically transactions involving venture capital take place under asymmetry of preferences and beliefs. It is presumed that the capitalist is more risk averse than the entrepreneur, and with some loss of generality the analysis is simplified by assuming that is the entrepreneur is risk neutral while the capitalist is risk-averse, that $U'' = 0$ and $V'' < 0$.[8]

Thus, the entrepreneur's objective function is now simplified to:

$$\bar{r}y + (1 - \gamma)\bar{r}x_1 + (\bar{r} - \beta)x_2 - c(y) \tag{5}$$

where $\bar{r}$ is the subjective expected rate of return of the venture.

The second facet of this analysis concerns the heterogeneous beliefs of the two parties to the venture. The capitalist assumes that the entrepreneur enjoys superior information about the firm. We assume that he/she has superior, but not perfect, information. That is, his/her subjective expected return, $\bar{r}$ is positively correlated with the true rate of return, but is not necessarily equal to it.[9] In this analysis it is presumed that the capitalist becomes more optimistic about the venture's prospects

---

[8] This is a simplifying assumption. Our results will not change as long as the entrepreneur remains less risk averse than the capitalist.

[9] The relationship between the true, but unknown rate of return on the venture, r_0 and the entrepreneur's evaluation of this rate of return r, may be assumed to be of a linear regression type:

$$r = a + br_0 + u$$

and

$$\bar{r} = a + b\bar{r}_0.$$

Where u, the error term, reflects the intrinsic uncertainty of the venture. The parameters reflect the entrepreneur's subjective biases. Over time or with experience, as the game is repeated, the entrepreneur will correct his biases in a kind of recurring process of the form:

$$\bar{r}_{t+1} = \frac{\bar{r}_t - \hat{a}}{\hat{b}_t}$$

where t is a time or experience index.

as the entrepreneur's contributions increase. More specifically we assume that the capitalist's subjective distribution function (cumulative probability function) exhibits first-degree stochastic dominance, i.e. $G(r|y_1) \leq G(r|y_2)$ for every value of r if $y_1 > y_2$, where $G(r) = \int_{-\infty}^{r} g(s)\mathrm{d}s$ is the capitalist's subjective cumulative probability function. In other words, for every r, the capitalist assigns a higher probability to more favorable state of the world under y_1 than under y_2. The capitalist's expected utility, then, is conditioned on y; his optimal solutions, x_1 and x_2, are functions of y as well as β, γ.

Define the first two moments of the capitalist's subjective density function of r: $\hat{r}(y) = \int rg(r|y)\mathrm{d}r$ as the expected rate of return, and $\hat{\sigma}(y) = \int [r - \hat{r}(y)]^2 g(r|y)\mathrm{d}r$ as the variance of the rate of return. The random variable $r(y)$ is now decomposed as follows:

$$r(y) = \hat{r}(y) + \hat{\sigma}(y)\varepsilon(y) \tag{6}$$

where $\varepsilon(y)$ is a standardized random variable with $E\varepsilon(y) = 0$ and $E\varepsilon^2(y) = 1$ for every y. With these assumptions and the concomitant decomposition, the entrepreneur's optimization problem can now be written as:

$$\begin{aligned} &\mathrm{Max}_{y,\beta,\gamma}\{\bar{r}y + (1 - \gamma)\bar{r}x_1 + (\bar{r} - \beta)x_2 - c(y)\} \\ &\text{s.t.} \quad x_1 x_2 \in \mathrm{Argmax} \int V\{\gamma\hat{r}(y)x_1 + \beta x_2 - d(x) + \gamma\hat{\sigma}(y)\varepsilon(y)x_1\} g(r|y)\mathrm{d}r. \end{aligned} \tag{7}$$

In this problem the entrepreneur is a Stackelberg leader in choosing his capital input y, and the contract parameters (β, γ), and the capitalist is the follower choosing her control variables given the entrepreneur's incentives.

3. CHARACTERISTICS OF THE SOLUTION

3.1. The capitalist

For every given value of y, β, γ the first order condition of the capitalist's optimization problem w.r.t. x_2 is given by

$$[\beta - d'(x)]EV' = 0 \tag{8}$$

where $EV' = \int V' g(r|y)\mathrm{d}r$ is the positive expected marginal utility. It follows from Eq. (8) that $\beta = d'(x)$. That is, in equilibrium the marginal borrowing rate of the capitalist equals his lending rate, the important implication of Eq. (8) is that the total investment for the capitalist, $x = x_1 + x_2$, is determined by β and $d'(x)$ and is independent of his attitudes toward risk, the parameter of the density function of r,

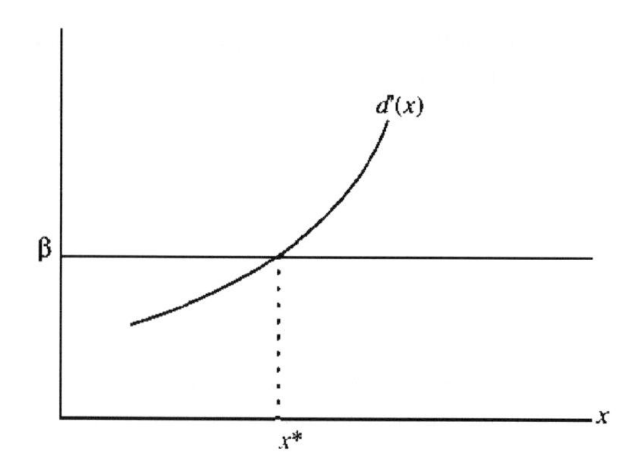

Figure 1.

the sharing rule γ and the action of the entrepreneur y (Figure 1).[10] The explanation of this novel result is that since x_2 is riskless with a positive marginal revenue, it is a residual which will be increased up to a point where marginal cost equals marginal revenue. Thus, the total investment is independent of risk considerations and parameters of the contract while the amount of equity is not.

The first order condition w.r.t. x_1 the capitalist's equity investment, is

$$[\gamma\hat{r} - d'(x)]EV'\gamma\hat{\sigma}E(V'\varepsilon) = 0 \tag{9}$$

where $E(V'\varepsilon) = \int[V'\varepsilon(y)]g(r/y)\mathrm{d}r$ is the expected value of the product of the marginal utility and the random element of the rate of return. Because $E(V'\varepsilon) = \mathrm{cov}(V', \varepsilon) + E(V')E(\varepsilon)$ and $E(\varepsilon) = 0$ the expected value of the product, $E(V'\varepsilon)$, is equal to the covariance of V' and ε. This covariance is negative: $E(V'\varepsilon) = \mathrm{cov}(V'\varepsilon) < 0$ since the capitalist is assumed to be risk averse, $V'' < 0$. More specifically the higher the value of ε, following (6), also the return on the project and consequently the value of the utility function V will be higher. But due to the capitalist's risk aversion, her marginal utility V' diminishes ($V'' < 0$) and thus the covariance between V' and ε is negative. The assumptions $d'' > 0$ and $V'' < 0$ guarantee that the second order condition holds globally.

Substituting β for $d'(x)$ in Eq. (9) and rewriting yields:

$$\frac{\gamma\hat{r} - \beta}{\gamma\hat{\sigma}} = A(x_1) \quad \text{or} \quad \gamma r - \beta = \gamma\hat{\sigma}A(x_1) \tag{10}$$

[10] This result is similar to the solution of a general portfolio selection problem, in which the investor faces a risk-free and risky asset and increasing costs. His total investment will be determined only by the risk-free return, and by the cost. His attitude toward risk and the riskiness of the assets will determine the composition of his portfolio; proportions of the risky and the risk-free assets.

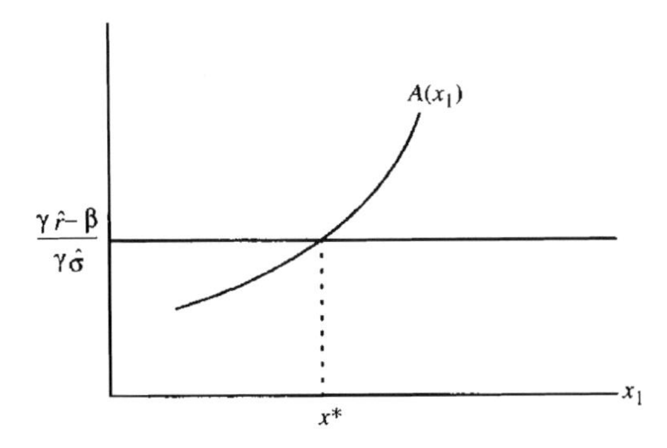

Figure 2.

where $A(x_1) = -E(V'\varepsilon)/EV' > 0$ is a factor representing the capitalist's attitude toward risk. This factor is related to the Pratt-Arrow measure of absolute risk aversion.[11] Equation (10) is compatible with a well known result in the finance literature. It states at the outset that the price (per unit) of risk $((\gamma\hat{r} - \beta)/\gamma\hat{\sigma})$, is determined in equilibrium by the investor's risk aversion. The product of this risk aversion factor and the risk of the investment $\gamma\hat{\sigma}$ determine the risk premium $\gamma\hat{r} - \beta$ in equilibrium.

Equations (8) and (9) or (10) describe a two-stage decision by the capitalist: in the first stage he or she determines the total investment in the venture, according to the lending rate and the financing cost. In the second stage he or she determines the allocation between the riskless debt and risky equity. This allocation is determined by characteristics of the investor: first by her risk aversion $A(x_1)$ and then by the parameters of her subjective density function: $\hat{r}(y)$ and $\hat{\sigma}(y)$ and thus in turn by the action of the entrepreneur, and the incentives γ and β (Figure 2).

---

[11] The relationship between risk aversion factor A, and the Pratt-Arrow measure of absolute risk aversion $\hat{A} = -V''/V'$ is obtained by a Taylor expansion around the mean:

$$A \approx \gamma\hat{\sigma}x_1\hat{A}(w) \tag{11}$$

where $w = (\gamma\hat{r} - \beta)x_1 + \beta x - d(x)$ is the wealth at the mean and $x_1 + x_2 = x$ is a constant

$$\frac{\partial A}{\partial x_1} = \gamma\hat{\sigma}\hat{A}(w) + \gamma\sigma x_1 \frac{\partial\hat{A}}{\partial w}\frac{\partial w}{\partial x_1} = \gamma\sigma\hat{A}(w) + \gamma\sigma x_1(\gamma\hat{r} - \beta)\frac{\partial\hat{A}}{\partial w} > 0 \tag{12}$$

That is, the expression is positive either under constant or increasing absolute risk aversion *or* under increasing relative risk aversion where the measure of relative risk aversion:

$$\frac{\gamma\hat{r} - \beta}{\gamma^2\hat{\sigma}^2} = x_1\hat{A}(w)$$

The effect of the entrepreneur's action y on the capitalist may now be defined. An increase in y will render the capitalist "more optimistic" in his beliefs about the distribution of r. This will be taken to mean either an increase in the subjective expected return, $\hat{r}(y)$ and/or a decrease in risk $\hat{\sigma}(y)$ that is,[12]

$$\frac{\partial \hat{r}(y)}{\partial y} \geq 0 \quad \text{and} \quad \frac{\partial \hat{\sigma}}{\partial y} \leq 0 \tag{13}$$

where at least one of the strong inequalities prevails.

It is important to recall that, in Eq. (8), y does not affect the total capital contributed by the capitalist. Thus, y affects the "capital structure" alone or equity vs. debt contributed by the capitalist. Following Eq. (10), it can be determined that, given an increasing absolute risk aversion function ($\partial A/\partial x_1 > 0$) either an increase of $\hat{r}(y)$ or a decline in $\hat{\sigma}(y)$ as specified in Eq. (13), will increase the price of risk and thereby increase equity capital x_1. Thus, as the capitalist becomes more optimistic, he will contribute more of his capital in the form of risky equity.

Investigation of the effects of the actions of the entrepreneur may now be completed. These effects include signal y and the incentives γ and β on the actions of the capitalist: total capital contributed x and its structure, that is, equity x_1 vs. debt x_2.

Total differentiation of Eq. (8) w.r.t. β, γ, and y yields,

$$\frac{\partial x}{\partial \beta} = \frac{1}{d''} > 0; \quad \frac{\partial x}{\partial \gamma} = 0; \quad \frac{\partial x}{\partial y} = 0. \tag{14}$$

Total differentiation of Eq. (10) w.r.t. β, γ, and y yields,

$$\begin{aligned} \frac{\partial x_1}{\partial \beta} &= -\frac{1}{\gamma\hat{\sigma}(\partial A/\partial x_1)} < 0 \\ \frac{\partial x_1}{\partial \gamma} &= \frac{\beta}{\gamma^2\hat{\sigma}(\partial A/\partial x_1)} > 0 \\ \frac{\partial x_1}{\partial y} &= \frac{\gamma\hat{\sigma}(\partial\hat{r}/\partial y) - (\gamma\hat{r} - \beta)(\partial\hat{\sigma}/\partial y)}{\gamma\sigma^2(\partial A/\partial x_1)} = \frac{(\partial\hat{r}/\partial y) - A(\partial\hat{\sigma}/\partial y)}{\hat{\sigma}(\partial A/\partial x_1)} > 0. \end{aligned} \tag{15}$$

[12] In general, "more optimistic" means $\partial G(r|y)/dy < 0$, i.e. the distribution becomes stochastically more dominated which can also be consistent with the case:

$$\frac{\partial \hat{r}(y)}{\partial y} < 0 \quad \text{and} \quad \frac{\partial \hat{\sigma}(y)}{\partial y} > 0.$$

We however, have limited ourselves to "mean-variance optimism."

The crucial role played by "optimism" in investment-financing decisions was emphasized by Keynes, "... a large proportion of our positive activities depend on spontaneous optimism rather than on a mathematical expectation. ... our decisions can only be taken as a result of *animal spirit*—as a spontaneous urge to action rather than inaction" (1961: 161).

Combining Eqs. (14) and (15) yields

$$\frac{\partial x_2}{\partial \beta} > 0 \quad \frac{\partial x_1}{\partial \beta} < 0$$
$$\frac{\partial x_2}{\partial \gamma} = -\frac{\partial x_1}{\partial \gamma} < 0 \tag{16}$$
$$\frac{\partial x_2}{\partial y} = -\frac{\partial x_1}{\partial y} < 0$$

The main findings are:

(1) The capitalist's total investment in the firm, x, is determined by β and is independent of γ, and A;
(2) The equity investment of the capitalist, x_1, is affected positively by both the return on equity, γ, and the investment of the entrepreneur, y that increases his optimism (given his risk aversion) and, thus, his risky investment in the firm. An increase in β will increase his lending at the expense of his equity investment;
(3) An increase of the risk aversion factor, A, will not affect total investment of the capitalist, but will reduce his equity (risky) investment.

3.2. The entrepreneur

The entrepreneur's optimal decisions regarding his financing y, and the incentives γ, β may now be determined. His objective function, as specified in Eq. (5) is:

$$\underset{y,\beta,\gamma}{\text{Max}} D(y, \beta, \gamma)$$

where

$$D = \bar{r}(1-\gamma)x_1(y, \beta, \gamma) + (\bar{r} - \beta)x_2(y, \beta, \gamma) + \bar{r}y - c(y). \tag{17}$$

This is a typical objective function of a leader who takes into account the follower's reaction.

The first order conditions are:

$$D_y = \bar{r} + (\beta - \gamma\bar{r})\frac{\partial x_1}{\partial y} - c'(y) = 0 \tag{18}$$

$$D_\gamma = (\beta - \gamma\bar{r})\frac{\partial x_1}{\partial \gamma} - \bar{r}x_1 = 0 \tag{19}$$

$$D_\beta = \bar{r}(1-\gamma)\frac{\partial x_1}{\partial \beta}(\bar{r} - \beta)\frac{\partial x_2}{\partial \beta} - x_2 = 0 \tag{20}$$

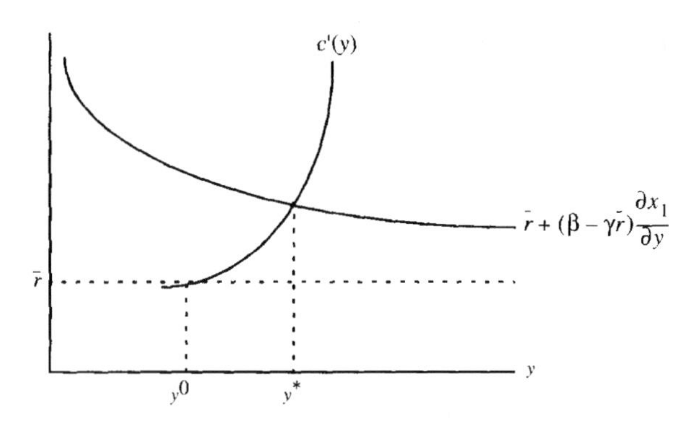

Figure 3.

Each of these equations, in fact, describes a relationship between marginal revenue (the left hand side) and marginal cost (the right hand side). A diagrammatic illustration of these equations is presented in Figures 3–5.

The entrepreneur's own capital optimal value, y^*, is reached at the point where his marginal cost $c'(y)$ equals the marginal revenue, $\bar{r} + (\beta - \gamma\bar{r})\partial x_1/\partial y$, as in Eq. (18). Marginal revenue consists of direct marginal revenue $\bar{r}$ and indirect marginal revenue $(\beta - \gamma\bar{r})\partial x_1/\partial y$. Indirect marginal revenue emerges as a result of additional external capital, which is motivated by y, $\partial x_1/\partial y$. The term $\beta - \gamma\bar{r} = \bar{r}(1 - \gamma) - (\bar{r} - \beta)$ is the net expected gain per dollar from the raising of external equity capital. This point is illustrated in Figure 3 in which the positive difference $y^* - y_0$ serves as a motivation for the additional external equity capital. The increment $y^* - y_0$ is the signaling part of the entrepreneur's investment. Its magnitude depends on the

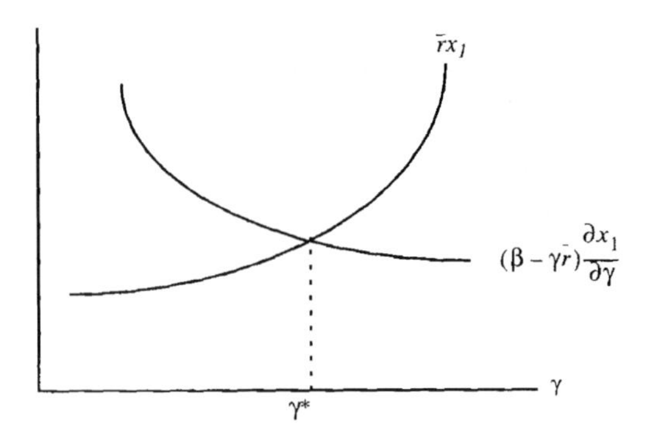

Figure 4.

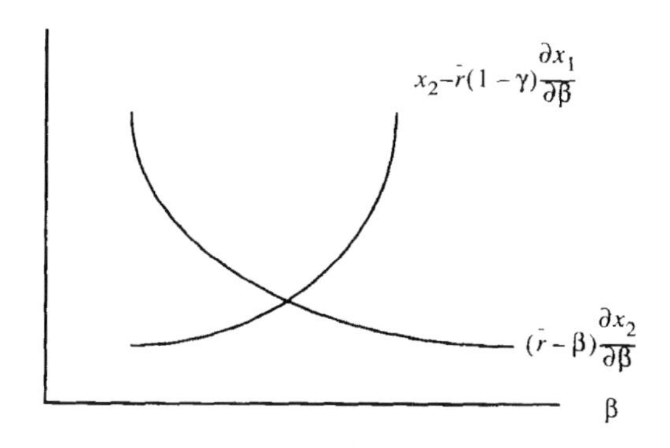

Figure 5.

capitalist's characteristics $\partial x_1/\partial y$ in Eq. (15). The second order conditions required for the stability of the solution are $c'' > 0$ and $\partial^2 x_1/\partial y^2 < 0$.

Equation (19) establishes the equilibrium condition for γ; Here, the marginal cost is $\bar{r}x_1$ the payment to external equity; marginal revenue is the gain from the additional external capital $(\beta - \gamma\bar{r})\partial x_1/\partial y$.

Since $r > \beta$, Eq. (19) implies that $\gamma < \eta/(1+\eta) < 1$ where $\eta = (\partial x_1/\partial\gamma)(\gamma/x_1)$ is the elasticity of x_1 with respect to γ. As in Figure 4, the second order condition necessary for the marginal revenue to decline is $\partial^2 x_1/\partial y^2 < 0$. This condition must be prevalent at the optimum for the stability of the solution.

Finally, the optimal incentive β, is determined by Eq. (20) and presented in Figure 5. Marginal cost is $x_2 - \bar{r}(1-\gamma)\partial x_1/\partial\beta$ where the first term, x_2, is a direct marginal cost and the second term, $\bar{r}(1-\gamma)\partial x_1/\partial\beta$ is an indirect marginal cost due to the inverse effect of β on x_1. Marginal revenue is $(\bar{r}-\beta)\partial x_2/\partial\beta$. The second order conditions necessary for the solution are $\partial^2 x_2/\partial\beta^2 < 0$ and $\partial^2 x_1/\partial\beta^2 < 0$.

An interesting result concerning the subjective beliefs of the entrepreneur compared with those of the capitalist is obtained by combining first order conditions of the two parties.

From Eq. (19), first order conditions of the entrepreneur, we obtain $\beta > \gamma r$, similarly this first order condition of the capitalist, Eq. (10), yields, $\gamma\hat{r}(y) > \beta$. Thus at the optimum we obtain:

$$\hat{r}(y) > \bar{r}. \tag{21}$$

In other words, to be a regular internal solution, γ must be large enough so that $\hat{r}(y)$ will exceed $\bar{r}$. The rationale of this result is clear. Since the capitalist is assumed to be more risk averse than the entrepreneur, the latter must encourage the capitalist to be more optimistic, if he has to induce him to supply external equity capital.[13] This may be a serious obstacle in raising external capital for a new venture. It should be noted that we are dealing with the first round of financing with no prior observations about the project and therefore with no indication about the true rate of return. At this state, it is unclear which of the two rates of return is closer to the true rate r_0. The result of Eq. (21) is a necessary condition that holds until the return is realized for the first time. Of course in the longer run with additional rounds of financing the game is repeated and observations about the true rate of return are obtained, the gap between the two subjective beliefs converges and the difference between the two perceived rates $\hat{r} - \bar{r}$ disappears. Additional rounds of financing will take place and the behavior of the firm may be described by a model such as Leland and Pyle (1977), where the equilibrium condition would be $\hat{r} - \bar{r}$.

4. MULTIPLE CAPITALISTS (AGENTS)

Thus far, only a single source of external financing (the capitalist) has been considered. In this section, two specialized sources of external financing are introduced: one providing only equity capital (venture capital) and the second only debt financing ("bank").

It is assumed that these are two independent sources, with independent financing cost functions. That is, d_i is a function of x_i only, while previously d was a function of both x_1, x_2. Denote by x_1 the equity capital contributed by Capitalist 1 and by x_2 the debt financing provided by Capitalist 2 and $d_1(x_1)$ and $d_2(x_2)$ are the financing opportunity cost of the two capitalists, respectively.

The objective function is:

$$\underset{x_1}{\text{Max}} \int V[\gamma r x_1 - d_1(x_1) g(r|y)] \mathrm{d}r \tag{22}$$

For the first capitalist, and

$$\underset{x_2}{\text{Max}}[\beta x_2 - d_2(x_2)] \tag{23}$$

for the second. The second capitalist makes, by assumptions, a riskless investment (through lending) and maximizes his profit. Since only the first capitalist faces uncertainty, only his beliefs will be affected by the signal of the entrepreneur, y.

[13] This result is consistent with that obtained by Paroush and Rubinstein (1982).

4.1. Solutions for the capitalists

The first order condition of the first capitalist yields

$$\frac{\gamma\hat{r} - d_1'}{\gamma\hat{\sigma}} = A \quad \text{or} \quad \gamma\hat{r} - d_1' = \gamma\hat{\sigma}A \tag{24}$$

This is a generalization of Eq. (10). A result similar to that of Eq. (10) occurs if $d_1'' = 0$, with the caveat that in Eq. (10), d' is a constant equal to β, but now d_1' is a function of x_1.

The first order condition for the second capitalist yields:

$$\beta = d_1'(x_2) \tag{25}$$

Total differentiation of Eqs. (24) and (25) w.r.t. β, γ and y yields

$$\begin{aligned} \frac{\partial x_1}{\partial \beta} &= 0 \\ \frac{\partial x_1}{\partial \gamma} &= \frac{d_1'(x_1)}{\gamma(d_1(x_1) + \gamma\hat{\sigma}(\mathrm{d}A/\mathrm{d}x_1))} > 0 \\ \frac{\partial x_1}{\partial y} &= \frac{\gamma((\mathrm{d}\hat{r}/\mathrm{d}y) - (\mathrm{d}\hat{\sigma}/\mathrm{d}y)A)}{d_1''(x_1) + \gamma\hat{\sigma}(\mathrm{d}A/\mathrm{d}x_1)} > 0 \end{aligned} \tag{26}$$

And

$$\begin{aligned} \frac{\partial x_2}{\partial \beta} &= \frac{1}{d_2''(x_2)} > 0 \\ \frac{\partial x_2}{\partial \gamma} &= \frac{\partial x_2}{\partial y} = 0 \end{aligned} \tag{27}$$

Note that Eq. (15) is a special case of Eq. (26), in which $d''(x_1) = 0$.

Thus, the existence of two separate financing sources generates several new results. Each incentive has only a direct effect on a single source of financing, γ on x_1 and β on x_2, such that an increase in either incentive will increase the total external investment. In the previous case, with its single source of financing, each incentive has a direct, positive effect on its own form of financing and a negative effect on the substitute form. As a result, only β had an effect on total investment while both β and γ affected the composition of external capital.

The signal provided by the entrepreneur, y, affects only the equity capitalist, but has no effect on the lender (second capitalist); thus, an increase in y increases the external investment. In the previous case, an increase in y had two offsetting effects on the two financing forms, so that the entrepreneur's action had no effect on the total external funds raised, but influenced capital structure only.

4.2. Solution for the entrepreneur

The Entrepreneur's objective function is assumed to be:

$$\underset{y,\beta,\gamma}{\text{Max}} F(y, \beta, \gamma) \tag{28}$$

where

$$F = (1-\gamma)\bar{r}x_1(y,\gamma) + (\bar{r}-\beta)x_2(\beta) + \bar{r}y - c(y).$$

The First order conditions are:

$$F_y = (1-\gamma)\bar{r}\frac{\partial x_1}{\partial y} + \bar{r} = c'(y) \tag{29}$$

$$F_\gamma = (1-\gamma)\bar{r}\frac{\partial x_1}{\partial \gamma} = \bar{r}x_1 \tag{30}$$

$$F_\beta = (\bar{r}-\beta)\frac{\partial x_2}{\partial \beta} = x_2 \tag{31}$$

That is, marginal revenue in equilibrium equals marginal cost with respect to each of the parameters. The net marginal gain per dollar of external equity is $(1-\gamma)\bar{r}$. Interestingly, unlike the previous case where $\gamma < \eta/(1+\eta)$ following (19), here Eq. (30) yields the equality $\gamma = \eta/(1+\eta)$.

5. CONCLUSIONS

This paper has analyzed the behavior of an entrepreneur and the response of a capitalist in a venture capital context. In our problem we consider first round financing for a new firm where no prior observations about its return exists. The entrepreneur is assumed to have superior but not perfect information concerning the outcome of the venture. In a feedback system the entrepreneur, who is a Stackelberg type leader, uses two types of controls: an incentive scheme and signaling to affect the action of the capitalist. The signal is the equity investment of the owner in the venture that is used to change the capitalist's beliefs, rendering him more optimistic. Thus, the owner's capital serves a dual purpose: as an input generating income and as a signal. The interesting result is that since the capitalist is assumed to be more risk-averse, at the optimum he is made to be more optimistic than the entrepreneur.

In our analysis we considered two models. In the first model (Model I), a single capitalist supplies equity and debt financing. The entrepreneur supplies "entrepreneurship," for which he or she is compensated by a fee paid from the

Table 1. Objective and responses

Model	I			II		
Entrepreneur						
Objective function	$\bar{r}y + (1+\gamma)\bar{r}x_1 + (\bar{r}-\beta)x_2 - c(y)$					
Capitalist(s)						
Objective function	$E\{V[\gamma r x_1 + \beta x_2 - d(x_1 + x_2)]\}$			$E\{V[\gamma r x_1 - d_1(x_1)]\}$		$\beta x_2 - d_2(x_2)$
Response capitalist	x	x_1	x_2	x	x_1	x_2
y signal	0	+	−	+	+	0
β debt incentive	+	−	+	+	0	+
γ equity incentive	0	+	−	+	+	0
Decisions (FOC) entrepreneur						
y	$[(1-\gamma)\bar{r} - (\bar{r}-\beta)]\dfrac{\partial x}{\partial y} + \bar{r} = c'(y)$			$[(1-\gamma)\bar{r}]\dfrac{\partial x_1}{\partial y} + \bar{r} = c'(y)$		
γ	$[(1-\gamma)\bar{r} - (\bar{r}-\beta)]\dfrac{\partial x_1}{\partial \gamma} = \bar{r}x_1$			$[(1-\gamma)\bar{r}]\dfrac{\partial x_1}{\partial \gamma} = \bar{r}x_1$		
β	$[(1-\gamma)\bar{r}]\dfrac{\partial x_1}{\partial \beta} + (\bar{r}-\beta)\dfrac{\partial x_2}{\partial \beta} = x_2$			$(\bar{r}-\beta)\dfrac{\partial x_2}{\partial \beta} = x_2$		

capitalist's equity income. The entrepreneur also supplies equity capital, which affects the equity investment of the capitalist; however his total investment in the firm, determined by the interest rate, is unaffected.

In the second model (Model II) specializing capitalists are considered, thus equity and debt capital is assumed to be independent. Here the signal affects both total external capital and its composition. The debt incentive (β) in both models determines total external investment, and has a direct positive effect on debt financing (x_2) and negative or zero effect on equity financing (x_1). The equity incentive, γ, affects total investment only in the second model; It has a direct positive effect on x_1 and negative or zero effect on x_2. Note that, in the first model, the positive effect of debt incentive β, on x_2, is greater than the negative effect on x_1, while the positive effect of the equity incentive γ, on x_1 is exactly offset by its negative effect on x_2. To summarize our results and clarify the differences between the models we present the Table 1.

REFERENCES

Berglof, E. (1994). A control theory of venture capital finance, *Journal of Law, Economics and Organization*, *10*, 247–267.

Chan, Y. S., Siegel, D., & Thakor, A. (1990). Learning, corporate control and performance requirements in venture capital contracts, *International Economic Review*, *31*, 281–365.

Downes, D. H., & Heinkel, R. (1982). Signaling and the valuation of unseasoned new issues, *Journal of Finance*, *37*, 1–10.

Gompers, P. (1995). Optimal investment, monitoring and the staging of venture capital, *Journal of Finance*, *50*, 1461–1489.

Gompers, P., & Lerner, J. (1999). *The venture capital cycle*, MIT Press.

Hellman, T. F. (1998). The allocation of control rights in venture capital contracts, *Rand Journal of Economics*, *29*, 57–76.

Keynes, J. M. (1961). *The general theory of employment, interest and money*, New York: Macmillian.

Leland, H. E., & Pyle, D. H. (1977). Informational asymmetries, financial structure, and financial intermediation, *Journal of Finance*, *32*, 371–387.

Paroush, J., & Rubinstein, A. (1982). Pessimism and risk-aversion in principal-agent relationship. *Research Report 145*, Department of Economics, Hebrew University of Jerusalem, Israel.

Ross, S. A. (1977). The determination of financial structure: The incentive signaling approach, *Bell Journal of Economics*, *8*, 23–40.

Sahlman, W. A. (1990). The structure and governance of venture capital organizations, *Journal of Financial Economics*, *27*, 473–521.

Stiglitz, J. E. (1974). Incentives and risk sharing in sharecropping, *Review of Economic Studies*, *61*, 219–256.

New Venture Investment: Choices and Consequences
A. Ginsberg and I. Hasan (editors)

Chapter 2

Venture Capital Investing and the "Calcutta Auction"

SRIDHAR SESHADRI[a], ZUR SHAPIRA[a]
and CHRISTOPHER L. TUCCI[b,*]

[a] *Stern School of Business, New York University, New York, USA*
[b] *Swiss Federal Institute of Technology, EPFL-CDH-ILEMT-CSI, Laussanne, Switzerland*

ABSTRACT

Rational models of firm valuation hold that the value of a firm derives from its future earnings. These valuations are grossly exceeded under certain circumstances such as when a potential market is first uncovered and there is only a short time frame for entrants to get into this market. The entrants are fully aware that as few as one (or even zero) firms might survive in this market. Further, investments by different firms tend to work together to develop the market, since entry, investment, and expansion help give the market credibility and bring in the first customers. The wave of investments in Internet-related businesses in the late 1990s is such an example.

In this chapter, we argue that the investment decision in such circumstances can be modeled as a "Calcutta auction." The Calcutta auction is a mechanism that fits nicely with venture capital investment dynamics and that leads to some non-obvious results. Based on this model, investments by venture capitalists and other professional investors follow a relatively simple formula that might be considered to be the "optimal bid" in a Calcutta auction. We derive several theoretical propositions on the nature of bidding in electronic

* Corresponding author. Tel.: +41-21-693-2463; fax: +41-21-693-5060.
E-mail address: christopher.tucci@epfl.ch (C. L. Tucci).

markets and illustrate the mechanism using data from 10 industries involved in Internet-based businesses. A consistent pattern of behavior emerges from the application of the model indicating that investors may have been much more rational than previously thought.

1. INTRODUCTION

Rational models of firm valuation hold that the value of a firm derives from its future earnings. These valuations are grossly exceeded under certain circumstances. Probably, most significantly when a potential market is first uncovered and there is only a short time frame for entrants to get into this market. The entrants are fully aware that as few as one (or even zero) firms might survive in this market. Further, investments by different firms tend to work together to develop the market, since entry, investment, and expansion help give the market credibility and bring in the first customers. The wave of investments in Internet-related businesses in the late 1990s is such an example.

Consider, for example, the investment in the Internet-based pet supply industry during the late 1990s. Total investments in firms offering pet supplies over the Internet were \$352 million and \$45 million in 1999 and 2000, respectively. The sales-to-capital ratio in the pet supply industry was approximately 30 in the late 1990s. Furthermore, total industry sales during the same period, based on Compustat data, was approximately \$1.6 billion per year. In calculating the incremental sales expected to justify the 1999 investment, it appears that Internet-based sales should equal, \$352M $\times$ 30 or \$10.6 billion, implying total industry sales of about \$12.1 billion rather than the current \$1.6 billion. The projected increase in sales of \$10.6 billion is a conservative estimate (if all the investments bear fruit) for several reasons. First, the venture capital (VC) investment in an industry does not represent the entire capital invested in it.[1] Nor does it include prospective future investments. It also does not account even for the capital invested by the start-up "owners" of a firm because the firm's senior management must already have invested money and sunk effort before going to the VC. Second, an upward trend in the sales-to-capital ratio implies that productivity has improved over time (more sales generated per dollar of investment in plant and equipment, for example).[2] As investments take time to

[1] For example, pets.com raised \$82.5 million in their IPO in February, 2000. This number is not included in our calculation. If one fifth of venture-backed firms go public (Gompers and Lerner, 1999), one gets a sense for how conservative this estimate is.

[2] One counter argument is that the trend in sales is also sharply higher. However, if one fits a line to the trend, even five years out (2004 estimate), the total industry sales would be expected to be only \$5.2 billion.

come to fruition (generate more sales revenue), a "rational" investor might expect even higher sales than the current multiple of 30.

Another possibility with a similar conclusion is from the point of view of profitability. If there must be commensurate return on investment for the risk in the industry, then the sales must grow (assuming the margins remain more or less the same) or the margins should widen substantially. Given few radical breakthroughs in cost reduction due to supply chain reconfiguration (notably Dell Computers, Cisco, Fedex, and eBay) because of the use of digital technology, it seems safe to conjecture that cost savings at the margin that will result in tripling or quadrupling of the gross profit are unlikely. Even if they were likely, these gains could come under attack due to competition.

Similar (but less dramatic) patterns hold for some industries as well as Internet markets as a whole, see Table 1. For example, using the same calculation, the sales expected to justify the 1999 investment in e-commerce software is $5.8 billion against total industry sales of approximately $3.5 billion in 1999. In Internet-based sporting goods, the sales expected to justify the approximately $179 million investment would be a $1.1 billion increment against industry sales of $335 million. In Internet-based biotech/pharmaceuticals, however, there appears to be more rational investment based on this method of calculation: $5 billion increment on total industry sales of approximately $183 billion.

In the economy as a whole, $90.5 billion was invested by venture capitalists (the majority of which was invested in Internet markets) in 2000 (CNN, 2001). Further, the domestic sales-to-capital ratio was approximately 8.6 and total gross domestic product was $8.9 trillion (U.S. Bureau of the Census, 2000; U.S. Department of Defense, 2000) in 1999. Thus the increase in GDP expected from the total venture capital invested would be about $800 billion, or almost 9%. Past explanations for this phenomenon have centered on market myopia, technology racing, herding and free cash flow theories. We briefly review each of these explanations before presenting our model.

Market myopia. Salhman and Stevenson (1986) describe the venture capitalists' (VCs') foray into the Winchester disk drive industry. During the late 1970s and early 1980s, nineteen disk drive companies received venture capital financing. Two-thirds of these investments came between 1982 and 1984, the period of rapid expansion of the venture capital industry. Many disk drive companies also went public during this time. While industry growth was rapid (sales increased from $27 million in 1978 to $1.3 billion in 1983), Salhman and Stevenson question whether the scale of investment was rational given any reasonable expectations of industry growth and future economic trends. They suggest that "market myopia" (irrationally short-term view of the market) affected the VC investments. They claim that individual investments may have been rational but individual

Table 1. Incremental sales necessary to justify investment in ten industries ($ millions except sales-to-capital and percentage increase)

Industry	VC Internet investment, 1999	Sales to capital ratio, 1999	Total industry sales, 1999	Incremental sales expected	Estimated new sales	Percentage increase
Pet supply	352	30	1550	10560	12110	681
Biotech/pharmaceuticals	179	28	183021	5012	188033	2.74
Books	194	24	24136	4656	28792	19.3
Broad retail	257	38	8387	9766	18153	116
E-software	193	30	3487	5790	9277	166
Food + garden	682	48	48273	32736	81009	68
Sporting goods	29	39	335	1131	1466	338
Telecoms	818	4.5	795340	3681	799021	0.49
Vehicle parts	1.5	33	8619	49.5	8668.5	0.57
Wireless	106	23	96249	2438	98687	2.53

investments conditioned on everyone else's investments were short-sighted or irrational.

Technology races. Lerner (1997) suggests that disk drive manufacturers may have displayed behavior consistent with strategic models of "technology races," in which firms that are behind expend more effort to overtake rivals and claim the indivisible, winner-take-all prize. One interpretation of this is that as firms had the option to exit the competition to develop a new disk drive, it may have been rational for venture capitalists to fund a substantial number of disk drive manufacturers. If the VCs saw that the firms were clearly losing, the VCs would have stopped funding the project. Similar interpretations have been made regarding investments in software, biotechnology, and the Internet. The phrase "too much money chasing too few deals" is a common refrain in the venture capital market during periods of rapid growth.

Herding. Several researchers argue that institutional investors frequently engage in "herding": making investments that are too similar to one another (Devenow and Welch, 1996). These models suggest that a variety of factors—such as when the assessment of performance is on a relative rather than an absolute basis—can lead investors to obtain poor performance when making too many similar investments. As a result, "social welfare may suffer because value-creating instruments in less popular technological areas may have been ignored" (Gompers and Lerner, 1999, p. 137).

Devenow and Welch propose three main reasons for "rational" herding behavior: (1) payoff externalities, in which payoffs to an action increase in the number of other people who also adopt the same action; (2) principal-agent issues, which refers to managers attempting to avoid evaluation by "standing out"; and (3) cascade models, where managers override their own private information and assume that prior actions by others reveals more valuable information in the absence of others' private information. It might be argued that each of these explanations has a bit of irrationality in it.

Cash flows. Jensen (1986) proposes that managers prefer to keep cash windfalls within the firm to retain control of the firm. He presents data from the oil industry that indicates that firms invested large cash surpluses excessively in exploration and development activities in the early 1980s. Jensen's approach implies that VCs who are flush with cash from earlier successful investments might be tempted to make further investments instead of keeping the cash in the bank. Note however that the free cash flow theory does not hold as strongly when managers are the owners of the firm.

1.1. The present approach

In this chapter, we argue that the investment decision in the circumstances outlined in the introduction can be modeled as a "Calcutta auction" (see below). The Calcutta auction is a mechanism that fits nicely with venture capital investment dynamics and that leads to some non-obvious results. Based on this model, investments by venture capitalists and other professional investors follow a relatively simple formula that might be considered to be the "optimal bid" in a Calcutta auction. We illustrate this mechanism using data from 10 industries and 81 firms involved in Internet-based businesses. A consistent pattern of behavior emerges from the application of the model indicating that investors may have been much more rational compared to the explanations offered above.

We also identify a new role for very early stage investors which we call "seeding the pool." Even though early stage investors (such as angel investors [wealthy individuals that invest their own funds in the early stages of ventures] or early-stage VC funds) may simply try to maximize their own company's chance of winning in the market, by their actions, they may be attracting VC funds to the industry as a whole as the market develops. This may or may not help the original startup but helps fuel the growth of that industry. One of our key conclusions is that higher valuation relative to the classical methods of firm valuation may not actually be irrational. We develop new measures for understanding when there is under- or over-investment based on the total investment in an industry.

The chapter is structured as follows. Some background on the Calcutta Auction is provided in the next section. We then describe a model of the auction, including an analysis of bids. We show that our model helps predict the flow of capital into ten Internet-based industries using data from the VentureSource database on startup companies and their venture capital funding. We conclude the paper with a discussion of rationality and exuberance.

1.2. The Calcutta auction

According to the Webster's Unabridged Dictionary (2000: 296), a "Calcutta" or "Calcutta pool" is "a form of betting pool for a competition or tournament, as golf or auto racing, in which gamblers bid for participating contestants in an auction, the proceeds from which are put into a pool for distribution, according to a prearranged scale of percentages, to those who selected winners." In a Calcutta auction, players compete in a tournament, such as a backgammon tournament. Each player in the tournament is "auctioned off" to the highest bidder. Each bid is then added up to a prize "pool" which is won by the "investor(s)" in the winning player, minus a commission for the organizer. The Appendix contains a detailed history of this auction.

For example, a backgammon tournament might have sixteen players. In addition to the players, there are many bidders (spectators, investors) whose number exceeds the number of players. Before the tournament begins, each of the sixteen players is "put up for auction" where the highest bidder wins the rights to the pool in case her player wins (in an actual tournament, players can also participate in their own auction). Thus each player is "owned" by one bidder. For example, Player 1 could be auctioned off to Jane, with a winning bid of $5000, followed by Player 2 with a bid of $5500 to James, and so on. All of the winning bids (the $5000 plus the $5500 plus the rest) are combined into a pool. If for example, the pool totals to $50,000 and Player 1 wins in the ensuing tournament, then Jane receives $50,000 minus a (say) 10% commission. Similarly, if Player 2 wins the tournament James wins $45,000. From the viewpoint of Jane, in the case in which she wins she receives $45,000, minus her initial bid of $5000 for a net gain of $40,000 otherwise she stands to lose $5,000.

Surprisingly, this widely used format of wagering has not been previously analyzed. Today this auction is used mainly in wagering on the winner(s) in games of skill (and not in games of chance—see Appendix for more details).

1.3. The analogy to venture capital investing

There are many similarities between the Calcutta auction and venture capital investing. The startups are analogous to the players that are about to compete. The venture capitalists are analogous to the bidders, while competing for market share in the common (industry specific) marketplace is equivalent to playing in the tournament. A startup firm's initial public offering is analogous to winning the tournament, as it allows the VCs to realize gains from their investments.[3] In reality, there are often few winners and the development of the market can occur quite quickly with many startups simultaneously developing the technologies or products that they hope will enable them to win once they hit the market (Gompers and Lerner, 1999).

Investments are made based on the probability that the player will win the tournament and the total estimated size of the pool. Just as a gambler attempts to assess the player's skills at backgammon, the VC attempts to ascertain the chance that a particular management team has the skills, experience, and ambition to win in the market. Complicating the analogy is the fact that VCs may not know the number of players

[3] We realize that VCs may not be able to (or want to) liquidate their entire holding after an IPO. However, they may be able to sell some and may be able to hedge their position with respect to the rest, allowing them to realize much of the gain. Likewise one might argue that "winning" is more analogous to sustained market leadership for years after an IPO. We agree, but we feel that for the early-stage investors, the IPO is a landmark that more directly corresponds to their "payoff." Acquisition is a distant second in terms of payoff (Gompers and Lerner, 1999).

in advance. In addition, they may take into consideration how the odds of success might be affected by not only their financial investment, but their managerial and industrial sector expertise. However, these complications do not necessarily change the optimal bid. To complete the analogy, in several industries, the amount "won" by the winning investor is approximately equal to the amount invested by all the rest.

1.4. Attractiveness of the Calcutta auction

One can ask, "why would anyone want to participate in such an auction?" We postulate that the Calcutta auction is more efficient: it is a market-based system for pricing information asymmetries in markets where the number of people examining the market is small. In a typical traditional professional football pool, bookies change their odds as people bet, but this requires in depth knowledge of the teams and many people betting. In contrast, it would be very difficult for an odds-maker to make a market in less publicized tournaments, such as a backgammon or amateur golf tournament, where information on different players is less widely known.

In addition, from the bidders' point of view, bidders might feel that they have an informational advantage (knowledge of the players) that they hope to exploit. That is, the bidder might have an a priori probability that a certain player might win and may hope to bid for the player at a price lower than what would be indicated by his chance of winning. In the VC world, this suggests that the bidders (the VCs) think that they may be able to influence the chances of the chosen player's victory.

2. THE MODEL

2.1. Analysis of the Calcutta auction mechanism

Assume that n players participate in a tournament. For simplicity assume also that only one player can win this tournament. Prior to the start of the tournament the n players are auctioned off, typically in a sequential manner. Label the players as one through n. Assume that they are auctioned off in the same sequence. A bidder can bid on any number of players. The auctions are of the English type, and we assume that in each auction the winning bidder pays the bid amount. Denote the winning bid for player i or equivalently the dollar amount paid for player i as x_i.

Assume that (according to the rules of the auction) there is a unique winner in each auction for a player, thus the same player can not be "won" by more than one bidder. Moreover, a bidder cannot win in more than one auction. Thus, the number of bidders is assumed to be strictly greater than n. The bid amounts are pooled and the total is sometimes simply called the Calcutta pool. In our notation, the Calcutta pool amount is given by: $X = x_1 + x_2 + \cdots + x_n$. We assume that there is a fixed fee as well as a variable administrative fee charged for running the auction. Thus we describe the administrative fee as: $a + bX$, where a and b are pre-specified real numbers.

The tournament begins after all players have been auctioned. The tournament could be an event in a bridge tournament, a backgammon tournament, a golf match, a fishing tourney, etc. If player i wins the tournament then the bidder who won the auction for player i, say, B_i, is awarded the amount: $X - (a + bX)$.

To keep the analysis simple, assume that player i is likely to win the tournament with probability p_i and that this probability does not depend on the bid amounts or the size of the pool. The bidders are aware of and agreed upon these probabilities. Assume that the amounts bid until the beginning of the auction for player i are known to all the bidders. All bidders are expected value maximizers and they do not collude with one another.

Given the sequence described above, We now address the following questions. What should be the optimal bid amount for player i? Which parameters of the auction determine the size of the Calcutta pool? What happens if the sequence of players auctioned is different? Is the knowledge of the number of players essential? Is the sequential auction format essential? What happens if the number of players is random? What is the impact of an auction for some player, say $i < n$, that ends with the bid amount less than or greater than the optimal bid amount? In order to relate the auction outcome to VC investment, we also ask whether and how the outcome is affected if there is a non-zero probability known to every player that the pool will not be distributed?

To facilitate the analysis, denote $X_i = x_1 + x_2 + \cdots + x_i$. In words, X_i is the total of the amounts bid for players one through i.

Proposition 1. The optimal bid for player i is given by

$$x_i = \frac{p_i}{1 - (1 - b)(p_i + p_{i+1} + \cdots + p_n)}((1 - b)X_{i-1} - a). \tag{1}$$

Proof: The auction can be modeled as a game of complete and perfect information. Thus, the optimal bid amounts can be computed by backwards induction. Consider the last auction prior to the beginning of the tournament. The bidders are aware that the amount so far collected equals X_{n-1}. Thus, the bid amount for the n-th player should just equal the expected value of winning the pool. After the auction the Calcutta pool equals $(X_{n-1} + x_n)$. The payout is not equal to this amount but an amount adjusted for the administrative fees and is given by:

$$X_{n-1} + x_n - a - b(X_{n-1} + x_n).$$

The successful bidder in the n-th auction wins the pool if the n-th player wins the tournament, an event that has probability equal to p_n. Thus, we require

$$x_n = p_n(X_{n-1} + x_n - a - b(X_{n-1} + x_n))$$

or

$$x_n = \frac{p_n}{1-(1-b)p_n}((1-b)X_{n-1} - a).$$

Now assume that the optimal bid amount is given by Eq. (1) for $i = j+1, j+2, \ldots, n$. Moreover, for the same values of i, assume that the total of the optimal amounts bid for players i through n is given by

$$x_i + x_{i+1} + \cdots + x_n = \frac{p_i + p_{i+1} + \cdots + p_n}{1-(1-b)(p_i + p_{i+1} + \cdots + p_n)}((1-b)X_{i-1} - a). \tag{2}$$

Thus, there are now two backward induction assumptions. We have established these to be true for $i = n$. Consider the j-th auction. The amount in the pool at the beginning of this auction is X_{j-1}. As before, we equate the bid amount x_j with the expected value of the payout from the pool. From (2), if the winning bid is x_j then the Calcutta pool is given by

$$X = X_{j-1} + x_j + \frac{p_{j+1} + \cdots + p_n}{1-(1-b)(p_{j+1} + \cdots + p_n)}((1-b)(X_{j-1} + x_j) - a).$$

Thus, we should expect

$$x_j = p_j\left((1-b)\left(X_{j-1} + x_j + \frac{p_{j+1} + \cdots + p_n}{1-(1-b)(p_{j+1} + \cdots + p_n)}\right.\right.$$
$$\left.\left.\times ((1-b)(X_{j-1} + x_j) - a)\right) - a\right)$$

or

$$x_j = \frac{p_j}{1-(1-b)(p_j + p_{j+1} + \cdots + p_n)}((1-b)(X_{j-1} + a)$$

Therefore, (1) holds for the j-th auction. Using this expression for x_j and plugging it into (2) for $(j+1)$, leads to:

$$\sum_{i=j}^{n} x_i = \frac{p_j}{1-(1-b)(p_j + p_{j+1} + \cdots + p_n)}((1-b)X_{j-1} - a)$$

$$+\sum_{i=j+1}^{n} x_i = \frac{p_j}{1-(1-b)(p_j+p_{j+1}+\cdots+p_n)}((1-b)X_{j-1}-a)$$

$$+\frac{p_{j+1}+\cdots+p_n}{1-(1-b)(p_j+p_{j+1}+\cdots+p_n)}((1-b)X_{j-1}-a)$$

or

$$x_j+x_{j+1}+\cdots+x_n = \frac{p_j+p_{j+1}+\cdots+p_n}{1-(1-b)(p_j+p_{j+1}+\cdots+p_n)}((1-b)X_{j-1}-a).$$

Thus, (2) holds for the j-th auction. This completes the proof of the induction step. □

Proposition 2. The optimal bid for player i is unaffected by the sequence in which players are auctioned off.

Proof: Due to (1) and (2) it follows that an interchange in the sequence for players i and $(i+1)$ does not affect the optimal bids for the remaining players. We show that such an interchange does not affect the bids for these two players as well. Let:

$$x_i = \frac{p_i}{1-(1-b)(p_i+p_{i+1}+\cdots+p_n)}((1-b)X_{i-1}-a)$$

$$y_i = \frac{p_i}{1-(1-b)(p_i+p_{i+2}+\cdots+p_n)}((1-b)(X_{i-1}+y_{i+1})-a)$$

$$x_{i+1} = \frac{p_{i+1}}{1-(1-b)(p_{i+1}+\cdots+p_n)}((1-b)(X_{i-1}+x_i)-a)$$

$$y_{i+1} = \frac{p_{i+1}}{1-(1-b)(p_i+p_{i+1}+\cdots+p_n)}((1-b)X_{i-1}-a).$$

That is, y_i and y_{i+1} stand for the optimal bids with the sequence of bidding interchanged. We have to show that $x_i = y_i$ and $x_{i+1} = y_{i+1}$. We can re-express,

$$\begin{aligned} x_i - y_i &= K(((1-b)X_{i-1}-a)(1-(1-b)(p_i+p_{i+2}+\cdots+p_n)) \\ &\quad -((1-b)(X_{i-1}+y_{i+1})-a)(1-(1-b)(p_i+p_{i+1}+\cdots+p_n))) \\ &= K(1-b)(((1-b)X_{i-1}-a)p_{i+1}-y_{i+1}(1-(1-b)(p_i+p_{i+1}+\cdots \\ &\quad +p_n))), \end{aligned}$$

where K does not involve the optimal bid quantities. Thus, x_i equals y_i if and only if $x_{i+1} = y_{i+1}$.

But,

$$
\begin{aligned}
x_{i+1} - y_{i+1} &= K_1(((1-b)(X_{i-1}+x_i) - a)(1-(1-b)(p_i + p_{i+2} + \cdots + p_n)) \\
&\quad - ((1-b)(X_{i-1} - a)(1-(1-b)(p_{i+1} + \cdots + p_n))) \\
&= K_1(1-b)(((1-b)X_{i-1} - a)p_i - x_i(1-(1-b)(p_i + p_{i+1} + \cdots \\
&\quad + p_n))) = 0,
\end{aligned}
$$

where K_1 does not involve the optimal bid quantities and the last equality follows from (1). □

This result leads to a somewhat surprising proposition.

Proposition 3. The key parameters that decide the value of the Calcutta pool are a and b. In fact, the value of the pool is given by: $X = -(a/b)$.[4]

Proof: From Proposition 2, we know that the sequence of the auction does not matter. Thus, we shall examine the cases that the first player is auctioned either first or last. In the former case, from (1) (and defining $X_0 = 0$)

$$
x_1 = \frac{p_1}{1-(1-b)(p_1 + p_2 + \cdots + p_n)}(-a).
$$

In the latter case, (by equating expected value of the payoff with the optimal bid amount)

$$
x_1 = p_1((1-b)X - a).
$$

Thus, equating the two expressions for x_1, we get the size of the Calcutta pool to be

$$
\begin{aligned}
&\frac{p_1}{1-(1-b)(p_1 + p_2 + \cdots + p_n)}(-a) = p_1((1-b)X - a) \\
&\Leftrightarrow X = -\frac{a}{(1-b)}\frac{(1-b)(p_1 + p_2 + \cdots + p_n)}{1-(1-b)(p_1 + p_2 + \cdots + p_n)} = -\frac{a}{b}.
\end{aligned}
$$

□

[4] In the originally envisioned Calcutta auction (see Appendix), 100 tickets were sold for 10 rupees each, thus creating a pool of 1000 rupees. Then tickets were drawn and matched against horses. The lucky holders of these tickets could then turn around and auction them off to the highest bidder before the race took place. Over the years, the initial pool concept has disappeared.

2.2. Seeding the pool

We term this result as "seeding the pool." The value of the pool is non-negative if the value of a is negative and the value of b is less than one. Thus, the promoter of the auction needs to entice the bidders with some initial pool amount. (Note that the promoter recovers the seed amount via the variable administrative fee, namely, $b \times X = b \times (-a/b) = -a$.) The fact that the organizers do not need to seed the pool in Calcutta auctions now-a-days is surprising. However, the excitement generated by the bidding as well as the interest in witnessing the tournament might explain why a is never negative in a real-world Calcutta auction. Of course, another explanation for why in reality pools do not need to be seeded might be that the estimates of the probability of winning add up to a quantity greater than one, i.e. bidders exhibit overconfidence. (If $(p_1 + p_2 + \cdots + p_n)$ exceeds one then the numerator in the second last expression in the last proof can become negative.) A third possibility for explaining why the pool does not have to be seeded in real-world Calcutta auctions is that the "winner's curse" (Thaler, 1988) might operate in overtime here if in each auction the winning bidder gets the winner's curse. Assume from this point forward that a is negative.

Proposition 4. The sequential bidding format is not necessary. The optimal bid in the auction for the i-th player is $x_i = -p_i a/b$ given bidders follows the same strategy in other auctions.

Proof: Follows from Propositions 2 and 3 above. □

Proposition 5. There is never a need to know the total number of players.

Proof: Follows from Proposition 4 as the optimal bid amount for any player depends only on the values of a and b.

Finally, we approach the last two questions we posed. □

Proposition 6. If the probability that the money in the pool is distributed according to the rules of the game is p_d (in other words the sponsors of the auction go bankrupt in the meanwhile with probability $1 - p_d$), then the optimal bid for player i is $x_i = -p_d(p_i a/b)$.

We can argue that if some player makes an error, say ε, in the sequential format of the auction and if the expected value of this error is zero then the net effect is for subsequent bidders to revise their bids by an amount equal to the value of the error divided by the probability of the player winning the tournament. In other words, errors affect subsequent bids, and in the same manner as the fixed fee a.

In summary, the three important conclusions are in Propositions 3, 4 and 6. We now apply them to new markets. The seeding of the pool $(-a)$ is equivalent to the value of early entry into a new potential profit site, which refers to a unique market or market segment in which a firm may compete and presumably reap profits (Afuah and Tucci, 2003). Firms that wish to enter and compete in this market make investments, knowing that others will be making similar investments and that these investments work together to develop the market. The extent to which there are market development externalities determines the value of b. If everybody's investment worked together to develop the market, then the winner realizes the benefits of everybody else's efforts. The amount gained relative to the total investment would therefore be large, implying a low variable administrative fee, b. Thus if these market development externalities are small, then b is large; otherwise if these externalities are large, then b is small. It is indeed surprising that it is the ratio of these two quantities (a and b) that determines investment behavior when there are market development externalities such as those that might be present during the development of new markets. We illustrate this in the next section.

2.3. Data

In this section, we illustrate several aspects of the model with archival data. The data are drawn from the VentureSource database, which contains information about most venture capital investments in companies during the last ten years. With more than 9,200 venture-backed companies, 26,000 financing transactions and 61,000 key executives in the U.S. venture capital industry, the database is considered one of the most comprehensive in the industry. We randomly selected ten industries within the "Internet Focus" section of the database. These ten industries have a wide range of total investment, average investment, entry timing, and IPO outcomes. The industries are: (1) telecommunications; (2) broad retailing; (3) books; (4) jewelry; (5) investment banking; (6) wireless; (7) pet supplies; (8) food; and (10) job placement services. Each industry has 5 to 10 startup companies in it, with an average of 8.1 companies.

We documented every stage, amount, and date of investment for the companies up to and including an IPO, if applicable. Thus we know the total VC investments in the industry and who made them. If a company eventually went public, we obtained from the SEC filings who owned it and matched the names of the earlier stage investors with those shareholders. We were thus able to compute a "payoff" to early stage investors by multiplying the value of the company at their IPO by the beneficial ownership of the company. While venture capitalists may not be able to sell at the IPO and thus the calculation might overstate their actual payoff, the VCs may be able to hedge against changes in the stock price post-IPO and thus realize most of their interest in the company. Furthermore, we assume that if the company does not go public, the payoff to the VCs on average is minimal (Gompers and Lerner, 1999).

We set the stage for the analysis of the Calcutta auction by displaying varying estimates produced by a traditional method of valuing firms. As explained in the introduction, one way of examining this is to look at VC investment in an industry and calculate how much an increase in sales in the industry would be expected to justify those investments (see Table 1). Notice the huge variation in sales growth expected across industries. When looking at this table one must ask whether information gleaned from any one subsector in electronic markets could possibly be useful in understanding any other subsector.

Our analysis of these investments is based on the predictions of the Calcutta auction model. If the auction model describes the investment phenomenon, we expect to verify at least the prediction of Proposition 3. We estimate the values of the fixed administrative fee a, and the size of the total investments (Calcutta pool) X. It seems logical that the Calcutta pool in any given industry can be set equal to the total venture investment in that industry. The estimate of a, or how the pool is seeded, might be done in several ways. For example, we could use (1) the first "spike" of investments in that industry as a proxy for the initial worth of the market; (2) the amount raised by a firm in the first IPO in that industry; (3) the total market value of the first IPO; or (4) the total market value of all IPOs in that industry.

We computed the "first spike" by graphing the value of the investments in ten industries and looking for an investment large relative to the prior investments. All industries had some small investments in the early stages and it was relatively easy to spot the first large investment (see for example Figure 1). Table 2 shows our estimates of a for each of these four methods. We evaluate these methods below.

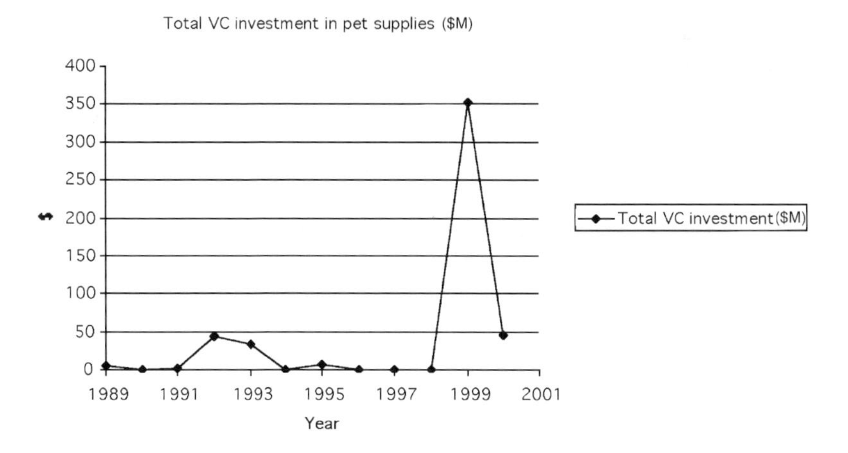

Figure 1. Total VC investment in pet supplies over time.

Table 2. Estimates of a and b in ten different industries

Industry	No. of firms	Total investment ($ million)	First spike		Amount raised in first IPO		Market cap of all IPOs		Market cap of first IPO	
			$-a$	b	$-a$	b	$-a$	b	$-a$	b
Wireless	10	171.3	96	0.56	96	0.56	202	1.18	202	1.18
Telecommunications	10	975	62	0.06	108.8	0.11	893	0.92	397	0.41
Pet supplies	10	488	130	0.27	130	0.27	165	0.34	165	0.34
Garden Supplies	5	70.5	20	0.28	49.2	0.70	130	1.84	130	1.84
Food	10	1036	120	0.12	64	0.06	2998	2.89	870	0.84
Broad retailing	6	343.8	25	0.07	129.6	0.38	36.4	0.11	36.4	0.11
Books	7	355.6	54	0.15	54	0.15	739	2.08	217	0.61
Jewelry	7	287	30	0.10	81.25	0.28	171	0.60	171	0.60
Investment banking	7	336	28	0.08	68.4	0.20	1025	3.05	513	1.53
Job placement	9	242.1	58	0.24	58.5	0.24	113	0.47	113	0.47
Average				0.19		0.30		1.35		0.79
Standard deviation				0.15		0.20		1.06		0.56

The values for the variable administrative fee b can now be computed as the ratio of the size of the pool divided the fixed administrative fee (see Table 2). The "first spike" method of estimating a produces estimates of b in the range of 0.06 to 0.56 with an average of 0.19. That is, on average, the winning firm expected to realize 81% of all investment in the industry from its IPO. If we assume that seeding of the pool was accomplished by the amount raised in the first IPO, we find that b ranges in the interval [0.07, 0.70] with an average of 0.29. For seeding based on all IPOs' market cap, we find b in the range [0.11, 3.05] with an average of 1.35 (!), while for seeding based on the market cap of the first IPO, we find b in the range of [0.11, 1.84] and an average of 0.79. The standard deviation tracks exactly the average (i.e. standard deviation of b using the first spike method is the lowest [0.15] and using the all-IPO market cap method is the highest [1.06]). Recall that if the variable administrative fee is very small, the total investment is roughly the same as the payoff to winning investors. However, if the administrative fee is very large, the payoff to investors is much smaller than the total amount invested in the industry. Note that when b is greater than 1.0, it signifies that the firm realized less than the seeding of the pool. Two out of the four ways of estimating a appear to produce values of b that are inconsistent with the model. This is not surprising as the latter two methods of estimating a refer to value of all firms accumulated over a long period of time and do not provide a good indication of "seeding of the pool," which occurs at the beginning of the "tournament."

3. DISCUSSION

In this chapter we propose a framework for analyzing seemingly non-rational investment behavior. The variety of examples we provided show that investments in Internet type businesses cannot be explained by traditional valuation models, which assume that the amount of investment is closely related to the expected cash flow. The numbers in our examples imply expected industry growth in excess of what would be predicted by any traditional valuation model. In examining this paradox we reviewed a variety of explanations that assume that investor behavior is not perfectly rational. In contrast, ideas such as herding and market myopia were proposed to account for such behavior.

We approach the problem in a different way. We model the VC investing as a Calcutta auction and propose that this mechanism can predict the amounts invested we observe. The Calcutta auction appears to be a good mechanism to explain behavior in markets that are characterized by uncertainty, information asymmetries and by small number of participants. Under these conditions behavior that departs from what would be typically prescribed by a "rational" model is more likely to be observed. However, the Calcutta auction mechanism is shown in this paper to provide a good explanation of the actual bids put up by VCs. Indeed, several bids that are characterized as departing from rationality according to traditional valuation models can be described more naturally by the Calcutta auction model.

Most of the models that were proposed to explain the "abnormal" investment in Internet industries included some element of irrationality. Such were the models that explained technology races and those that assumed herd behavior or investors' irrational exuberance. Our approach suggests that in considering investment under such conditions of high uncertainty, the behavior of investors can be explained without resorting to arguments about irrationality albeit including some behavioral components.

For instance, in the absence of past data on investment in such industries what value would one take as an indicator of the potential value of such investment? An investor can ask what is the correct signal: Is it the market value of the firm or the first spike of money raised by the VCs? These two signals provide information to subsequent investors. The fact that a firm goes to an IPO can be considered as analogous to a "victory" in the market. However, it may not be accurate to equate market valuation based on an IPO with the seeding of the pool. We can distinguish between two signals coming out of the first IPO. The first is the total market capitalization of the firm immediately following the IPO. As mentioned above, the IPO indicates the "victory" of the VCs who put the money in the particular company earlier. However, the total market cap may be an inflated estimate of the pool of money available. In contrast, the cash amount raised in the first IPO is a more natural measure of the seeding of the pool. It is also more natural in that it represents the investment by early investors and thus provides a measure of that part of a firm's value that accrues to early investors. The VC's are probably more interested in this cash amount than in the total value of the firm. The first spike of VC investment provides a similar signal.

In estimating the values of a and b we note that in some cases, as shown in Table 2, the value of b is higher than one. This occurs when we use the market value of the firm after IPOs (either the first one or the total of all of them) as an estimate for the seeding of the pool. On the other hand, both the first spike and the cash raised in the first IPO lead to sensible values of the administrative fee, b. For the case of the first spike, the values of b are also close together (mean of 0.19 and standard deviation of 0.15). The relatively low values of b for either of the first two methods indicate that the Calcutta auction accurately describes the behavior—a winner take all phenomenon, probably fueled by the externalities in market development efforts. Notice that the claim is not that individual investments affect the probability of success of a given firm but that they increase the conditional expectation of return when that firm is successful. In other words, when more firms enter everybody knows that the probability of winning is lower but the stakes are higher hence the conditional expectation upon winning is higher. Large values of b indicate that the return to a firm is not affected by the efforts of other firms, whereas small values of b imply spillover effects from one firm to another.

One can argue that an indicator such as the first spike may be unreliable or biased. Yet, behavioral studies suggest that in forecasting under uncertainty people often use heuristics rather than normative rules such as Bayes theorem. One of these heuristics, called anchoring and adjustment (Tversky and Kahneman, 1974), suggests that in making predictions under uncertainty people cling to an anchor that is provided to them and then they do no adjust enough (in comparison to Bayes rule) when new information arrives. Thus, the first spike may provide VCs with an anchor according to which they decide on their investment without adjusting it enough when new information becomes available.

Other heuristics that may play a role in affecting VCs decisions whether to invest in such industries are the availability and representativeness heuristics (Tversky and Kahneman, 1974). According to the former, when making a probabilistic judgment people often rely on information that is readily available and salient. For instance, after a plane crash people tend to overestimate the probability of airline accidents. According to the representativeness heuristic, people perceive certain events as a reliable source if they appear to be representative of the population of events that they are looking at to make a probabilistic estimate. Thus, in many cases people would base their estimates on a small (and unstable) sample for making such a forecast if it appears to be representative of the population upon which they make inferences. If the sample appears to be respresentative they may correct their estimates accordingly. For instance, many people expect a fair coin flipped randomly to alternate between heads and tails even in a small sample of coin flips. This phenomenon has long been known in statistical inference as the "gambler's fallacy." The use of these heuristics has often been interpreted as a demonstration of the fallibility of judgment; yet, as Hogarth (1981) noted, they sometime serve a function with respect to the effort people put into processing information. It is plausible that in the context we are examining such heuristics may affect the way VCs make their judgments and subsequent decisions.

Given that in many auction type situations people behave in a way that can be considered as irrational such as in the case of the winner's curse (Thaler, 1992), one can ask if the behavior in the context of the Calcutta auction is irrational. In discussing this issue, we examine two aspects of the decisions: First, is the particular investment considered an overpayment or an underpayment? Second, is the value of the variable administrative fee, b, small or large? We consider the two aspects jointly.

It can be argued that an investment that deviates significantly from the "true" value of the firm is not rational. Clearly an overpayment is irrational, while an

Future research should examine the applicability of the Calcutta auction to explain investment phenomenon in other newly discovered markets.

REFERENCES

Afuah, A., & Tucci, C. L. (2003). *Internet business models and strategies*, 2nd Edition. New York: McGraw-Hill.

CNN (2001). Venture capital dips again (November 1), http://money.cnn.com/2001/11/01/smbusiness/venture/

Devenow, A., & Welch, I. (1996). Rational herding in financial economics, *European Economic Review*, *40*, 603–615.

Gompers, P. A., & Lerner, J. (1999). *The Venture Capital Cycle*, Cambridge, MA: MIT Press.

Hobbs, H. (1930). *The Romance of the Calcutta Sweep*, Calcutta: Thacker's Press & Directories.

Hogarth, R. (1981). Beyond discrete biases: Functional and dysfunctional aspects of judgmental heuristics, *Psychological Bulletin*, *90*, 187–217.

Jensen, M. C. (1986). Agency costs of free cash flow, corporate finance, and takeovers, *The American Economic Review*, *76*(2), 323–329.

Lerner, J. (1997). An empirical examination of a technology race, *Rand Journal of Economics*, *25*, 319–333.

Pearson, R. (1954). *Eastern Interlude*, Calcutta: Thacker, Spink, & Co.

Perkins, E. B. (1950). *Gambling in English Life*, London: Epworth.

Salhman, W. A., & Stevenson, H. (1986). Capital market myopia, *Journal of Business Venturing*, *1*, 7–30.

Surita, P. H. R. (2002). *The Parades Gone By: 150 Years of Royal Calcutta Turf Club*, Calcutta: Royal Calcutta Turf Club.

Thaler, R. H. (1992). *The Winner's Curse*, New York: Free Press.

Tversky, A., & Kahneman, D. (1974). Judgment under uncertainty: Heuristics and biases, *Science*, *185*, 1124–1131.

U.S. Bureau of the Census (2000). http://www.census.gov/csd/ace/view/ace99.html

U.S. Department of Defense (2000). http://www.defenselink.mil/pubs/allied_contrib2000/E-1.html

Webster's Unabridged Dictionary of the English Language (2000).

APPENDIX 1. HISTORY OF THE CALCUTTA AUCTION[5]

In the early 1800s when horse racing was introduced in Calcutta an auction form came to be used for betting on the outcome in the following way. Tickets were sold to the public for ten rupees each. Typically, there were 100 tickets, which was more than the number of horses in the race. After all the tickets were sold, a lottery was held for each horse in the race out of these 100 tickets. The ticket associated with the horse had an exclusive claim on the winnings for that horse. For example, after all 100 tickets were put into drum, the organizer would say, the next ticket we draw will be for "Valentine" for a race on February 1st, and that ticket would pay if Valentine won the race (or came in second or third as we will see below). Thus if there were fifteen horses in the race there would be 15 lucky winners. After the lottery, each of the fifteen tickets would be auctioned off to the highest bidder and the winning bid amounts were collected and pooled as described below. The winner of the ticket in these auctions would have exclusive rights to the winnings of that particular horse. Thus, favorites in the races would command higher auction prices than long-shots. The money collected was split between the original holder of the ticket and the pool. For example, if the ticket for Valentine on February 1st was auctioned for 100,000 rupees, then 50,000 would go to the lottery winner and the other 50,000 would be added to the original 1000 in the pool. The lottery winner would not have to auction off the ticket but could also hold onto it, or sell off only a partial share in the ticket (Hobbs, 1930). This process would continue for all fifteen of the horses. When the race was over, the tickets paid the following: winner, 40% of the pool; second place, 20% of the pool, third place, 10% of the pool, and unplaced (divided equally among unplaced horses), 20% of the pool. The organizer kept 10% for its own use and for expenses related to the process.

Eventually, it seems as if an inability to pay the auction bids by bidders led to the formation of an official club in 1847, the Calcutta Turf Club, which regulated all aspects of the lottery, subsequent auction, and the race itself (Surita, 2002). The lottery was only open to members of the club (who could also act as agents and purchase tickets for others). In its heyday of the late 1920s, the pool reached levels of the equivalent of one million British pounds. After 1930 the

[5] It is not surprising that such an auction format evolved opportunistically in Calcutta. For example, the East India Company sold its cargoes using the "public outcry" method early on, but soon a number of private firms sprang up in the Dharamtola-Lal Bazar area to liquidate the holdings of wealthy retiring merchants or those who died in Bengal without heirs (Pearson, 1954). Disease and natural disasters claimed the lives of many in the city every year, swelling the number of goods available at auction and increasing the scale of the auction area. Soon it was well known that almost any type of good—new or used—could be purchased at a "Calcutta auction."

popularity of the Calcutta auction diminished due to the introduction of pari-mutuel betting.

The Calcutta auction contrasts with traditional horse racing where anyone who bids on a winning horse has a claim on that horse's winnings. Perkins (1950) claims that the outcomes of some of these Calcutta auctions could make the winners wealthy—and by the way, perhaps ruin their lives much as we suspect these days that winning the lottery might do.

From there it seems natural that the auction technique could be extended to those participating in tournaments or sporting events.

A search on Google for the phrase "Calcutta Auction" resulted in over 250 hits. The summary of the first 12 pages of hits is as follows (name of sport or type of page followed by the number of hits—total hits are 67): Backgammon (18), golf (9), game fishing (7), boat racing (6), blackjack/poker (4), horse racing/jumping etc (4), horse show (3), tennis (3), dog racing (2), squash (2), gaming laws etc (2), and one each to bridge, car racing, cattle show, miniature horse show, pigeon racing, snow machine racing. As for geographic representation, results came from the following countries: Australia, Canada, The Netherlands, New Zealand, and the U.S. For example, a form from New Zealand elaborately sets out the information needed prior to granting a "Licence to Conduct a Calcutta."

Regarding the law, the Attorney General of Idaho responds to an inquiry and states: "In conclusion, it is the opinion of this office that calcutta wagering on events such as golf tournaments is not a permitted exception to Idaho's public policy prohibiting gambling as articulated in art. 3, § 20 of the Idaho Constitution and is prohibited by Idaho Code § 18–3801. Further, betting at a calcutta auction is criminally punishable as a misdemeanor pursuant to Idaho Code § 18–3802." September 17, 1993. It is fascinating to note that the same opinion states: "Germane to our analysis is art. 3, § 20(1)(b), which does permit a form of "betting" as opposed to a particular gaming activity. It is important to note that pari-mutuel betting is distinguishable from calcutta wagering or auction pools. Pari-mutuel betting allows patrons to select a contestant and place a wager upon that contestant, generally a horse or dog. Rather than one patron bidding against the other for the right to wager on a particular contestant, every patron is allowed to wager on the contestant of choice. The money wagered is pooled and odds are computed based upon the amount of money wagered on one contestant in relation to the other contestants in the race. The odds then determine how much money is paid to successful patrons. See Oneida County Fair Board v. Smiley, 86 Idaho 341, 386 P.2d 374 (1963)."

New Venture Investment: Choices and Consequences
A. Ginsberg and I. Hasan (editors)

Chapter 3

The Entrepreneur's Initial Contact with a Venture Capitalist. Why Good Projects may Choose to Wait

TOM BERGLUND[a,*] and EDVARD JOHANSSON[b]

[a] *Department of Finance, Swedish School of Economics, P.O. Box 479, 00101 Helsinki, Finland*
[b] *The Research Institute of the Finnish Economy, Helsinki, Finland*

ABSTRACT

We show that profit-maximizing entrepreneurs may choose to avoid venture capitalists in the very first stages of a project's life. By developing the project on his own up to a certain stage the entrepreneur will enhance his bargaining position, and capture a larger share of the total surplus. We derive the condition for the optimal delay from the birth of the project until the initial contact with the venture capitalist. This delay depends on the competitive pressure in the market, and the entrepreneur's skills. Our model implies that the market for a venture capitalist's services at the start-up stage is likely to fail.

1. INTRODUCTION

Recent work in economics has identified the overcoming of asymmetric information, the provision of expert advice and the allocation of property rights as

* Corresponding author. Tel.: +358-9-431331; fax: +358-9-43133382.
E-mail addresses: tom.berglund@shh.fi (T. Berglund); ejohanss@lehman.com (E. Johansson).

important reasons for venture capital firms to provide finance for risky start-up and development projects.

Venture capital firms provide advice, contacts and reputation on issues such as the selection of key personnel and the dealing with customers and suppliers. The venture capitalist's role as key advisor also provides a cost efficient way of solving problems of asymmetric information that typically arise when a firm that lacks an established reputation requires outside financing. On one hand, the venture capitalist is well informed about the prospects of the firm, and on the other hand the venture capitalist can raise required external funds from outside investors by using reputation acquired in earlier cases.

There are several papers in the literature, both theoretical and empirical, that deal with the various agency problems that occur in a venture capital setting. See Barry (1994) for a survey, and for theoretical contributions see, for example, Chan (1983), Sahlman (1990), Admati and Pfleiderer (1994), Berglöf (1994) and Repullo and Suarez (1998).

The paper that most closely resembles our chapter is by Anton and Yao (1994). That paper discusses the problem faced by an inventor who for the lack of personal wealth is forced to sell his invention to an existing firm. The paper shows that when verification of the value of the invention is straightforward the inventor can reveal the idea to the firm and still obtain a fair price simply by threatening to take the idea to a competitor. In our chapter, the entrepreneur cannot pursue this strategy since maximizing the value of the project is assumed to require some strategic knowledge possessed by the venture capitalist as well. In that situation, the venture capitalist may not find it worthwhile to compensate the entrepreneur once the idea has been revealed.

Empirical papers that explain how agency problems are dealt with by venture capitalists in small, rapidly growing firms include Gompers (1995), Gompers (1996), Gompers and Lerner (1996) and Lerner (1994). An interesting finding revealed in survey evidence (e.g. Freear and Wetzel, 1990) is that venture capital finance is not the dominant form of equity financing for start-up firms. Business angels and other investors invest considerably more in those firms than do venture capitalists. Also, venture capitalists seldom provide their advisory services at an early stage of a project's life. Typically investors which are involved at the start-up phase place funds in a venture without taking on the active involvement in the project that characterizes a venture capitalist.

Why is this occurring? Why do we not see venture capitalists providing finance and expert advice in the very first stages of a project's life? According to Lerner (1998), the literature on capital constraints,[1] provides evidence that the inability

[1] See Hubbard (1998).

to obtain external financing limits many forms of business investment. This could indicate that the "lack" of venture capital is simply due to a shortfall of venture capitalists. This argument became less valid in the 1990s, since the pool of venture capital funds grew dramatically, thus mitigating potential problems of constraints on capital provided through venture capitalists (Gompers, 1998). Secondly, the investment policy of venture capital firms might be inappropriate for starting firms, since venture capitalists prefer making substantial investments in their portfolio firms, rather than paying out small sums which typically are required in the initial stages of a firm's existence (Lerner, 1998). For example, Gompers (1995) reports that the mean venture investment in a start-up or early stage business between 1961 and 1992 was $2.0 million in 1996 dollars. The existence of a lower limit on investment size seems to be a consequence of the fact that institutions that invest in venture capital partnerships face monitoring costs that are to a large extent fixed for a target firm. Furthermore, governance and regulatory considerations lead institutions to limit the share of the fund that any one limited partner holds. This means that the minimum feasible size of an investment by a venture capital fund is substantial. Because each firm in the fund's portfolio must be closely scrutinized, the typical venture capitalist is normally responsible for no more than a dozen investments. Consequently, venture organizations are unwilling to invest in very young firms that require only small capital infusions.

Although this argument may explain why venture capital funds will not be involved in the early stages of a new venture it does not explain why a profit maximizing venture capitalist should not try to pick the best projects at an early stage of the projects life. By getting involved from the very beginning the venture capitalist would presumably be able to make the most of the project in terms of shareholder value. Thus, the question of why there seems to be a lack of venture capitalist involvement at the earlier stages of a company's growth phase is still largely unresolved.

In this study, we offer a simple explanation to this phenomenon. Our explanation is based on the assumption that a profit-maximizing entrepreneur selects the timing of his initial contact with a venture capitalist to maximize the value of his personal share of the total expected surplus of the project.

2. THE MODEL

The starting point of this paper is a new business opportunity that has been discovered by an entrepreneur. We employ a broad interpretation of what constitutes a business opportunity. Simply stated it is an idea of how to combine existing resources in order to produce something that can generate an economic surplus. Thus, we do not restrict attention to technological inventions which result from R&D or innovative cost reducing processes. The idea may simply be a new application of existing technologies, a product or service innovation, exploiting a location advantage, or introducing a new way of doing business.

At this early stage it is highly unlikely that the entrepreneur by acting alone would maximize the potential surplus from the business idea. Usually, there are specialized individuals and institutions, like venture capital firms, that possess strategic intelligence that could substantially enhance the value of the project. The first-best solution that would maximize social welfare would thus imply the immediate involvement of these outside "advisors". By providing the entrepreneur with know-how on important issues, contacts and reputation the size of the profits from the business idea would be maximized.

However, we argue that a profit maximizing entrepreneur deliberately may avoid approaching a venture capitalist in the very first stages of a projects life. This is because what matters for the entrepreneur is not the total surplus of the project but the value of the part he is able to secure for himself. When a venture capitalist is involved the returns of the project will be split between the venture capitalist and the entrepreneur. The outcome of this split will depend on the relative bargaining position of the two parties.

Formally, we assume that there are two types of agents in the model, entrepreneurs and the venture capitalists. The entrepreneur wants to maximize her wealth. The venture capitalists are the sole providers of advice. This follows from the cost advantage that venture capitalists as providers of a package consisting of advice as well as funding will have relative to firms that specialize in providing either financing or advice.[2] Competition will thus drive those consultants that cannot contribute to funding of the venture out the market.

We begin by considering the nature of the investment project. We assume that the entrepreneur has an idea which will yield a positive net present value (NPV). The magnitude of the NPV depend on the technical skill and effort of the entrepreneur, denoted by s, the level of advice, and other resources provided by the venture capitalist, denoted by a and the delay, t until contact is first established with the venture capitalist. Formally, let $\text{NPV} = f(a, s, t)$, where $(\partial f/\partial a) > 0$, $(\partial^2 f/\partial^2 a) < 0$, $(\partial f/\partial s) > 0$, and $(\partial^2 f/\partial^2 s) < 0$. We also assume that the NPV is a decreasing function of the delay, i.e. $(\partial f/\partial t) < 0$. This is due to the fact that imitation or competition will erode the NPV over time. By taking the project

[2] This is formalized in Repullo and Suarez (1998). Empirical research by Hellmann and Puri (2000) on a sample of U.S. start-up firms that have been assisted by venture capital firms indicate that there is substantial active involvement on behalf of the venture capital firm in the business development of the target firm.

to the advisor the entrepreneur is opening up a set of alternative actions not previously available to him. These alternatives can be regarded as real options. By delaying the contact the entrepreneur will reduce the time to maturity of those options and thus the value of the project. Regarding the role of skill s, we assume that a higher level of skill implies that the NPV will fall at a slower rate, i.e. $(\partial^2 f/\partial t \partial s) > 0$.

Next, we consider more carefully the role of advice a. In this study, we simplify by assuming that once having been engaged, the venture capitalist will provide the amount of advice that will maximize the total value of the project. This amount will naturally depend on when the initial contact with the venture capitalist is being established.[3] Therefore, we can write the NPV in the following way: $\text{NPV} = f(a^*(t), s, t) = f_{a=a^*}(s, t)$, where $a^*(t)$ is the value of a that maximizes the NPV at time t.

Finally, we consider the entrepreneur's maximization problem. What the entrepreneur is interested in is not the entire NPV but the share that he can secure for himself. When the venture capitalist has been engaged in the project, the entrepreneur has to give up a part of the total expected NPV to the venture capitalist. The fraction that the entrepreneur can secure for himself is thus of crucial importance. In cases where the business idea involves a patent or something else on which the entrepreneur has a specific property right the entrepreneur will probably be able to secure a substantial part of the surplus simply by allowing competing bids from different venture capitalists. However, if the entrepreneur lacks property rights to his business idea and he doesn't own any specific assets required for the project, the situation will be different. In the worst case, if the entrepreneur explains his idea to an efficient venture capitalist, the venture capitalist may simply reject this entrepreneur and put useful parts of the business idea to work with his established client firms.[4]

To avoid this moral hazard problem, the entrepreneur may invest in setting up a production unit, or in R&D aiming at a patentable version of the original idea. We thus assume that the investor is not strictly cash constrained.[5] The investment in contractibility will improve the bargaining position of the entrepreneur and make it more difficult for the venture capitalist to resort to opportunistic behavior. The entrepreneur will thus delay the initial contact with the venture capitalist in order

[3] This is simply Coase's (1960) theorem saying that the value maximizing solution will be chosen if the parties can bargain freely.

[4] This situation is also described in Kortum and Lerner (1998).

[5] A cash constraint that would limit the entrepreneur's option to continue on his own can be interpreted as a rising cost of capital that rapidly will reduce the NPV of the project if the entrepreneur continues on his own. With this interpretation a cash constraint fits well into the present model.

to invest in improving his possibilities to extract a larger share of the total profits of the project.

From the venture capitalist's point of view the important question is the expected value of joining forces with this entrepreneur compared to the alternative of exploiting the essence of the idea in collaboration with someone else. The longer the head start of the entrepreneur the more difficult it will be for the advisor to find a competing outlet for the idea, that would match joining forces with this entrepreneur. Thus, the longer the head start of the entrepreneur the better the deal that he can extract from the venture capitalist without a serious threat in the form of an attractive alternative for the venture capitalist.

The fraction that the entrepreneur can expect to secure for himself δ ($\delta \in [0, 1]$), at the point in time where the venture capitalist is first contacted will accordingly increase with the delay. Furthermore, we assume that the impact of the effort invested by the entrepreneur in improving his bargaining position will be subject to decreasing marginal returns. This follows if the entrepreneur chooses between available actions in the order of decreasing efficiency to improve contractibility. Consequently, the entrepreneur's share of the NPV as a function of the delay until he initially approaches the venture capitalist has the following properties: $(\partial\delta/\partial t) > 0$, $(\partial^2\delta/\partial t) < 0$, $\delta(0) = 0$, $\partial(\infty) = 1$.[6]

Graphically the entrepreneur's maximization problem is illustrated in Figure 1. The total value of the project decreases according to the dashed line in the upper diagram but the share that goes to the entrepreneur is maximized at t^*.

Formally, the maximization problem of the entrepreneur is the following:

$$\max_t \pi[\delta(t)\text{NPV}(s, t)] \tag{1}$$

The first-order condition for this maximum is:

$$\frac{\partial\pi}{\partial t} = \left(\frac{\partial\delta}{\partial t}\right)\text{NPV} + \delta\left(\frac{\partial\text{NPV}}{\partial t}\right) = 0 \tag{2}$$

Rearranging we arrive at the following equality:

$$\frac{\partial\ln(\text{NPV})}{\partial t} = -\frac{\partial\ln(\delta)}{\partial t} \tag{3}$$

The NPV of the project will decrease over time, but the share of the cash flow from the project that the entrepreneur can expect to receive increases over time. That there usually will be a value t^* which optimizes the value for the entrepreneur, an optimal delay, is seen as follows: The expected decrease in the NPV will

[6] Please note that δ may also depend on the skill of the entrepreneur. For the sake of simplicity, however, we have not taken these contracting skills into account.

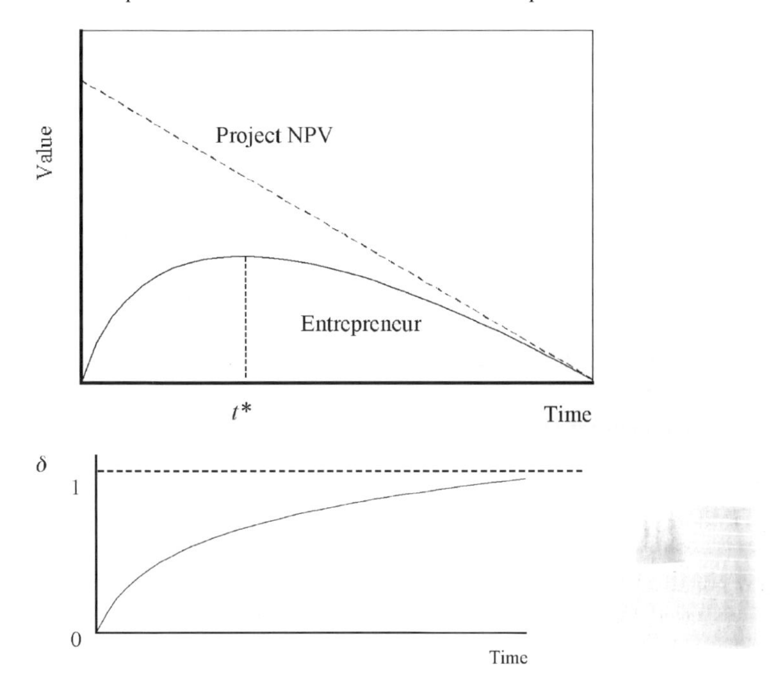

Figure 1. The value of the entrepreneur's part of the project given that the value of the NPV deteriorates according to the dashed line in the upper diagram and that the part he can secure for himself develops according to the lower diagram.

be approximately constant over time. Since we know that the rate of increase in contractibility will be relatively high in the beginning and converge towards zero as time goes on, there will be a value t^* for which the condition in Eq. (3) holds. However, there are cases where the rate of decay in NPV is very slow or non-existent, for instance if the entrepreneur is highly skilled. In these cases, the condition in Eq. (3) will not be satisfied for finite values of time t. This corresponds to the case where the optimal strategy for the entrepreneur is to never approach a venture capitalist.

Simple comparative statics using condition (3) reveal that the optimal waiting time to involve a venture capitalist will be negatively correlated with the rate at which the net present value of the project decreases and negatively correlated with the rate at which the contractibility of the project improves over time.

The above model can also be used assess the effect of an increase in skill on the optimal time for the entrepreneur to involve a venture capitalist. Totally differentiating expression (3) and rearranging we arrive at the following:

$$\frac{\mathrm{d}t}{\mathrm{d}s} = -\frac{(\partial\delta/\partial t)(\partial \mathrm{NPV}/\partial s) + \delta(\partial^2 \mathrm{NPV}/\partial t \partial s)}{2[(\partial s/\partial t)(\partial \mathrm{NPV}/\partial t)] + \mathrm{NPV}(\partial^2\delta/\partial^2 t) + \delta(\partial^2 \mathrm{NPV}/\partial^2 t)} \tag{4}$$

Clearly, the numerator of the expression is positive. The two first terms of the denominator are clearly negative. Regarding the last term of the denominator, it is not clear what the sign is, but we can safely say that it should be close to zero[7] and therefore dominated by the first two terms. The whole expression will thus be negative and therefore, we can conclude that an increase in entrepreneurial skill will make the entrepreneur wait longer to approach a venture capitalist.

The above result can be compared with the findings of Amit, Glosten and Muller (1990). The result of Amit et al. indicate that the most skilled entrepreneurs will develop their projects without the involvement of venture capitalists, whereas less skilled entrepreneurs will end up with venture capitalists' assistance. That result can be seen as a special case of the results of the analysis in this paper. If we assume a high enough entrepreneurial skill condition in Eq. (3) will not hold because it will be optimal for the entrepreneur to wait infinitely to approach a venture capitalist.

Based on this model we can also conclude that providing start-up financing should normally not constitute an attractive business for venture capital firms. The reason is the severe adverse selection problem that is implied by the model. The universe of entrepreneurs that will contact a VC is by no means restricted to entrepreneurs with positive NPV projects, like the ones we have assumed above. Any entrepreneur with a negative or zero NPV-project may find it profitable to go to a venture capitalist if he believes that the venture capitalist is willing to put up enough resources to more than cover the losses from the project. If entrepreneurs with positive NPV projects, following the logic above, tend to postpone their initial contact, negative NPV-projects will be overrepresented in the group of entrepreneurs which do not hesitate to make the initial contact immediately.

Another interesting implication concerns the design of publicly funded institutions tailored to support entrepreneurs at the start up stage of their business venture. Considerable care should be taken to design those institutions to convince potential clients that profitable ideas will not end up in the pockets of potential competitors. Strict employment conditions limiting outside activities and options of key employees, which hopefully provides for a gradual accumulation of reputation is probably the only available alternative for these institutions.

[7] Exactly zero with linear expected decay in the NPV.

3. CONCLUSIONS

This chapter provides an explanation for why we do not see more venture capitalists providing their services to client firms at an early stage of the firm's life. There are two cornerstones of our model. Firstly, the complementarity of financing and advice, which as pointed out by Repullo and Suarez (1999), provide the venture capitalist with an advantage compared to firms specialized in either or. Secondly, an entrepreneur is not interested in the total value of an investment project, but rather in the part that he can secure for himself. Being aware that a venture capitalist may find it more profitable to reject the project and transfer useful parts of the idea to other clients than to join forces with the entrepreneur, the profit-maximizing entrepreneur may find it in his best interests to wait and develop the project himself before involving the venture capitalist. By developing the idea without the involvement of a venture capitalist, the entrepreneur improves his bargaining position against the venture capitalist. This allows the entrepreneur to secure a larger share of the expected profits of the business idea once the venture capitalist is being involved. The optimal length of the delay will be determined by the point where the marginal increase in contractibility exactly balances the marginal deterioration in total value of the project due to competitive pressure. Using comparative statics we show that the delay in involving a venture capitalist will be longer the less rapidly the NPV deteriorates, the more skilled the entrepreneur is, and the slower the rate of increase in contractibility is.

The chapter also provides an explanation for why the market for a venture capitalist's services at the start-up stage is likely to fail. Since entrepreneurs with projects that have above average expected profitability will postpone their initial contact with the venture capitalist, entrepreneurs that do contact a venture capital firm at the start-up stage tend to have an expected profitability below average.

REFERENCES

Admati, A., & Pfleiderer, P. (1994). Robust financial contracting and the role of venture capitalists, *Journal of Finance*, *49*, 371–402.

Amit, R., Glosten, L., & Muller, E. (1990). Entrepreneurial ability, venture investments and risk sharing, *Management Science*, *10*, 1232–1245.

Anton, J., & Yao, D. (1994). Expropriation and inventions: Appropriable rents in the absence of property rights, *American Economic Review*, *84*, 190–209.

Barry, C. (1994). New directions in research on venture capital finance, *Financial Management*, *23*, 3–15.

Berglöf, E. (1994). A control theory of venture capital, *Journal of Law, Economics and Organization*, *10*, 247–267.

Chan, Y. (1983). On the positive role of financial intermediation in allocations of venture capital in a market with imperfect information, *Journal of Finance*, *38*, 1543–1561.

Coase, R. (1960). The problem of social cost, *Journal of Law and Economics*, *3*, 1–44.

Freear, J., & Wetzel, W. (1990). Who bankrolls high-tech entrepreneurs? *Journal of Business Venturing*, *5*, 77–89.

Gompers, P. (1995). Optimal investment, monitoring, and the staging of venture capital, *Journal of Finance*, *50*, 1461–1489.

Gompers, P. (1996). Grandstanding in the venture capital industry, *Journal of Financial Economics*, *42*, 133–156.

Gompers, P. (1998). Venture capital growing pains: Should the market diet? *Journal of Banking and Finance*, *22*, 1089–1104.

Gompers, P., & Lerner, J. (1996). The use of covenants: An empirical analysis of venture partnership agreements, *Journal of Law and Economics*, *34*, 463–498.

Hellmann, T., & Puri, M. (2000). Venture capital and the professionalization of start-up firms: Empirical evidence, *GSB Research Paper No. 1661*.

Hubbard, R. (1998). Capital-market imperfections and investment, *Journal of Economic Literature*, *36*, 193–225.

Kortum, S., & Lerner, J. (1998). Does venture capital spur innovation? NBER Working Paper 6846.

Lerner, J. (1994). Venture capitalists and the decision to go public, *Journal of Financial Economics*, *35*, 293–318.

Lerner, J. (1998). Angel financing and public policy: An overview, *Journal of Banking and Finance*, *22*, 773–783.

Repullo, R., & Suarez, J. (1998). Venture capital finance: A security design approach, Working Paper 9804, CEMFI.

Sahlman, W. (1990). The structure and governance of venture-capital organizations, *Journal of Financial Economics*, *27*, 473–521.

New Venture Investment: Choices and Consequences
A. Ginsberg and I. Hasan (editors)

Chapter 4

How Should Entrepreneurs Choose Their Investors?

DIMA LESHCHINSKII*

Department of Finance and Economics, Groupe HEC, 78351 Jouy-en-Josas, France

ABSTRACT

In this work, we describe what value different investors can bring to the entrepreneurial company apart from providing financing of the project. We consider such aspects of investors' activity as project screening, stage financing and investment into a portfolio of several projects. We analyze costs and benefits for the entrepreneur and compare investors to each other. The entrepreneur's choice of financing depends on the additional value to the project brought by the investors' abilities to resolve the uncertainties about the project and by the actions they can take, such as replacing the manager or cutting the investment.

1. INTRODUCTION

Some entrepreneurial companies get money from individual investors (angels), others get them from venture capitalists (VCs), third manage to get bank loans to start their operations. It is often thought that the sequence of angel investment, VC funding, and IPO with diffuse atomistic investors is a natural succession that a young company goes through in the course of its development. This is what we normally

* Corresponding author. Tel.: +33-1-39-67-94-10; fax: +33-1-39-67-70-85.
E-mail address: leshchinskii@hec.fr (D. Leshchinskii).

observe in practice, but we could ask ourselves—how was this sequence formed, what is the reason for this order? Do entrepreneurs get angel financing first, because they have no other choice, or is it because of other factors related to the angel financing? Why would VCs invest more into some industries than into others? Why do some companies become public relatively fats, while others remain private for a long time?

Sources of capital vary and we admit that their choice is often dictated by its cost and availability. However, these are not the only factors that entrepreneurs consider (or should consider) when looking for sources of funding, because investors bring to the company more than just the money. Investors' influence and specific investor-entrepreneur relationship can affect both the company value and the company future development.

Therefore, when choosing an investor for their projects, entrepreneurs should consider additional issues. Is entrepreneur confident enough in the project's quality or would he like to get an independent opinion? How much does he know about the market, its size and growth potential, about customers and competitors? Does the entrepreneur have knowledge and business experience of running a company? Is he planning to stay with the company for a very long time, or would he rather leave to pursue other agenda, once the project's results materialize? In other words, is this venture a possibly life-long commitment or is it just a speck in his career? What is more important for the manager—the project's success or the company's independence?

Evidently, a "serial" entrepreneur, who is starting his fourth or fifth venture, would answer these questions very differently, than a scientist, who decides to quit academia in order to commercialize her invention, or than a French actor, who wants to use his name to raise the money in order to buy a vineyard in Bordeaux and start selling wine.

In this work, we will discuss what value different investors can bring to the company. Among others, we will focus on such aspects as project screening, stage financing and investment into a portfolio of several projects.

For example, we will show, how project screening by business angels and VCs increases the project's value and makes further financing cheaper for the entrepreneur; how stage financing coupled with monitoring and possibility of managerial replacement makes VCs indispensable investors for some entrepreneurs and their ventures. We also discuss why entrepreneurs should take into account what other companies "their" VC has in his investment portfolio.

Different types of uncertainty can be efficiently resolved over different periods of time, thus determining the importance of specific information production technologies over the project's lifetime. For example, the technological uncertainty may be a very important factor at the beginning and very valuable information about the project's technological feasibility can be obtained from expert analysis of the project (project "screening"), whereas the market uncertainty becomes more important at the later stages. An investor, who has a superior information production technology about a source of uncertainty, crucial at the current phase of project development, should be the leading investor at this stage. This leads to the observed fact that a project's sources of finance evolve and change over the lifetime of the project.

An important characteristic of innovative projects is stage financing, whereas investments are made gradually, in stages. Gompers (1995) describes in detail the stage financing process that allows venture capitalists to gather information and monitor the progress of firms whilst maintaining the abandonment option. The paper tests empirically the agency and monitoring cost predictions on a sample of 794 venture capital-backed companies. VCs concentrate investments in early stage companies and high-tech industries, where information asymmetries are significant and monitoring is valuable. Gompers shows that an increase in asset tangibility (thus, lower uncertainty) increases financing duration (timing between different investment stages) and reduces monitoring intensity. Time between investment stages declines as the potential for future investment opportunities increases (higher market-to-book ratios) and in industries with higher R&D. Berk et al. (1998) develop a quite sophisticated and comprehensive model of a multi-stage investment project. The project's profitability is revealed to the firm over its lifetime because technological uncertainty about required R&D effort can be resolved only through additional investment, whereas other uncertainties remain unresolved until the last period. Continuation decisions are made conditional on the resolution of systematic as well as unsystematic uncertainty.

Portfolio approach is another mean to create additional value for portfolio firms. Bhattacharya and Chiesa (1995) study the interaction between financial decisions and the disclosure of interim research results to competing firms. Technological knowledge revealed to a firm's financier(s) need not also flow to its R&D and product market competitors. The authors show that the choice of financing source can serve as a precommitment device for pursuing ex-ante efficient strategies in knowledge-intensive environments.

Potential investors into entrepreneurial projects include parent companies, banks, independent corporate entities (strategic investors), business angels, VCs and traditional dispersed small (atomistic) investors. In the next section, we will talk about some of these categories in more detail. Since we consider entrepreneurial projects with a high degree of uncertainty about the required investment and about the payoff, we will focus mainly on business angels and VCs as the most suitable investors for these projects. We call them "active investors," because of their ongoing involvement with the project. Both types have access to information production technologies and have the experience necessary to screen the projects.

However, there exist important differences between VCs and business angels and these differences can play a key role in the entrepreneur's choice of investor. While business angels invest their own money, VCs act as managers of other people's money and require a fee for their professional advice. Due to cash constraints, angels typically invest in a single project, while VCs invest into a portfolio of several projects within the same industry. Another crucial difference is that only the VCs have the contractual right to replace the manager.

2. DIFFERENT TYPES OF INVESTORS—WHAT DISTINGUISHES THEM?

This paper mostly focuses on information production capabilities of different groups of investors. We can divide investors with abilities to produce information about the company into the following groups: existing parent company; banks; dispersed atomistic investors; business angels; venture capitalists (VCs) and strategic investors.

We can give the following characteristics to these groups:

2.1. Existing parent company (internal investment)

Staying with his own parent company, i.e. his employer, might seem to be an obvious choice as a source of financing for an employed manager-entrepreneur with a novel idea. However, corporate industrial investment accounted for only a tiny fraction of capital committed into new ventures. Hellmann (2002) shows that the entrepreneur may prefer to go to an independent venture capitalist, because corporate investors, including the entrepreneur's the parent company, may be unable to credibly commit ex ante not to shirk with support, not to exercise self-interested control or not to invest in a rival internal venture. The outcomes crucially depend on the extent of complementarity/supplementarity (cannibalization) of the proposed new project with the profits of the corporate investor's core business, because the new venture generates an externality to it.

Information asymmetry between the parent company and the manager is quite low, especially concerning existing projects and the manager's abilities. There is little danger that the asymmetry will increase in the future. The future information production costs depend on the nature of the new project and can be quite high if the new project lies outside the company's core competence. The company can exercise complete control over the project—shut it down, replace the manager, or inject additional investment. If the manager-entrepreneur sells his project to the company, then his compensation becomes part of the project's cost and is determined by the original contract. In this case the parent company gets all the benefits from the project and assumes all the costs. As we have already mentioned, the interests of the parent company and the entrepreneur do not necessary coincide.

2.2. Banks

Bankers are experienced investors, normally with a good knowledge of new business developments. Their expertise would be helpful to entrepreneurs, if bank were interested in financing entrepreneurial projects. However, due to the specifics of their business, sych as low limits to the allowed risk exposure and debt financing, normally they are reluctant to give credits for risky projects without a good collateral. They do give credits to wealthy entrepreneurs or finance some projects with tangible assets. Some banks also established venture capital divisions, which act relatively independently and should be considered as VCs.

2.3. Traditional dispersed investors

Traditional dispersed investors, like bond- and equity holders, have quite limited capabilities to monitor and influence the start-up company's activities. With some rare exceptions, they usually stay away from investment into entrepreneurial projects.

2.4. Business angels

Business angels play an extremely important role in financing new start-ups. The lack of a reliable data source makes it difficult to give precise figures, but according to some estimates reported by Hellmann (2002) angel capital provides from two to seven times the funding provided by "organized" venture capital. However, a typical individual angel's investment is smaller than that of a VC and varies in the range from $50,000 to $1 million.

In real life, business angels are heterogeneous investors, some of them are very sophisticated, while others are quite naive. However, in this chapter we consider the angels who are experts and professionals, thus called "professional angels". Information asymmetry between them and the manager-entrepreneur is quite low, because usually angels invest into the projects in their area of expertise and give money to people whom they know (Prowse, 1998). There is little danger that the asymmetry will increase over time.

Although active angels are usually very involved in the company's operations, they rarely use employment contracts that penalize poor performance. Their role is more advisory. Therefore, we can assume that they do not make strategic decisions themselves. For example, they cannot fire a manager-entrepreneur.

Usually angel investors do not have deep pockets to finance several projects at the same time. Also, the time they can spend with the companies is quite limited, because investment is not their full-time job. Normally, they are single-company investors.

2.5. Venture capitalists

Sahlman (1990) describes the role and structure of venture capital. Venture capital organizations raise money for investment in early stage businesses with high potential and high risks. Their forms of organization vary from corporations and captive subsidiaries of banks and corporations to small business investment corporations (SBIC) and limited partnerships, with the latter being the most popular form of venture capital organization. By VCs we mean the managing partners of these partnerships. VCs are usually contribute about 1% of total partnership funds raised and receive compensation—the management fee—equal to 2.5% of funds raised and a portion (about 20%) of carried interest. Partnerships have a finite life (10 years, extendable for another 3). Most VCs specialize in a particular stage of development and/or in a particular industry. Experience is a key resource that the venture capitalist brings to the table in helping to nurture a developing firm. Agency conflicts are usually remedied by having entrepreneurs being paid in the form of equity interest, which is proportionally larger for more successful performance ("earn-out"). Staged capital financing lessens potential asymmetric information moral hazard problems by creating an abandonment option and thus increasing the value of the project.

VCs and angels are involved in similar types of activities, namely, funding, monitoring, advising, and formulating business strategy. Enrlich et al. (1994) find that in comparison with business angels, VCs are more involved in the management of portfolio companies. They set higher performance standards, but give better feedback in return. The more formalized approach of VCs is often better for entrepreneurs with technical or scientific background and who have limited managerial experience.

Several papers study the involvement of venture capitalists in the strategic decisions of their portfolio companies. Gorman and Salman (1989) report that venture capital firms in their sample have an average portfolio of nine entrepreneurial companies. The venture capitalists spend half of their time monitoring the portfolio companies and typical respondents said that during their affiliation with a venture capital firm, they had replaced three CEOs in portfolio companies. From the venture capitalist's point of view, weak management is the dominant cause of start-up failure. This fact motivates our model assumption that manager quality is an important variable and that the potential to change the manager can be important.

From the VC's point of view, weak management is the dominant cause of start-up failure and decisions to replace CEOs in portfolio companies are not rare—Hannan et al. (1996) give the following attrition rates for companies' CEOs: in the first 20 months of a company's life, the likelihood that a non-founder is appointed as a CEO is 10%, after 40 months it is 40% and after 80 months it is 80%. Lerner (1995) finds that venture capitalists' involvement is more intense when the need for oversight is greater—venture capitalists' representation on the board increases around the time of CEO turnover, while the number of other outsiders remains constant.

Even if, at the initial stage, VCs may not possess all the information that is available to the entrepreneur, the high degree of monitoring of the company's progress and involvement in its activities make them well informed at later stages. Moreover, as we have mentioned, both business angels and venture capitalists bring to the company their own past experience and knowledge, thus producing information important to the company's success.

2.6. Strategic investors

Strategic investors are other companies, who invest capital into entrepreneurial projects. As investors, they have characteristics in many respects similar to VCs or angels. Nevertheless, in addition to purely financial interests, they also can have strategic interests in the entrepreneurial company. We have already discussed Hellmann's (2002) model, where he shows why entrepreneurs are sometimes

Table 1.

Investors	Precision of information	Cost of information	Level of control
Parent company	Same as the manager-entrepreneur	Same as the manager-entrepreneur	High
Banks	Low/medium	Low/medium	Low/medium
Atomistic investors	Usually low	Medium/high	Low/medium
Angels	Medium/high	Low/medium	Medium
VCs	Medium/high	Low/medium	High
Strategic investors	Medium/high	Low/medium	Medium/high

reluctant to get funding from strategic investors. In this paper, we will consider strategic investors as a subgroup of VCs.

To summarize, we characterize these groups according to the factors just described in Table 1.

3. PROJECT SCREENING

In this section, we show how investor's ability to screen the project increases its value.

Suppose that entrepreneurial projects can be either workable or not. If the project is not workable then its payoff is zero, if it is potentially workable then its payoff is positive provided that it would have received enough funding and market conditions would have turned out to be good. Let's call the workable type "W".

Entrepreneur and different investors observe signals about the project's type. These are the binary signals, with possible values of high ("H") or low ("L") and they can be observed before an actual investment is made. For simplicity, we assume that an "L" signal, indicating that the project is not workable, is perfect. That is, if the signal is "L", then its observer knows for sure that the project will never work. On the other hand, if the signal is "H", then there is still a (non-zero) probability that the project won't work.

It is natural to assume that more experienced and sophisticated investors, such as business angels and VC, who were actively involved with projects they financed in the past, have better screening capabilities, than less experienced ones and, sometimes, even the entrepreneur, if he does not have much business experience at the time when he starts his project. Suppose that if the entrepreneur or unsophisticated investors observe "H" (we denote the high signal observed by the entrepreneur as "H_E"), then this signal is "correct" with probability $p_E = \Pr\{W|H_E\}$. If a business angel or a VC, whom we call active investors, observe "H" (we denote the high signal observed by the angel or VC as "H_A"), then this signal is "correct" with probability $p_A = \Pr\{W|H_A\}$. Active investors get more precise signals than the inexperienced entrepreneurs do. Superior screening capability is characterized by $p_A > p_E$. Furthermore, we assume that the signals are nested, that is $\Pr(H_E|H_A) = 1$.

If the project requires investment K, then after the entrepreneur observes signal H_E, he knows that the project's expected net payoff is

$$E[\text{NPV}|H_E] = p_E\mu - K$$

where μ is the expected payoff of the workable project. This is the value of the project if financed by unsophisticated investors. Investment by active investors, such as angels or VCs, creates additional value.

Proposition. If the entrepreneur observes signal H_E, then by asking active investors to provide financing, the entrepreneur creates additional value

$$\Delta \text{NPV} = \frac{p_A - p_E}{p_A} K > 0;$$
$$0 \leq \Delta \text{NPV} \leq (1 - p_E)K.$$

Proof: Project financed by sophisticated investor has an NPV

$$E[\text{NPV}|H_A] = p_A \mu - K.$$

Before going to the sophisticated investor, the entrepreneur does not know, what signal the investor will observe. From the entrepreneur's point of view, the ex-ante probability of H_A is

$$\Pr(H_A|H_E) = \frac{\Pr(H_E|H_A)\Pr(H_A)}{\Pr(H_E)} = \frac{p_E}{p_A}$$

If the active investor observes "L", then the project will not get funding and its NPV is zero. Therefore, *even before* the entrepreneur knows what signal the active investor observes, the NPV becomes

$$E[\text{NPV}|H_E, \text{ investor } A \text{ invited}]$$
$$= \Pr(H_A|H_E)(p_A \mu - K) + \Pr(L_A|H_E) \times 0 = p_E \mu - \frac{p_E}{p_A} K$$

From which we get

$$\Delta \text{NPV} = \frac{p_A - p_E}{p_A} K$$

From $p_E < p_A$ we get the inequality sought. □

The net expected payoff increases because of decrease in the expected changes in investment and not because of the increase in the expected project payoff. All other things being equal the higher the precision of the information that the investor possesses the bigger the increase in the expected NPV is. However, this increase is bounded above by $(1 - p_E)K$. Notice also that absolute difference in signal precision does not matter as much as the relative difference. If $p_E = 0.95$ and $p_A = 1$, then the gain is only 5% of investment. However, if $p_E = 0.05$ and $p_A = 0.1$ then the gain is 50% of invested capital! Of course, with so low precision, the project's upside should be *very* high to get the funding in the first place!

To illustrate this proposition, consider the following example. Suppose that you are the entrepreneur and you have a project, which, *you think*, is workable. Its expected payoff is $\mu = \$8$ M and it requires $K = \$5$ M. However, you understand that you might be wrong (too optimistic) in your assessment of the project feasibility and that, in fact, there is only 50% chance that the project is *actually* workable. If you try to raise money from investors as unsophisticated as you, you will never succeed, because the project has a negative NPV: $0.5 \times 8 - 5 = -1 < 0$. If you decide to ask for funding from more experienced and sophisticated investors, for whom the probability that the project is good if they are thinking favorably about it is 0.9, then the NPV will increase by \$2.2 and it will become positive! Note, that this happens *before* the investor actually screens the project! After the screening, the expected NPV will be either \$2.2 M, following approval, or 0 the rejection.

On this example, we see that for some risky entrepreneurial projects active investors could be the only financiers, who would agree to finance them. Active investors have other means to increase the project's NPV and one of them is the stage financing.

4. STAGE FINANCING: PROJECT MONITORING AND MANAGERIAL REPLACEMENT

Let us have a look at how the mechanism of stage financing practiced by VC and some angel investors, increases the value of the project. Under the stage financing, investments into the project are made gradually, in stages, and the new round (stage) of financing depends on the outcome of the previous round and on the new information, available by this time. This contingency increases the project value adding to it the value of real option.

This value increase comes in at least three forms: (1) due to the diminished entrepreneur-manager's incentives to screw-up investors and misappropriate the disbursed funds; (2) through information revelation about the project value; and (3) through information revelation about the managerial capabilities of the entrepreneur coupled with possibility to replace him as a manager. In the last two, the stage financing works as a real option, adding the value of the abandonment option and the value of the manager replacement option to the project. Abandonment option of VC stage financing was studied in several papers, for example in Berk et al. (1998), Bergemann and Hege (2002) and Gompers (1995). The value of managerial replacement option is studied in Brisley and Leshchinskii (2002). In this section, we will report a brief summary of their results.

Stage financing is not a VC invention. For example, it has been practiced to finance construction projects for long time. It is used in construction industry to put a tight grip on how contractors spend the money, thus allowing investors to avoid cost overruns and embezzlement of funds by contractors. With some exceptions, new information is not important in this case and it would not play a significant role in the project continuation decision.

For projects of risky and innovative nature, which VC investors are dealing with, new information arrival is very important and can be crucial for the project continuation/abandonment decision. Stage financing is one of the mechanisms, which allows VCs and angels to add value to the firms they invest in. Nevertheless, without investor having received new information (observing a signal) before proceeding to the new round of investment, the stage financing by itself simply would not work. Generally, the higher the precision of the interim information, the higher the value of stage financing. The first-best decision rule remains the same—all projects with positive expected NPV should get financing. However, as in the screening example, the stage financing coupled with observing signal Ψ about the project payoff V at the first stage increases the project's ex-ante NPV. This increase does not come from the increase in the expected project's payoff—before the signal is observed, the investor's expectation about the future payoff remains the same—but incomes from the investment side—the active investor expects to continue financing of only good projects.

Suppose that after making initial investment K_0 the active investors get signal Ψ about the project's payoff V. There is no direct cost to observe the signal, but it can be observed only if K_0 has been invested in the project at stage 1. Without possibility to observe the signal the project should be undertaken iff $\mu - K > 0$, where $\mu = E[V]$ is the expected project payoff and K is the total investment into the project. If the investor knows that he will observe signal Ψ after K_0 was invested at stage 1 and before making the investment decision for stage 2, the expected ex-ante NPV at the *beginning* of stage 1 becomes

$$E_\Psi\{\Pr(\mu_\Psi > K - K_0) \times [E\{V|\mu_\Psi > K - K_0\}] - K_1\} - K_0 > \mu - K$$

where $\mu_\Psi = E[V|\Psi]$.

This expression takes into account the fact that after observing Ψ the investor continues financing for all projects with the positive NPV positive at the second stage, including some projects for which

$$K - K_0 < \mu_\Psi < K_0$$

Clearly, if Ψ were observed at the very beginning, these projects would not be undertaken, but since investment K_0 has already been made and it has become the sunk cost, the financing will continue at stage 2 in order to minimize the losses. This helps us to understand that K_0 is not equivalent to the signal cost. The cost of stage financing is the sum of two components—investment of K_0 into bad projects and continuation of the projects, which would not be undertaken, if Ψ were observed at the very beginning.

Notice that all projects undertaken with one-stage financing, would be undertaken with the two-stage financing as well (absent agency problems, the value of an abandonment option cannot be negative).

Let us illustrate on an example the value of information learned during the first stage and the expected cost of learning.

Example. Consider the project, which requires the total investment, $K = \$8$ M. Ex-ante, the distribution of payoffs is the following:

$$V = \begin{cases} 17, & \text{with probability } 0.25 \\ 7, & \text{with probability } 0.25 \\ 0, & \text{with probability } 0.50 \end{cases}$$

$\mu = \$6$M and by straightforward calculation, the single stage financed project has the expected NPV $= \$ - 2$M. The project will not get financing. However, if after investing $K_0 = \$2$M the active investor learns the exact project payoff, then the *ex-ante* expected NPV becomes $E(\text{NPV}) = 0.25(7-6) + 0.25(17-6) - 2 = \$1\text{M} > 0$ and the project will be undertaken. Notice, that if the true value observed before the second stage financing is $V = \$7$M, then the project will be continued, because $K_0 = \$2$M will have become a sunk cost and should be disregarded at the time of the project continuation decision.

The cost of stage financing is \$1.25M. It is equal to the difference between \$2.25M (the project NPV given the perfect information available before financing starts) and the expected project NPV of \$1 M under the stage financing. We can consider this cost as the sum of expected wasted investment of \$1M (with probability 0.5 there will be no second-stage investment) and expected continuation if the realized value is \$7M, which gives the ex-ante negative NPV of −\$0.25M.

Learning the project value has the following empirical implications for stage financing: (1) the project abandonment rate should be higher around the time when new information is produced; and (2) at the more advanced stages of financing one should observe higher share of projects that are getting funding despite the fact that they are loss-making from the ex-ante perspective.

We did not say much about the nature of signal Ψ about the project's payoff. These signals are coming from new developments in the industry, events, happening inside the rival companies, changes with the consumer base etc. Since active investors are the industry insiders, they learn these signals faster and interpret them better, than other investors. We may say that although there is not much information asymmetry between investors and companies they invest in, there might exist information asymmetry between different groups of investors. For outsiders it is not easy to analyze new industries with only few existing public companies or no public companies at all.

Putting right people into these companies is even harder. This is another source of value increasing from the stage financing—learning the manager's type and possibility to replace him.

4.1. Learning the manager's type

Quite often entrepreneurs, who start innovative companies, do not have past managerial experience, especially those of them, who are good scientists or engineers willing to start their new business. At the same time, majority of VCs would say that a good managerial team is at least as important for the success of a new venture as the entrepreneurial idea itself. From the VC's point of view, weak management is the dominant cause of start-up failure. Even with a good idea, the team quality determines the time and amount of resources spent to achieve the goal. Therefore, having a good manager is indispensable to guarantee an ultimate success of the venture.

One solution would be to hire an experienced management team to run the entrepreneurial company from the onset. However, despite its obvious attractiveness, this solution has several drawbacks. First, a good manager is normally very expensive and hiring him to run the company without knowing, whether the business idea itself works or not, could be a waste of money and scarce human capital. Second, at the early stage of company development, scientific expertise of the entrepreneur could be crucially important and sidestepping him might do more harm than good. Finally, the entrepreneur might turn out to be a good manager himself, and his replacement would become unnecessary ex-post.

Stage financing combined with VC's access to the pool of good managers creates a real option to replace the manager and thus increases the project value. Decisions to replace CEOs in portfolio companies are not rare—Hannan et al. (1996) give the following attrition rates for companies' CEOs: in the first 20 months of a company's life, the likelihood that a non-founder is appointed as a CEO is 10%, after 40 months it is 40% and after 80 months it is 80%. Lerner (1995) finds that venture capitalists' involvement is more intense when the need for oversight is greater—venture capitalists' representation on the board increases around the time of CEO turnover, while the number of other outsiders remains constant.

Brisley and Leshchinskii (2002) employ the following model. An entrepreneur is seeking investment for his project. If the project gets sufficient investment, then its payoff is a random variable V with mean μ (we have discussed the role of stage financing in observing the signal about payoff already). Investment required for the project success is a random variable and its value depends on the managerial quality of the entrepreneur. He can be a good manager with ex-ante probability q, or a bad manager with ex-ante probability $(1 - q)$. If he is a good manager, then the sufficient investment is I. If he is a good manager, then the sufficient investment is αI, where α characterizes improvement that the good manager can achieve, $\alpha < 1$. Nobody, including the entrepreneur himself, knows ex-ante his managerial quality.

VC can bring a replacement manager M, who is always good, but hiring him is costly. The replacement cost is proportional to the investment K_M that manager M oversees and is equal to βK_M. Unlike VCs, angel investors cannot hire the replacement manager. This difference is consistent with empirical observations by Prowse (1998), Enrlich et al. (1994) and Hannan, Burton and Barron (1996). One of possible explanations to this difference is that VCs know good managers from the past experience, while angel investors do not have neither this vast experience nor the time to find the replacement.

Suppose that the first stage investment is K_0 and the second stage investment is K_1. If K_0 is managed by a bad manager and the rest by a good manager, then the total required investment, including the replacement cost is $\alpha(1 + \beta)(I - K_0) + K_0$, where the part managed by the good manager is $K_1 = \alpha(I - K_0)$

The model assumes that active investors, both VC and angel investors, might learn the managerial quality of the entrepreneur. The probability of learning the entrepreneur's type is an increasing function of the capital managed by the entrepreneur, $f(K_0)$. Thus, the initial investment into the project can potentially reveal the manager's type before the final investment is made. If the entrepreneur is a bad manager, the VC can replace (or sideline) the manager-entrepreneur. If such a development takes place, then the new manager, who is known to be good, is brought in to help the entrepreneur to manage the project. Hiring the new manager saves investor's money on the remaining part of investment, but the new manager requires an extra fee to do the job.

Of course, it might happen that after the first stage the managerial quality of entrepreneur still remains unknown. Then, based upon the updated information about the project payoff that we have discussed before, the VC has to make a decision, whether to replace the manager or keep the incumbent entrepreneur as a manager or abandon the project. For the angel investor the choice is narrower: whether to continue with the incumbent or to abandon the project.

Notice that if for one reason or another the managerial replacement is never optimal, then the angel investor creates as much value as the VC. If at $t = 1$ there is new information at all, then atomistic investors with one-time investment will give financing terms as good as active investors, provided that we put disciplining and other considerations aside.

To illustrate advantages of stage financing for the managerial replacement, consider first a one-time investment scheme, when all capital is invested at $t = 0$ and no further investment is done.

If no replacement is made and the manager-entrepreneur manages the total investment K, then the optimal investment is the solution to the optimization problem

$$\max E\{\text{NPV}\} = \max\{q\mu - \alpha I, \mu - I, 0\}$$

giving

$$\max E\{\text{NPV}\} = \begin{cases} 0 \quad \text{if } \dfrac{I}{\mu} > \max\left(1, \dfrac{q}{\alpha}\right), \\ q\mu - \alpha I \quad \text{if } \dfrac{1-q}{1-\alpha} < \dfrac{I}{\mu} < \dfrac{q}{\alpha}, \\ \mu - I \quad \text{if } \dfrac{I}{\mu} < \min\left(1; \dfrac{1-q}{1-\alpha}\right), \end{cases} \qquad (*)$$

The upper line in $(*)$ corresponds to no investment at all, the second line—to taking the risk and investing αI in anticipation that the manager-entrepreneur is good, and finally, the third line corresponds to playing safe and investing I. In the latter case, with probability q the investor overspends $(1 - \alpha)I$.

If the VC investor decides to replace the manager-entrepreneur by the good manager immediately, then

$$\max E\{\text{NPV}\} = \mu - \alpha(1+\beta)I$$

Replacement is never optimal for $\alpha(1 + \beta) > 1$. If $\alpha(1 + \beta) < 1$ holds, then the replacement is optimal if $I/\mu < min(1; 1 - q/1 - \alpha)$ or $1 - q/1 - \alpha < I/\mu < 1 - q/\alpha\beta$. With probability q the investor is overspending $\alpha\beta I$ by replacing the good manager and with probability $(1 - q)$ he is saving $(1 - \alpha(1 + \beta))I$.

Now we consider the case of stage financing with two investment rounds. Assume that, at $t = 0$, the investor can invest K_0 and at $t = 1$ he observes signal Ψ about the project payoff V and learns with probability $f(K_0)$ the manager's type. At that time, the replacement decision and the further investment decision should be made. If K_1 is invested, it is managed either by the incumbent manager-entrepreneur or by the new manager.

Remember, that if the bad incumbent is replaced by the good manager, then the money will be saved only on the remaining part of investment, K_1. So, at the first stage, the investor is facing a trade-off—if he invests more money at the first round, then he will increase the probability of learning the manager's type, but there won't be much left to save. If K_0 is small, then a lot can be saved in the future, but the manager's type might not be revealed. If a housing contractor builds a foundation, it's hard to judge the contractor's quality, but after all floors are finished and the homeowner understands that the contractor is bad, how much can be saved on the roof-work?

At $t = 1$, the choice of decision for each of three possible observations of the incumbent manager's type (we can observe that he is good, bad, or his type remains unknown) is determined by investment K_0, replacement cost β and observed signal Ψ about the project payoff. To find optimal K_0 that maximizes the expected NPV at $t = 0$ under the two-stage investment scheme, we have to solve the problem backwards by starting from the decision choice at the second stage.

If at $t = 1$ the manager-entrepreneur turns out to be a good manager, then the investor invests more into the project if $K_0 < \alpha I$ and $\mu_\Psi - (\alpha I - K_0) > 0$. In this case, the amount invested at the second stage is $K_1 = \alpha I - K_0$ and the project NPV becomes $E\{\text{NPV}\} = \mu_\Psi - \alpha I$. Since K_0 is the sunk cost at $t = 1$, the total expected NPV can be negative, that is, we can have $\mu_\Psi - (\alpha I - K_0) > 0$, but $\mu_\Psi - \alpha I < 0$.

The situation with the bad manager is depicted on Figure 1.

If after observing signal Ψ, the expected value of the project's payoff, $\mu_\Psi = E\{V|\Psi\} > I$, then for $\alpha(1 + \beta) < 1$ the replacement is always optimal (a horizontal line in the upper part of Figure 1 is the expected NPV without replacement and

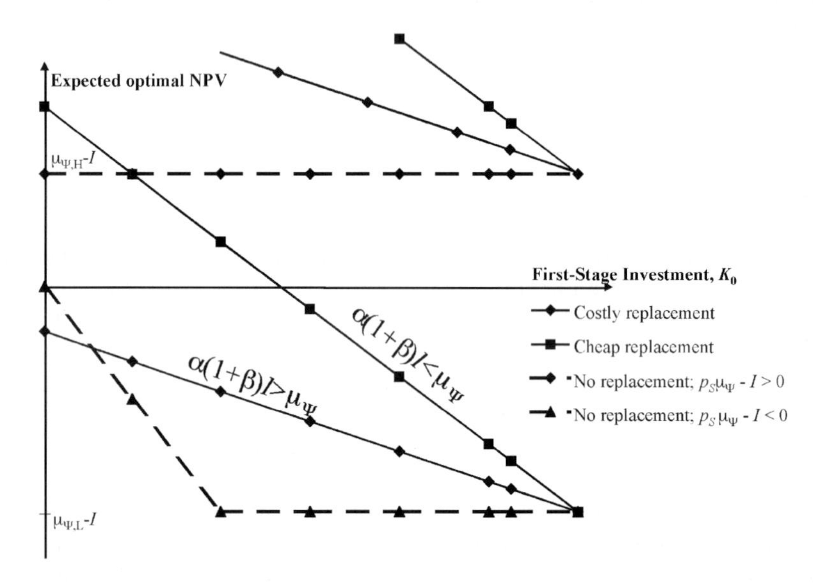

Figure 1. Net payoff when the manager's type is revealed as bad at $t = 1$. *Note*: Solid lines show the net payoff when the bad manager is replaced at $t = 1$. Dashed lines show the net payoff without replacement. Replacement is optimal, when the solid line is above the dashed line corresponding to the same signal. The upper dashed line corresponds to the positive net payoff (Ψ_H is high, $\mu_{\Psi,H}-I > 0$), if no replacement is made. In this case the replacement is optimal whenever $\alpha(1 + \beta) < 1$. The lower dashed line corresponds to the negative net payoff (Ψ_L is low, $\mu_{\Psi,L}-I < 0$). Replacement is always optimal, if $\alpha(1 + \beta) < (E\{V|\Psi\}/I - K_0) < 1$. If $K_0 < I-(E\{V|\Psi\}/\alpha(1 + \beta))$ then it is optimal to abandon the project.

negatively sloped lines coming from its right end are the NPVs with replacement for different replacement costs).

If $\mu_\Psi < I$, then the project has a negative NPV with the incumbent manager and everything depends on how expensive the replacement is, and on how much has been already spend. It turns out that for $\mu_\Psi - \alpha(1-\beta)I > 0$ the replacement is always optimal (a steeper negatively sloped line). If K_0 was small enough, then the project's NVP will become positive, but even for the negative NPV project, the new manager should be hired to minimize the total losses. For $\mu_\Psi - \alpha(1+\beta)I < 0$, the NPV is always negative and the replacement is optimal only if $K_0 > I - \mu_\Psi/\alpha(1+\beta)$. Otherwise, it is better to abandon the project, because savings from replacement will not justify the cost of continuing the project.

Before the project starts, it can be said that for $\alpha(1+\beta) < 1$ the bad manager is most likely to be replaced, unless the project's payoff will look so bad at $t = 1$ that it won't be worth continuing it at all.

If after investing K_0 the manager-entrepreneur's type remains unknown, then we have two possibilities. The first one is where $\mu_\Psi - I > q\mu_\Psi - \alpha I$. In this case, the situation is the same as in the case with the bad manager, which has been considered before. Therefore, for $\mu_\Psi - I > 0$, whenever $\alpha(1+\beta) < 1$, the manager of unknown quality should be replaced and investment $K_1 = \alpha(I - K_0)$ should be managed by new manager M. In order words, even when the project remains profitable, if run by a bad manager, it's better to replace the unknown incumbent, if the probability that he is bad is too high. For $\mu_\Psi - I < 0$, if $\mu_\Psi - \alpha(1+\beta)I > 0$, then the replacement is always optimal. If K_0 was small enough, then the project's NVP will become positive, but even for the negative NPV project the new manager will minimize the total losses. For $\mu_\Psi - \alpha(1+\beta)I < 0$, the NPV is always negative and the replacement is optimal only if $K_0 > I - \mu_\Psi/\alpha(1+\beta)$. Otherwise, it is better to abandon the project, because savings from replacement will not justify the cost of continuing the project. Before the first stage investment, the sign of inequality $\mu_\Psi - I >< q\mu_\Psi - \alpha I$ is known, is not known whether the manager is good or not.

The second possibility is that $q\mu_\Psi - \alpha I > \mu_\Psi - I$. Now the situation becomes more complicated and it is illustrated on Figure 2. This picture shows the expected NPV if the manager's type remains unknown at $t = 1$. Solid lines correspond to the managerial replacement at $t = 1$, and dashed lines correspond to the NPV of projects, in which the incumbent manager remains in control, including the situation when the project is discontinued, which corresponds to the negatively sloped segments of these lines. Cases, where the replacement is optimal, correspond to the areas, where the solid line is above the dashed line.

"Big quality gain" means the projects, for which $q\mu_\Psi - \alpha I > \max(0, \mu_\Psi - I)$ (the dashed line with diamond dots). For these projects, if the replacement were

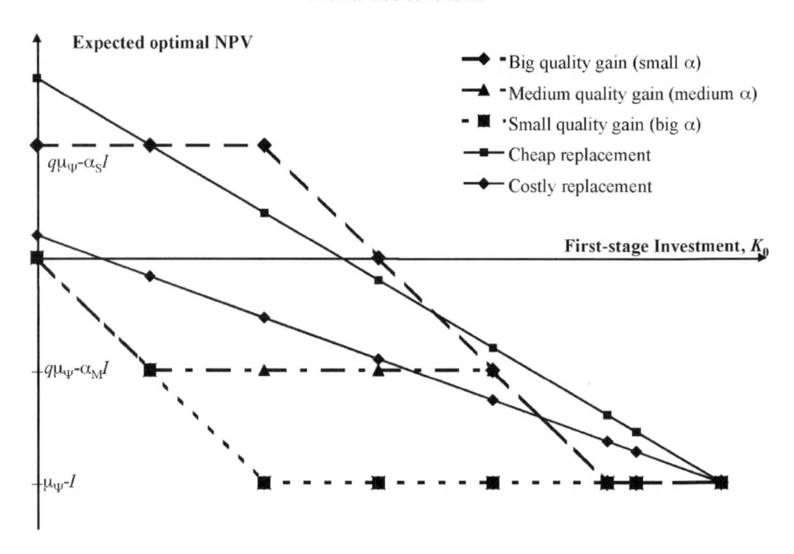

Figure 2. Net payoff when the manager's type remains unknown at $t = 1$. *Note*: Solid lines show net payoffs when the manager is replaced at $t = 1$. Dashed lines show the net payoff without replacement. Segments of dashed lines with negative slopes correspond to the cases when no second-stage investment, K_1, should be made under the incumbent manager. Replacement is optimal when the solid line is above the dashed line. By "big quality gain" we denote projects, for which $q\mu_\Psi - \alpha I > \max(0, \mu_\Psi - I)$. The "medium quality gain" projects have $q\mu_\Psi/I < \alpha < 1 - (1-q)\mu_\Psi/I$. Finally, the "small quality gain" projects are the ones with $1 - ((1-q)\mu_\Psi)/I < \alpha < 1$. If replacement were not possible, the high quality gain projects (the dashed line with diamond dots) should have got investment $K_1 = \alpha I - K_0$, unless $\alpha I < K_0$ (the horizontal section of the line). The medium gain projects (the dashed line with triangular dots) should get investment $K_1 = \alpha I - K_0$, if $\alpha I - q\mu_\Psi < K_0 < \alpha I$ (the horizontal section in the middle of the line), and investment $K_1 = I - K_0$, if $K_0 > I - (1-q)\mu_\Psi$ (the horizontal section at the right-hand end. For the small gain projects (the dashed line with starred dots), investment $K_1 = I - K_0$ should be made if $K_0 > I - \mu_\Psi$.

not possible, then they should have got investment $K_1 = \alpha I - K_0$ (the horizontal section of the line at the left end), unless the initial investment was too big, $K_0 > \alpha I$. With the VC investor, the replacement is possible, because we are still considering the situation, where $\alpha(1+\beta) < 1$. If the replacement is very cheap, meaning that $1 + \beta < (1-q)\mu_\Psi/\alpha(1-\alpha)I$, then the replacement of the incumbent manager of unknown quality is *always* optimal (this case is not shown on Figure 2). Notice, that this does not mean that the stage financing is useless, because we are talking about replacement at the second stage. By this time two events will happen: first, with probability $f(K_0)$ the investor will learn the exact type of the manager and, second, he will observe the signal Ψ about the project payoff. The higher the value of the signal, the "cheaper" the replacement is.

If the replacement is relatively cheap, i.e. $(1-q)\mu_\Psi/\alpha(1-\alpha)I - 1 < \beta < (1-q)\mu_\Psi/\alpha I$ (the solid line with square dots on Figure 2), then the replacement is optimal either where K_0 is very small, $K_0 < (1-q)\mu_\Psi - \alpha\beta I/1 - \alpha(1+\beta)$, or where it is very big, $I - (1-q)\mu_\Psi/\alpha(1+\beta) < K_0 < I$. For K_0 relatively close to αI, we observe the region, where the replacement is not optimal. Even for $K_0 < \alpha I$ it is better to take the chance with the incumbent manager, because the cost of the risk, $(1-q)\mu_\Psi$ is smaller than the replacement cost $\alpha\beta(I - K_0)$.

Finally, the relatively costly replacement with $(1-q)\mu_\Psi/\alpha I < \beta < 1/\alpha - 1$ (the solid line with diamond dots on Figure 2) is optimal only if K_0 is very big, $I - (1-q)\mu_\Psi/\alpha(1+\beta) < K_0 < I$.

Empirically, for "high quality gain" projects, we will observe more managerial replacements for projects with higher payoff.

The "medium quality gain" projects (the dashed line with triangular dots on Figure 2) have $q\mu_\Psi/I < \alpha < 1 - (1-q)\mu_\Psi/I$. These projects have the negative NPV with the incumbent manager and they should get loss-minimizing investment $K_1 = \alpha I - K_0$, if $\alpha I - q\mu_\Psi < K_0 < \alpha I$ (the horizontal section in the middle of the line), and investment $K_1 = I - K_0$, if $K_0 > I - (1-q)\mu_\Psi$ (the horizontal section at right). Following the logic of "high quality gain" projects one could think that in loss-minimizing scenarios, the managerial replacements should be observed less frequently. However, the replacement rules remain the same as in the "high quality gain" case.

Finally, the "small quality gain" projects (the dashed line with starred dots on Figure 2) are the ones with $1 - (1-q)\mu_\Psi/I < \alpha < 1$. For these projects, without replacing the incumbent manager, investment $K_1 = I - K_0$ should be made if $K_0 > I - \mu_\Psi$. With VC investors the replacement is possible. Given $\alpha(1+\beta) < 1$, the replacement is always optimal for $\mu_\Psi - \alpha(1+\beta)I > 0$. This inequality does not guarantee that the project's NPV will be positive, but it guarantees the optimality of replacement. For smaller values of μ_Ψ, with $\mu_\Psi - \alpha(1+\beta)I < 0$, the replacement is optimal only if $K_0 > I - \mu_\Psi/\alpha(1+\beta)$, i.e. if considerable initial investment has been made. Otherwise, it is better to abandon the project.

If $\alpha(1+\beta) > 1$, then the entrepreneur's replacement as the manager is never optimal. This means that VC investors do not have advantage in comparison with angel investors. When the replacement cost is too high or when a good manager won't make significant savings in comparison with a bad manager, angel investors are as good as VCs. At the second stage, the project should be either abandoned or continued with the incumbent manager-entrepreneur. Since this inequality does not depend on μ_Ψ, the fact that no replacement will be possible is known at time $t = 0$, but, nevertheless, the possibility to learn the manager's type and to get better information about the project's payoff increases the NPV of stage financing in comparison with the one-time investment. At time $t = 1$, the expected NPV is

$$E\{\text{NPV}\} = \begin{cases} -K_0 & \text{if } \dfrac{I - K_0}{\mu} > \max\left(1, \dfrac{q}{\alpha}\right) \\ q\mu - \alpha I & \text{if } \dfrac{I - q}{1-\alpha} < \dfrac{I - K_0}{\mu} < \dfrac{q}{\alpha}, \\ \mu - I & \text{if } \dfrac{I - K_0}{\mu} < \min\left(1; \dfrac{1 - q}{1-\alpha}\right), \end{cases}$$

If K_0 were equal to zero, then this expression would be identical to Expression (*) for the single-stage investment without the replacement. However, with $K_0 > 0$, the manager's type will remain unknown to the active investor only with probability $(1 - f(K_0))$. With probability $qf(K_0)$ it will be revealed as the good type and with probability $(1 - q)f(K_0)$ the entrepreneur's type will be observed as the bad one.

In order to find optimal investment K_0, which maximizes the expected NPV, specific assumptions about the functional form of $f(\cdot)$ and distribution of Ψ is necessary. Brisley and Leshchinskii (2002) find solutions for K_0 for some particular examples of $f(K_0)$ and Ψ, which we won't study here.

Here, for illustrative purposes, we were looking at the two-stage investment only. In reality, we often observe stage financing with more than two investment rounds. This happens because information keeps arriving and its better to make new financing decisions contingent on this new information. Berk et al. (2003) and Bergemann and Hege (1998, 2002) study multi-stage project financing. One of the results that Bergemann and Hege (2002) get is that as the project progresses over time, the amount invested at each stage becomes bigger and bigger and the time period between stages becomes longer and longer. We can conjecture that at some point the frequency of information arrival diminishes to the rate, at which stage financing is not advantageous anymore. This is the time for the investor's optimal exit. However, is the investor's exit decision always optimal from the entrepreneur's point of view?

5. INVESTOR'S EXITS

Most common ways of private investor's exit from investment in a relatively successful entrepreneurial company are (1) an initial public offering (IPO) of the company; (2) its trade sale trough a merger with or acquisition by another company (M&A); (3) buyout, when the entrepreneur or the manager or another private investor buy the company. If the company is not successful, we have to add a write-off to this list.

When making investment, investors normally anticipate the investment horizon, i.e. when and how they will exit from the investment project. For example, for bank loans this is normally the maturity of investment, for business angels this could be a long-term investment—if the company is performing well and becomes publicly traded, angel investors can keep their shares for quite an extended period. In contrast, VC investors have shorter investment horizon of 3 to 7 years, which is dictated by the vintage of the fund.

Shorter investment horizon of the VC investor can be a reason for the decision suboptimal to the firm value, such as a premature IPO or an underpriced trade sale of the company. One has to remember that an IPO is not an ultimate goal of the VC investor. It is simply a mean of achieving the goal of successful exit from the venture.

Of course, the track record of past glamorous IPOs by a VC brings with aura of success more business in the future. Nevertheless, this consideration does not play a role as important as the monetary return on investment. And with this respect an IPO has no advantage for the VC in comparison with a trade sale (M&A) or a buyout.

Moreover, quite often an IPO does not provide an immediate exit to the private investor. A lock-up agreement would require him to retain a significant amount (about 80%) if not the entire stake for a quite extended period of time from 6 months to 2 years (as in the case of Hawks Industries IPO). When market conditions are favorable, the IPO underwriter would allow investors to sell more shares and exit earlier, but this does not happen very often. Therefore, 2- and 3-digit rates of return at the time of IPO do not give a realistic picture of investment profitability, because at the end of the lock-up period the numbers look much more somber. After summer 2000, IPOs have become an even more problematic way of exit. For more discussion on bias in IPO returns, see Cohrane (2001).

That's why M&A is at least equally preferred exit for VCs. According to the National Venture Capital Association (NVCA) data, in 2001, 305 VC-backed M&As took place with a reported total value of $14.75 billion, vs. $3.23 billion raised in VC-backed IPOs. For example, four out of five exits by 3i, one of the biggest English VC, are done through M&As.

It should be mentioned that trade sales also sometimes include restrictions, such as warranties or deferred consideration. Nevertheless, trade sales are less restrictive than IPOs in terms of VC's quick access to cash and they have less uncertainty than an IPO.

If a trade sale generates more cash and it is faster than an IPO, why should the entrepreneur worry about this way of exit by investors? There are two main reasons for the entrepreneur possible concern: (1) the loss of company independence; and (2) the possible suboptimality of timing. Since at the end of a venture fund life, the VC has a lot of pressure to exit from investment, he is more willing to sell an undervalued company at a depressed price, rather than continue the project and wait for a few more years for better results or more favorable market conditions.

We know from empirical literature, e.g. from Kaplan and Stromberg (2002) that often VC investors contractually have control rights of underperforming companies. According to interviews with some VCs, when making the exit decision, they normally try to avoid direct confrontation with the company founders trying to be more persuasive instead. Nevertheless, usually they have enough arguments (and power) to convince entrepreneurs to follow their course of action, see, e.g. Hellmann and Puri (2000).

To illustrate the point, we will consider two hypothetical examples based on real-life situations. Suppose that after three years of receiving VC financing, an entrepreneur has developed a new drug, which still has to go through a series of tests in order to find its application. Potentially, it can cure diseases α or β or both, but maybe it will fail. Company A is willing to buy the entrepreneur's company, because of its potential to fight disease α. This firm does not care much about disease β, because this is not their area of expertise. Buying the entrepreneur's firm is risky and company A would buy it at a discount. IPO is not an option—the firm does not have a product yet. Optimally, the firm should wait until the results are ready and then in the case of at least partial success either do an IPO or sell to a the highest bidder. Would an investor agree to wait another year, if the venture fund is approaching maturity and an immediate exit though the trade sale is more attractive for the VC?

The second example is the following: a biologist came up with an idea of using some screening procedure as a diagnostic tool for a group of diseases. Relatively easy he got funding from a venture capitalist and surprisingly quickly found few diseases for which this procedure worked and developed a commercial version of the product. This discovery gave him new ideas for which other diseases this procedure might work. It also led him to some ideas about the nature of these diseases and their potential treatment. The scientist admits himself that these new ideas are quite risky (speculative) and that it would take much more time and a lot of money to test them. Of course, if undertaken, this work would bring to the scientist worldwide recognition and praise from other academics. Unfortunately for the scientist, his investor refuses to follow this route and he is pushing the entrepreneur for an IPO with existing product.

These two examples show that entrepreneur's and investors' interests might diverge over time. The second example shows that sometimes it is the investor, rather than the entrepreneur, who cares more about the company value.

In order to avoid unpleasant surprises in the future, at the very beginning the entrepreneur have to look into possible scenarios of events in anticipation of different outcomes and make a contingency planning. He should be aware of the difference of his and investor's interests. Sometimes investor's objectives are in conflict with entrepreneur's interests, but sometimes they act to the entrepreneur's benefit.

One of the sources of such difference of interests is the portfolio approach adopted by venture capitalists.

6. PORTFOLIO APPROACH

Normally, VCs apply portfolio approach to their investment. Instead of investing money into one company, they finance a portfolio of companies. Quite often, these companies belong to the same industry and they are at about the same stage of company development.

It is logical to think that investing into more than one company increases the investor's profits. First, there is a natural limit to how much can be invested into one company. In addition to that, other considerations can play a role in choosing portfolio approach. If VCs get rents from entrepreneurs for their advisory and monitoring role, then more firms in their portfolio means higher rents. This reasoning is used by Kanniainen and Keuschnigg (2000) and (2001) to find an optimal portfolio size. In their model, VC faces a trade-off between rents from bigger number of companies and diminishing quality of advice provided to each company. Since diluted quality diminishes probability of success for each portfolio company, the portfolio has a finite optimal size. In their model, success or failure of each individual project does not affect the outcome of other projects. Consequently, entrepreneurs would be indifferent in choosing between VC and angel investors, as long as they provide advising service of the same quality. Therefore, in this model the portfolio approach practiced by VCs should not give the VC investment advantage over the angel investment.

Intuitively, if VC could coordinate actions of portfolio firms and use information spillovers between them, then the portfolio investment could probably increase the overall portfolio value. One of the early examples of studying such approach is Bhattacharya and Chiesa (1995), who analyze advantages and disadvantages of bilateral financing vs. multilateral financing. In their model, two firms go through two stages of product development: the R&D stage and the market implementation stage. The outcome of the R&D stage is uncertain. Success of the R&D stage increases the probability of success of the second stage. If both firms are successful at the market implementation stage, then their competition drives profits to zero. In bilateral financing, each firm uses a separate financier (a bank in their model), and then the projects' outcomes are independent. The probability of each project's success at the R&D stage is low and, consequently, the probability of the market implementation is also low. The upside is that the probability of having both firms at the market is low as well. In multilateral financing, both firms use the same financier (bank), who can reveal the results of successful R&D stage of one firm to another firm. This information spillover increases the probability of individual success. The downside is that it increases the probability of a joint success and, consequently, zero profit. Bhattacharya and Chiesa find the set of parameters, for which the multilateral financing is preferable to the bilateral one. Technological knowledge revealed to a firm's financier(s) need not also flow to its R&D and product market competitors. The authors show that the choice of financing source can serve as a precommitment device for pursuing ex-ante efficient strategies in knowledge-intensive environments.

Hellmann (2002) and Ueda (2000) use somewhat similar models applied to VC financing. In these models an entrepreneur chooses an investors for his innovative project by taking into account the possibility that some investors can steal the entrepreneur's idea and apply it to other projects in their portfolio. In Hellmann's paper, this is the corporate investor, who can steal the idea and implement it in his own corporate project. In Ueda's paper, this is the VC investor, who can "steal." Unlike these papers, where the analysis is limited to the phenomenon of investors stealing from the entrepreneur, in Leshchinskii (2002) the information spillovers are two-directional. In this model, portfolio approach of the VC investor can bring more value to the portfolio firms thanks to two mechanisms: (1) entrepreneurs can potentially benefit from getting information about other VC's projects, and (2) the externalities effect, combined with VCs coordination of projects' continuation/termination decision, increases the future payoff. There are two types of externalities. First, the results of the R&D stage of one project affect the success/failure of the second stage of other projects. If at least one project is successful at the R&D stage, then all other firms on the market can use its results. In some sense, this is the extreme case of Bhattacharya and Chiesa (1995). The second externality directly affects the firms' profits at the market stage: the final payoff to each firm depends on how many projects are continued at the market stage, which generalizes Bhattacharya and Chiesa (1995) model. If the market externality is negative, then the total firms' value decreases with the number of firms (in the extreme case it can be zero). If the market externality is positive, then the overall value increases with the number of firms. If the externality is high and positive, then each firm is interested in a bigger number of firms at the market. The latter effect is similar to the one studied by Economides (1996), where an incumbent firm subsidizes the entry of other firms into a new industry with positive network externalities.

It is interesting that even positive externalities can serve as an impediment to getting funding from single-company investors, because these investors would prefer to wait indefinitely for the entrance of other entrepreneurial companies, reap the results of their R&D efforts and benefit from their presence on the market. VCs are able to internalize these externalities in the interest of the entire portfolio, rather than to the benefit of individual projects. Leshchinskii (2002) shows, that when the network externalities are positive, coordinated investment by VCs guarantees profitable investment into some projects that would otherwise have ex-ante negative NPV and fail to attract funding. When the network externalities are negative, coordinated investment allows early termination of some projects. Some positive NPV projects may even be sacrificed, in order to limit the number of surviving firms to increase the overall value of the VC portfolio. This early termination should not come as a surprise to the entrepreneurs, who must be aware of the possibility that their project will be terminated. The model finds the set of parameters, for which an entrepreneur is willing to take this risk and go after the VC financing, because the net benefits to the project's payoff will be much higher, than with non-coordinated individual investment of a business angel.

Taking into account only the "portfolio aspect" of VC financing, we can say that for entrepreneurs choosing the VC investor means: (1) higher probability of success thanks to getting some technological solutions and business ideas from the VC; (2) bigger number of competitors; (3) sometimes higher probability of early termination of a relatively successful project; and (4) high profits for the survived project.

If ex-ante the expected net benefits from getting funding from a portfolio investor (VC) are positive, then the entrepreneur will prefer this investor to a single-company investor (business angel).

In addition to establishing conditions, under which VCs successfully compete with angel investors, Leshchinskii (2002) makes predictions about characteristics of portfolio firms. These firms would have high growth potential, high profitability in the case of success and the size relatively similar to other portfolio companies. These predictions are consistent with known empirical observations. VCs should emerge as dominant investors in new industries with few existing companies and high market externalities (both negative and positive). The paper also predicts that in industries with negative market externalities, one should observe lower entrepreneur's ownership and higher returns to the VCs' investment into surviving firms. This happens, because entrepreneurs want to ensure continuation of their project and prefer smaller share of ownership to the project's termination.

7. CONCLUSION

Various groups of investors differ by their abilities to screen and monitor innovative projects, as well as to bring aboard alternative managers and to practice portfolio approach. We have shown how these differences change the expected net present value of an entrepreneurial project and, therefore, determine the choice of investor by the entrepreneur. Different factors can be critical for the project's payoff at each

Table 2. Entrepreneur's "checklist" of what different investors can do for the company. Although in the chapter we did not put strategic investors into a separate category, we put them here separately for illustrative purposes

	Parent company	Bank	Atomistic investors	Angel investor	VC	Strategic investor
Is investor's knowledge of industry same or greater than entrepreneur's?	Same or greater	Same or greater	Mostly not	Often greater	Greater	Often greater
Can investor easily understand the project and correctly evaluate it	Yes	Sometimes	Usually no	Usually yes	Yes	Usually yes
Can investor provide some insight about the project?	Yes	Quite often	No	Yes	Yes	Yes
Can investor help with business contacts?	Yes	Sometimes	No	Quite often	Yes	Quite often
Can investor bring the right people (new management) onboard?	Yes	Sometimes	Very rarely	Sometimes	Yes	Quite often
Would investor have a hidden agenda?	Yes	No	No	Very rarely	Yes	Quite often
If things go out of control can investor quickly understand the situation and interfere	Yes	Sometimes	No	Quite often	Yes	Quite often
Will investor care much about timing, maybe more than necessary	Sometimes	Yes	No	Sometimes	Yes	Sometimes

particular stage of its development. Therefore, for each stage the entrepreneur may attract investors with the highest level of expertise in the relevant area. For example, professional angels and venture capitalists obtain better information about a project's potential outcomes and emerge as providers of capital when this information is crucial. Venture capitalists and corporate investors, sometimes including the parent companies, are the industry's experts who have access to a pool of professional managers. Therefore, they can furnish least costly managerial replacement and they should be the main source of capital at the stage when the top-quality management is the decisive factor. The VC investment can mean higher probability of success of the R&D phase.

Table 2 provides a "checklist", which helps the entrepreneur to identify the investor most valuable for his company.

In this work, we did not focus on the specific form of contracts between entrepreneurs and investors. We took the existing contracts and investors characteristics as given and tried to explain what value these characteristics bring to the company.

ACKNOWLEDGMENTS

I would like to thank HEC students Bich-Tyên Diep, Ivan de Heeckeren d'Anthès and Sabine Mathis for their assistance in working Section 5.

REFERENCES

Berk, J. B., Green, R., & Naik, V. (2003). Valuation and return dynamics of new ventures. *Review of Financial Studies*, forthcoming.

Bergemann, D., & Hege, U. (1998). Venture capital financing, moral hazard and learning, *Journal of Banking and Finance*, *22*(5–6), 703–735.

Bergemann, D., & Hege, U. (2002). The value of benchmarking, Working paper.

Bhattacharya, S., & Chiesa, G. (1995). Proprietary information, financial intermediation and research incentives, *Journal of Financial Intermediation*, *4*, 328–357.

Brisley, N., & Leshchinskii, D. (2002). Stage financing, VCs' short-termism and managerial replacement, Working Paper.

Cohrane, J. H. (2001). The risk and return of venture capital, Working Paper.

Economides, N. (1996). Network externalities, complementarities and invitation to enter, *European Journal of Political Economy*, *12*.

Enrlich, S. B., de Noble, A. F., Moore, T., & Weaver, R. R. (1994). After the cash arrives: A comparative study of venture capital and private investor involvement in entrepreneurial firms, *Journal of Business Venturing*, *9*, 67–82.

Gompers, P. (1995). Optimal investment, monitoring and the staging of venture capital, *Journal of Finance*, *50*(5), 1461–1489.

Gorman, M., & Salman, W. (1989). What do venture capitalists do, *Journal of Business Venturing*, *4*, 231–248.

Hannan, M., Burton, D., & Barron, J. (1996). Inertia and change in the early years: Employment relations in young, high-technology firms, *Industrial and Corporate Change*, *5*, 503–535.

Hellmann, T. (2002). A theory of strategic venture investing, *Journal of Financial Economics*, *64*(2), 285–314.

Hellmann, T., & Puri, M. (2000). The interaction between product market and financing strategy: The role of venture capital, *Review of Financial Studies*, *13*(4), 959–984.

Kanniainen, V., & Keuschnigg, C. (2000). The optimal portfolio of start-up firms in venture capital finance, Unpublished.

Kanniainen, V., & Keuschnigg, C. (2001). Start-up investment with scarce venture capital support, Unpublished.

Kaplan, S., & Stromberg, P. (2002). Financial contracting theory meets the real world: An empirical analysis of venture capital contracts, *Journal of Economic Theory*, forthcoming.

Lerner, J. (1995). Venture capitalists and the oversight of private firms, *Journal of Finance*, *50*(1), 301–318.

Leshchinskii, D. (2002). Venture capitalists as benevolent vultures: The role of network externalities in financing choice, Unpublished.

Prowse, S. (1998). Angel investors and the market for angel investment, *Journal of Banking and Finance*, *22*, 785–792.

Sahlman, W. (1990). The structure and governance of venture capital organization, *Journal of Financial Economics*, *27*, 473–521.

Ueda, M. (2000). Bank vs. venture capital, Unpublished.

New Venture Investment: Choices and Consequences
A. Ginsberg and I. Hasan (editors)

Chapter 5

In Quest for Equity Partners: The Determinants of the Going Public-Large Blockholder Choice

MICHELE BAGELLA, LEONARDO BECCHETTI* and BARBARA MARTINI

Facoltà di Economia, Università Tor Vergata, Roma, Via di Tor Vergata snc, 00133 Roma, Italy

ABSTRACT

The chapter presents a theoretical analysis of the determinants affecting the controlling shareholders choice which include going public, remaining private and looking for a blockholder when they need external financiers. We show that the profitability of the going public choice is inversely related to monitoring costs of new stock exchange shareholders, and directly related to the presence of informational asymmetries between managers and controlling shareholders. The chapter also shows that a divergence between private and social optimum may arise when there is a weak institutional environment and when the private good of CSs control has high value.

* Corresponding author. Tel.: +39-6-7259-5723.
E-mail addresses: Bagella@economia.uniroma2.it (M. Bagella), Becchetti@economia.uniroma2.it (L. Becchetti), Martini@economia.uniroma2.it (B. Martini).

1. INTRODUCTION

Asymmetric information among managers, controlling shareholders and external investors and non competitiveness in the market of investors and financiers may significantly affect availability, conditions and the choice of external equity finance for firms which cannot entirely satisfy their investment needs with retained earnings. The quest for external equity partners may then be a crucial moment in which the financial sector affects real aggregate growth through its impact on the distribution of profits and therefore on incentives to run investment projects efficiently. The recent literature focuses on the determinants of the choice of equity financiers with several empirical and theoretical contributions (Bolton and Von Thadden, 1998; Chemmanur, 1993; Chemmanur and Fulghieri, 1999; Pagano et al., 1996; Pagano and Roell, 1988; Ransley, 1984).

Most of these papers focus on advantages and disavantages of going public vis-à-vis the choice of remaining private and choosing large blockholders. Bolton and Von Thadden (1998) emphasise that dispersed ownership is preferred, the lower the costs of monitoring the managers, the higher the transaction costs for secondary market trading and the higher the potential benefit from monitoring managers. Pagano and Roell (1988) stress listing costs as a disadvantage of the going public choice which, on the other hand, avoid excessive monitoring by large blockholders. In their perspective a large blockholder has more incentive than small minority shareholders to monitor managers as their effort has stronger effects on the market value of his firm share.

An important difference between the two choices is consequently that, under the going public choice, takeovers are a substitute for large blockholders monitoring when secondary markets are sufficiently liquid.

Lack of transparency and institutional weakness reduce the possibility of takeovers and are therefore a limit to the going public choice (Bolton and Von Thadden, 1998).

The contribution of our chapter to this literature is the explicit modelling of this choice in a bargaining framework between controlling shareholders (from now on also CSs) and the large blockholders in which going public represents the controlling shareholders' outside option. This framework has significant consequences on factors affecting the relative convenience of the two choices and their contribution to social welfare. In an extension of the model we also consider the impact of institutional environment on this choice under a framework which considers two different perspectives: the existence of an efficient market for corporate control and the possibility of illegal collusion between CSs and the large blockholder.

In our model CSs consider two alternative choices of external finance when financing their project. Under the first they maximize investment profits after compensating minority shareholders. The share of the latter depends on their financial contribution to the investment corrected for a dilution factor which is expression of the equilibrium value of external finance on the stock market.

Under the second choice they remain private and bargain their share of investment project with a large blockholder in a framework in which CSs expected returns under the going public choice represent their "outside option" in case of agreement failure and therefore affect the bargaining outcome.

After outlining this framework the model analyses the effects of project intrinsic value, small shareholders monitoring costs, discount rates, bearish or bullish stock market conditions, ex ante available cash flow and investment size on the equilibrium CSs profit shares under the two different options (Section 2, Proposition 1). It then draws some testable implications on the effects of changes in the above variables on the relative profitability of the two choices (Section 2, Proposition 2) and shows which conditions may generate a divergence between the controlling shareholder and the socially optimal choice between the two options (Section 2, Proposition 3). In the following sections it presents some extensions of the base model by analyzing how weak institutional framework allowing illegal diversion of funds (Section 3) and a more efficient market for corporate control (Section 4) may affect the going public blockholder capital choice.

2. THE BASE MODEL UNDER FEASIBILITY OF THE GOING PUBLIC FINANCING CHOICE

We consider a firm with cash flow CF and an investment project which costs I and can not be entirely internally financed ($I > \mathrm{CF}$). The investment project is profitable and yields $(q + e)$, where q is the project intrinsic value minus managerial remuneration and e is CSs effort in controlling the manager. Since existing financial slack is not sufficient to finance the new investment the control group must decide whether to finance it by going public or by choosing a large blockholder.

The model analyses the interaction among five types of actors. Three are internal to the firm (the manager, the controlling shareholders, the minority shareholders already existing before the decision to finance the new investment is taken), two are external (the large blockholder, if the firm opts for this choice, and the new minority shareholders if the firm opts for going public). The manager is in charge of operating the firm and we assume for simplicity that managerial labour market is tight so that his equilibrium wage is equal to zero.

Controlling shareholders may finance a share $\alpha_0 = \mathrm{CF}/I$ of the new investment while they need from external financiers $(1 - \alpha_0) = (I - \mathrm{CF})/I$.

When the CSs decide to go public new minority shareholders are compensated with a share of firm profits which is proportional to their contribution to the investment corrected for a dilution factor which is determined in equilibrium by the crossing of aggregate demand and supply of equity finance on the stock market. With perfect information their share would therefore be equal to $(1-\alpha_0)(1-x)[q+e-I]$ where $x \in [0, 1]$ is the dilution factor.[1] On the contrary, when external financiers are imperfectly informed over the quality of the investment and must pay a monitoring cost c_{M} proportional[2] to the project value to obtain this information, their share becomes: $(1-\alpha_0-c_{\mathrm{M}})(1-x)[q+e-I]$.

By taking into account new shareholders profits, ex post utility of the controlling shareholders may be written as:

$$W_{\mathrm{CS}} = (1-C)[q+e-I] - \psi(e) - \mathrm{CF} \tag{1}$$

where $C = (1-\alpha_0-c_{\mathrm{M}})(1-x)$ is the cost of raising equity funds in the stock market and $\psi(e)$ is the CSs cost of monitoring the manager. CSs then choose the optimal effort in monitoring the manager so that, in equilibrium, their marginal benefit be equal to their marginal cost from monitoring. If we conveniently formalize the effort of monitoring the manager as $\psi(e) = e^2, e \in [0, 1]$, we obtain the desirable property that $\psi'(e) > 0$ and $\psi''(e) > 0$ and an ex post equilibrium level of effort which may be written as:

$$e_{\mathrm{p}}^{*} = \frac{1-C}{2} \tag{2}$$

By replacing the optimal effort, the utility of controlling shareholders when they choose to go public becomes:

$$W_{\mathrm{CS}}^{*} = (1-C)\left[\left(q + \frac{1-C}{2}\right) - I\right] - \left(\frac{1-C}{2}\right)^{2} - \mathrm{CF} \tag{3}$$

[1] The demand (supply) of equity finance is obviously upward (downward) sloping in x. Therefore a bullish stock market, by shifting up the supply of equity finance increases the equilibrium dilution factor. This feature of the model is consistent with the empirical finding of seasoned IPOs (Ritter, 1987).

[2] The monitoring cost is assumed to be proportional to the project value as we imagine that the value of the project also proxies its complexity. For instance, small shareholders have relatively higher returns but also higher informational asymmetries in financing high tech projects versus traditional projects. The removal of this assumption does not change substantially the content of propositions presented in this chapter.

where CSs costs from monitoring managerial effort at optimum under the going public is[3]

$$\psi_{p}^{*}(e) = \left(\frac{1-C}{2}\right)^2. \tag{4}$$

Consider what may happen instead if the controlling shareholders decide to obtain external finance from a large blockholder. In this case we must consider that the ex post controlling shareholders share of profits must be bargained between the two counterparts. We may therefore write the following generic Nash maximand:[4]

$$\max \Omega = (V_{CS} - \bar{V}_{CS})^{\beta}(V_{BH} - \bar{V}_{BH}) \tag{5}$$

where V_{CS} and $\bar{V}_{CS}$ are respectively the income in case of agreement and the fallback income of the controlling shareholders, while V_{BH} and $\bar{V}_{BH}$ are respectively the income in case of agreement and the fallback income of the blockholder partner. β is the index of relative impatience of the two counterparts or the ratio between the blockholder (r_{BH}) and the controlling shareholders (r_{CS}) discount rates. We assume that $r_{BH}, r_{CS} \in [r, \infty]$, where the lower bound r is the rate of return of the riskless asset. To define the fallback income of the CSs we consider that, if the agreement is not reached, CSs may go public if they obtain positive ex post profits from this choice. Otherwise the firm is liquidated and the existing cash flow is invested in the risk free asset. In this simple case the firm and blockholders are both monopolist so that, if the agreement is not reached, there is not any possibility of finding any other blockholder partner. Therefore, the agreement and fallback incomes for the two counterparts may be written as:

$$V_{CS} = \alpha(q + e) - \psi_{CS} - \mathrm{CF} \tag{6}$$

$$\bar{V}_{CS} = \max[(1 - C)(q + e_{p}^{*} - I) - \psi_{p}^{*} - \mathrm{CF}, r_{CS}\mathrm{CF}] \tag{7}$$

$$V_{BH} = (1 - \alpha)(q + e) - M \tag{8}$$

$$\bar{V}_{BH} = r_{BH}M \tag{9}$$

where M is $I(1 - \alpha_0)$ is the investment share financed by external finance.

Notice that the value of the benefit from remaining in control of the firm does not change if the CSs choose the BH agreement as this one is just considered as an

[3] In the two limiting cases, $e_{p}^{*} = 1/2$ as CSs (first best optimum), if $x = 1$ (bull market), while $e_{p}^{*} = (\alpha_0 + c_M)/2$ if $x = 0$.

[4] Theoretical references on the simple approach to bilateral bargaining followed here may be found on Rubinstein (1982), Binmore et al. (1986) and Sutton (1986).

agreement on the share of current project profits and does not generate dilution on future profits.

To find the equilibrium CSs' profit share we maximise the log of the Nash maximand with respect to the ex post controlling shareholders share:

$$\max\log\Omega = \beta\log\{[\alpha(q+e) - \psi_{CS} - CF] - \max[(1-C)(q+e_p^* - I) - \psi_p^* - CF, r_{CS}CF]\} + \log[(1-\alpha)(q+e) - (1+r_{BH})M] \quad (10)$$

By examining this Nash maximand we may see that there are two factors affecting relative bargaining power. The first is β, which we call "relative impatience".[5] The second is the relative size of gains from bargaining for the two counterparts which are represented by the difference between the value of the agreement and the value of the fallback income. When $V_{CS} > \bar{V}_{CS}$ the fallback income for the controlling shareholders is given by the decision to go public (as this is more convenient than firm liquidation). Differentiating Eq. (10) with respect to α and calculating the first order condition gives:

$$\alpha = \frac{\beta(q+e-(1+r_{BH})M) + (1-C)(q+e_p^* - I) + \psi_{CS} - \psi_p^*}{(1+\beta)(q+e)} \quad (11)$$

The equilibrium value of α and e is given by the solution of the system including (11) and the following first order condition of the controlling shareholders maximizing problem when they choose blockholder financing. In this case the optimal contractual solution for the optimal share becomes an incentive mechanism for CSs as their share is increasing in their effort:

$$2e = \frac{\partial\alpha}{\partial e}(q+e) + \alpha \quad (12)$$

By solving the system we find that the solution for the optimal effort under blockholder financing is

$$e_{BH}^* = \frac{1}{2} \quad (13)$$

Proposition 1. When going public is relatively more profitable than investing in a risk free asset, the controlling shareholders share under the blockholder financing choice is increasing in β (index of CSs relatively lower impatience), in the blockholder discount rate, in the CSs share under the going public choice, in the internal finance/investment ratio, in the dilution factor and in the project intrinsic

[5] Relative impatience may depend, for instance, from the fact that the investment needs to be started within a limited time window to realize expected profits.

value (if beta is not too low). It is decreasing in the interest rate of the riskless asset, in the CSs discount rate, in the external finance/investment ratio, in the investment cost and in the monitoring cost of new shareholders if the firm goes public.

Proof: See Appendix 1. □

The partial derivative on β has positive sign if the sum of the potential gains from the agreement is positive and that reduction in benefits from control is not too high under the going public choice.

When we look at the project intrinsic value we find that this expression is higher than zero when beta is not too low. The intuition behind the partial derivative result on q is that a higher intrinsic value of the investment project affects both agreement and fallback income of the controlling shareholders but only the agreement income of the blockholder. This increases gains from the agreement more for the blockholder than for the controlling shareholders. The blockholder therefore has more interest in the agreement not to be left out from a project with high intrinsic value.

Proposition 2 (effect of changes in exogenous variables on the going public-blockholder choice for CSs). For reasonable levels of the dilution factor ($x < 1$), the relative profitability of the going public choice for controlling shareholders is increasing when CSs discount rate increases, BH discount rate decreases, project intrinsic value decreases, CSs bargaining power decreases, investment costs decrease. The same relative profitability is decreasing when monitoring costs increase and interest rate decreases.

Proof: See Appendix 2. □

Intuition for these results is straightforward. A project requiring higher investment increases, *coeteris paribus*, the amount of external finance needed, reducing both CSs share and equilibrium effort under the going public option, while it increases blockholder outside option and bargaining power, therefore reducing CSs share but not equilibrium effort under the BH option. Therefore the BH choice becomes more convenient for CSs unless the stock market is extremely bullish ($x = 1$). The opposite reasoning needs to be done for an increase in CSs cash flow. On the other side, higher project intrinsic value does not affect equilibrium share and effort under the going public option, but weakens the bargaining power of the blockholder (as shown in Proposition 1) and therefore increases CSs equilibrium share under the blockholder choice. The relatively lower propensity to go public in presence of high monitoring cost is the typical Bolton-Von Thadden (1998) effect. The same happens with the dilution factor which gives a result which is analogous to that of stock market liquidity in the Bolton-Von Thadden (1998) model. What this paper adds, though, is that higher monitoring cost and a lower dilution factor increase also the cost of choosing a large blockholder due to the bargaining structure postulated by the model.

Finally, a rise in the interest rate increases the outside option and the bargaining power of the blockholder therefore increasing the relative profitability of the going public choice for CSs.

An interesting result of the paper is that we also obtain the Pagano-Roell (1988) result demonstrating that the decision to go public is relatively more convenient when the amount of external finance needed is higher. In the Pagano-Roell (1988) model this result depends on the fact that large blockholder overmonitoring is increasing in the amount of external finance needed. In our model it depends on the fact that a higher amount of external finance increases the bargaining power of the large blockholder but not that of dispersed and uncordinated small shareholders.

Proposition 3. The blockholder solution always achieves social optimum when the CSs have nonzero bargaining power. The go public solution never achieves it unless the dilution factor reaches its maximum level. Social welfare under the going public choice may be improved upon by reducing small shareholders monitoring costs or when the stock market expectations are bullish.

To find the social optimum in our model we neglect the difference between agents and maximize effort as if the firm were operated by the same agent:

$$\mathrm{SW} = (q + e) - \psi(e) - I \tag{14}$$

The socially optimal level of effort is therefore $e^*_{\mathrm{SW}} = 1/2$ when $\psi(e) = e^2$. It is easy to check that this level of effort is always reached under the blockholder choice (if $\beta > 0$), while it is achieved under the going public option, only when $x = 1$.

It is intuitively clear that there may be a divergence between private and social optimum, when, for instance, $e^*_{\mathrm{BH}} = 1/2$ and CSs find it more convenient to go public.

It is also clear that reductions of small shareholders monitoring costs, upward shift in the supply curve of equity finance, and therefore in the dilution factor and higher availability of internal finance all have positive effects on welfare by increasing effort and total output of CSs who opt for the going public option. With this respect, model features support the hypothesis that a bullish stock exchange may positively affect output and that traders optimism may be self-fulfilling.[6]

The welfare superiority of the blockholder solution may seem puzzling, and strongly depends on our assumption on the absence of positive informational

[6] A positive impact on welfare would also be generated by someone who could have private benefits from producing information to reduce small shareholders monitoring costs. This may be the case of a privatised stock exchange.

externalities from the going public decision. We must also consider, though, that the likelihood that the blockholder partner brings a productive contribution has equally not been modelled. The decisive point in favour of the blockholder solution is that, both the nature of the contract and the possibility of the blockholder partner to verify CSs effort, push the latter to exert optimal effort.

We must also consider that, from an aggregate point of view, the feasibility of the going public choice may have significant positive welfare effects as it reinforces the CSs bargaining position and brings into existence projects which were not privately convenient for the CSs because of their scarce bargaining power. Equilibrium effort is always the first best optimum in the blockholder agreement and not when the firm goes public. This result depends from the fact that the first order condition in the BH case has an extra term. This term is positive and expresses the CSs' marginal benefit from increasing his effort in terms of higher profit share. This means that the CSs have an interest in increasing their effort as effort reduces their gains in the bargaining and therefore increases their bargaining power. This marginal benefit is higher than the marginal cost of effort up to the socially optimal effort. This result holds only for $\beta \neq 0$ since, for $\beta = 0$ controlling shareholders do not bargain. The result crucially depends on the fact that CSs monitoring effort is verifiable by the BH but is not observable by new minority shareholders.

Comparative statics results on the choice of the equity partner are somewhat different when the going public choice is not profitable as explained in the following proposition.

Proposition 4. (a) If the going public choice is not profitable for CSs (yields a return which is lower than the opportunity cost), we may fall in a case in which the CSs share under the BH financing choice is decreasing in the project intrinsic value (or less increasing than in the situation in which going public is profitable). Therefore non feasibility of the going public choice may weaken CSs incentives to pursue BH financed projects of high intrinsic value. (b) Optimal effort under the BH choice is unchanged with respect to the base model, therefore it is not affected by the feasibility of the going public solution.

Suppose that there is not agreement: $V_{\mathrm{CS}} > \bar{V}_{\mathrm{CS}}$. Going public is not feasible for the CSs as it yields less than the opportunity cost of internal finance. The equilibrium CSs share under blockholder financing becomes:

$$\alpha = \frac{(-\beta(1 - r_{\mathrm{BH}})M - \psi_{\mathrm{CS}}) + \beta(q + e) + (1 + r_{\mathrm{CS}})I}{(1 + \beta)(q + e)} \tag{15}$$

Again, the sign of the derivative is positive as the sum of the potential gains from the agreement is positive. It is easy to check that optimal effort under blockholder financing is unchanged:

$$e^*_{\mathrm{BH}} = \frac{1}{2}$$

A main difference with respect to the previous comparative static analysis is then that the impossibility of going public eliminates the impact of an increase in the intrinsic value of the project on the CSs fallback income. The increase in CSs gains from the agreement is therefore larger and their bargaining position is weaker.

The feasibility of the going public choice has therefore another important welfare effect as it increases the marginal CSs gains from investing in projects of higher intrinsic value.

3. THE EFFECTS OF POOR SHAREHOLDER PROTECTION IN A WEAK INSTITUTIONAL ENVIRONMENT

In many developing (and developed) countries the weakness of the institutional system is an important variable affecting the decision of going public vs. that of choosing a large blockholder. In this case remaining private (or going public in a market where small shareholder protection is insufficient) makes it easier for CSs to find forms of legal or illegal collusion with the manager or influential external financiers at the expense of existing small minority shareholders. In this extension of the model therefore consider the presence of minority shareholders with a property right share of $\alpha_{\mathrm{MS}} > 0$. Total available internal finance is provided by both control and minority shareholders. Therefore $\alpha_{\mathrm{CS}}/(\alpha_{\mathrm{CS}} + \alpha_{\mathrm{MS}})$ is the ex ante share of CSs, $\alpha_{\mathrm{MS}}/(\alpha_{\mathrm{CS}} + \alpha_{\mathrm{MS}})$ the ex ante share of minority shareholders, $\alpha_0 = \mathrm{CF}/I$ the ratio between internal finance and investment costs and $\mathrm{CF}' = [\alpha_{\mathrm{CS}}/(\alpha_{\mathrm{CS}} + \alpha_{\mathrm{MS}})]\mathrm{CF}$ the cash flow of existing shareholders.

We may conceive different forms of legal or illegal collusion. Illegal collusion may be easier when the firm is private as the possibility of ex post hidden information may in this case be enhanced. Legal collusion may involve practices which are not legally forbidden but which obtain as a result the reduction of the profit share of minority shareholders. An example may be the discounted sale of firm assets to another company in which controlling shareholders shares are less diluted. Or, alternatively, the purchase of overvalued assets of the other company with firm profits. These forms of legal collusion would be possible also if the firm decides to go public. This extension of the model tries to address these issues in a simple framework.

We assume here that controlling shareholders, the manager and external financier may decide to collude to hide project profits to minority shareholders. Illegal profits are equally shared between CSs and the blockholder financier. This strategy will provide each of them an additional share $\delta\alpha_{\mathrm{MS}}/2$, of firm profit, where α_{MS} is the

Table 1. Synthesis of results on collusion

	Illegal with infinite penalty	Legal collusion
Collusive profit split	1/3 each among CSs, the manager and the blockholder	Bargained between CSs and the blockholder if the firm opts for BH financing. All for CSs under the going public choice
Legal profit split	1/3 each among CSs, the manager and the blockholder	Bargained between CSs and the blockholder if the firm opts for BH financing. All for CSs under the going public choice
Ex post CSs share under the going public choice	Illegal collusion is not possible	Higher
Equilibrium effort under the going public choice	Illegal collusion is not possible	Higher than in the base case
Ex post CSs share under the blockholder choice	$1/3[1-(1-\delta)\alpha_{MS}]$	Higher
Equilibrium effort under the blockholder choice	$1/6[1-(1-\delta)\alpha_{MS}]$	Reduced
Relative profitability of the going public vs. the blockholder choice for the CSs	Uncertain	Reduced

profit share of existing minority shareholders and $\delta \in [0, 1]$ is the measure of the weakness of the institutional environment. We assume that going public involves transparency vs. minority shareholders and therefore that collusion is possible only under blockholder capital choice (Table 1).[7]

> **Proposition 5.** Under illegal collusion with infinite penalty an egalitarian split among the three colluding agents occurs at the expense of minority shareholders. In this case effort under the blockholder agreement is lower than the social optimum and may be lower than optimal effort under the going public choice.

The egalitarian split of collusive profits is the result when prosecution of the crime leads to infinite penalty. In this case if one of the three agents is left with less than 1/3 he may credibly threaten the other two agents that he will go to court if he is left out of the agreement. This means that in a bargain between him and the other two agents, the other two agents outside option is minus infinity and their equilibrium

[7] This is obviously a restrictive assumption but its justification is that going public requires more severe informational requirements and therefore involves higher transparency than remaining private, even though in reality it does not eliminate the possibility of illegal collusion.

share is zero. Therefore there are no profitable deviations from the egalitarian split.[8] The bargaining on legal profits is subject to the same rule so that each of the three agents gets 1/3. The maximizing function of controlling shareholders under the blockholder choice is:

$$V_{\mathrm{CS}} = \frac{2}{3}(q+e)[1-(1-\delta)\alpha_{\mathrm{MS}}] - \psi(e) - \frac{1}{3}[1-(1-\delta)\alpha_{\mathrm{MS}}](q+e) \quad (16)$$

or

$$V_{\mathrm{CS}} = \frac{1}{3}(q+e)[1-(1-\delta)\alpha_{\mathrm{MS}}] - \psi(e) \quad (17)$$

Then

$$e^*_{\mathrm{BH}} = \frac{1}{6}[1-(1-\delta)\alpha_{\mathrm{MS}}] \quad (18)$$

This effort is lower than social optimum (in fact the CSs have positive marginal costs and zero marginal benefits from increasing effort) and may be lower than optimal effort under the going public choice if, for instance, the stock market is bullish and the dilution factor is high.[9] The condition of social preference for the going public choice is:

$$\frac{1}{6}[1-(1-\delta)\alpha_{\mathrm{MS}}] > [1-(1-\alpha_0+c_{\mathrm{M}})(1-x)](1-\alpha_{\mathrm{MS}}) \quad (19)$$

This result suggests that one of the social costs of the blockholder choice is that lack of transparency (assumed relatively higher since going public requires more severe disclosure requirements) may foster illegal collusion among insiders, bringing effort below social optimum. In that case the going public choice may happen to be socially preferred.

Proposition 6. With perfect information on the stock market legal collusion increases effort more under the BH choice than under the going public choice and increases the relative profitability of the BH choice for CSs when project intrinsic value is sufficiently high.

Imagine that controlling shareholders have an ownership share in a second firm which is higher than that held in the observed firm. Collusion may then be legal by selling assets of the first firm to the second at a discounted price.

[8] The egalitarian share solution is the outcome of a bargaining in which outside option is $-\infty$ (infinite penalty) and β is equal to one.

[9] Note that in this case optimal effort and optimal CSs share are not necessarily higher with than without collusion. This is because the advantage of a share of minority shareholder profits must be traded off with the disadvantage of a higher participation of the manager to firm profits.

In this case imagine that the CSs invest all expected profits of the first firm in capital goods and then sell the capital goods to the second firm with a discount of $(1-\delta)$. The first firm has now a debt of $\delta(q+e-I)$ before making profits, while the second firm has a profit of $\delta(q+e-I)$ which may arise when selling the capital goods at their market value. If the first firm goes public new minority shareholders are informed and increase their reservation value needed to become equity financiers. After the intragroup sale future minority shareholders anticipate the effect of the news on the firm market value and need a compensation of $\delta(q+e-I)(1-\alpha_0+c_{\mathrm{M}})$. Therefore total CSs wealth with collusion under the going public choice will be

$$W_{\mathrm{CS}} = \alpha_1'(q+e-I)(1-\delta) + (q+e-I)\delta - \delta(q+e-I)C - \mathrm{CF}' - \psi(e) \tag{20}$$

with

$$\alpha_1' = (1-C)(1-\alpha_{\mathrm{MS}}) \tag{21}$$

The new equilibrium effort will be

$$e_{\mathrm{p}}^{*\prime} = \frac{\alpha_1'}{2} + \frac{\alpha_{\mathrm{MS}}'}{2} \tag{22}$$

where

$$\alpha_{\mathrm{MS}}' = 1 - \alpha_1' - C \tag{23}$$

is the after investment share of old minority shareholders under the going public choice. The intuition behind this result is that gains from collusion which motivate extra effort are proportional to the share of that class of shareholders (old minority shareholders) which are deceived by collusion.

If the CSs opt for BH financing they have to bargain profits of both the first and the second firm with the BH partner. CSs will get from the original company:

$$\alpha(q+e)(1-\alpha_{\mathrm{MS}})(1-\delta) - \psi_{\mathrm{BH}} - I + \alpha(q+e)\delta \tag{24}$$

or

$$\alpha(q+e)(1-(1-\delta)\alpha_{\mathrm{MS}}) - \psi_{\mathrm{BH}} - I \tag{25}$$

and:

$$\alpha = \frac{[1+\alpha_{\mathrm{MS}}((1-\beta)(1-\delta)) + \beta(1-\delta)](\psi_{\mathrm{BH}}+I)}{(1+\beta)(q+e)(1-\alpha_{\mathrm{MS}}(1-\delta))(-1+\delta)(-1+\alpha_{\mathrm{MS}})} \tag{26}$$

Notice that a higher degree of institutional weakness increases effort more under the BH than under the going public choice as old minority shareholders have to partially compensate new minority shareholders for their monitoring costs. The marginal increase in CSs welfare with legal collusion under the blockholder choice is higher than zero.

4. THE EFFECTS OF A MORE EFFICIENT MARKET FOR CORPORATE CONTROL

Proposition 3 showed that the blockholder solution seems socially preferred under assumptions of the base model. This conclusion may be reversed if we consider the effects of the market for corporate control in the model as specified in the proposition below.

> **Proposition 7.** Under reasonable assumptions the possibility of takeovers in the stock exchange (implying an efficient market for corporate control) generates an increase in effort under the going public choice if: (i) the project has high intrinsic value; (ii) monitoring effort is not too costly; (iii) the new productive idea is not too superior to that of existing CSs; (iv) the ex ante share of the existing minority shareholders is high; and (v) the value of the benefits from control is high.

For high levels of outsiders innovation and probability of takeover, the going public choice is socially preferred to the BH choice. If the CSs highly value the benefit from remaining in control they prefer the BH choice and therefore their privately optimal choice does not coincide with the socially optimal choice.

Consider that, under an efficient market for corporate control, existing minority shareholders may, with probability $\lambda(e) < 1$, come out with a project with higher intrinsic value $\nu_{\text{MS}} = q + i$ (where $i > e^*_{\text{SW}} = 1/2$ is higher than the socially optimal effort and represents the superior ability of the new management or the "outsiders innovation" brought into by the minority shareholder) and take over the firm. The hypothesis is that the CS initial share α_{cs} is lower than 50%, and that the existing (before investment) minority shareholders can acquire the control share on the market when the firm goes public. We also assume that $\lambda'(e) < 0$ as higher effort from the CSs may reduce the probability of takeover. Incorporating in a simple framework the Shleifer-Vishny (1986) idea we state that, the higher the ex ante property right share of minority shareholders, the lower the winner's curse effect and therefore the compensation of the CSs for releasing control over the firm. The winner's curse rationale is that CSs will refuse to tender until the price of the shares they are selling to minority shareholders which are taking over the firm includes the positive contribution to market value of the future new owners. Therefore, the lower the ex ante share of minority shareholders, the higher the offer they need to make to existing CSs to take over the firm.

The winner's curse effect occurs as CSs actions are individually taken implying that control group members can refuse to tender if the compensation is not adequate (individual CSs may free ride). The control group under the going public option therefore maximises:

$$\begin{aligned} W_{\mathrm{CS}}(1-\lambda(e))[\alpha_1'(q+e-I)-\psi(e)-\mathrm{CF}'] \\ +\lambda(e)(1-\alpha_{\mathrm{MS}})[\alpha_1'(q+e_{\mathrm{N}}+i-I)-B-\mathrm{CF}'] \end{aligned} \tag{27}$$

The fist term is the revenue when the controlling shareholders realize their project. The second term is the revenue in case someone else will make the project, e_{N} is the effort contribution to output given by the effort of new CSs in monitoring the manager, $(1-\alpha_{\mathrm{MS}})$ is the *winner course* effect, $\alpha_1' = (1-C)(1-\alpha_{\mathrm{MS}})$ and $\mathrm{CF}' = [\alpha_{\mathrm{CS}}/(\alpha_{\mathrm{CS}}+\alpha_{\mathrm{MS}})]$.

We assume here that, if the takeover occurs, controlling effort will be exerted by the new control group $\lambda(e)(1-\alpha_{\mathrm{MS}})[\alpha_1'(q+e_{\mathrm{N}}+i-I)-\mathrm{CF}']$ is CSs compensation in case of takeover. Note that $-\psi(e)$ is not part of this compensation as, if CSs refuse to tender, they become minority shareholders and do not have effort costs while they loose all benefits of control (B).

The ex post CSs profits is unchanged as only CSs and not new minority shareholders have disadvantages from takeover risk. Therefore takeover risk has no effects on the reservation value of new minority shareholders. The equilibrium effort changes as the new first order condition is:

$$\begin{aligned} \frac{\partial W_{\mathrm{CS}}}{\partial e} = (1-\lambda(e))[\alpha_1'-\psi(e)] - \lambda'(e)[\alpha_1'(q+e_{\mathrm{N}}+i-I)-\mathrm{CF}'] \\ +\lambda'(e)(1-\alpha_{\mathrm{MS}})[\alpha_1'(q+e_{\mathrm{N}}+i-I)-B-\mathrm{CF}'] \end{aligned} \tag{28}$$

The first term is the discouragement effect. With the positive probability of takeover more effort means lower expected profits. The second term is the incentive effect, more effort means reducing the probability of takeover. The third term is a different discouragement effect, more effort means reducing the benefit for CSs (if B is not too high) of becoming minority shareholders in a company with higher value project.

We have an interior maximum if $W(\cdot)$ is convex in e. Consider that $\psi'(e) = 0$ and $-\lambda(e)[\alpha_1' - \psi(e)]$ at optimum. Therefore, by reasonably assuming that marginal returns from effort in avoiding the takeover are decreasing—$\lambda''(e) > 0$—, second order condition for an interior optimum requires that

$$[\alpha_1'(q+e)-\psi(e)-\mathrm{CF}'] > (1-\alpha_{\mathrm{MS}})[\alpha_1'(q+e_{\mathrm{N}}+i-I)-B-\mathrm{CF}']$$

or benefits from the takeover for the existing CSs are lower than profits from remaining in control.

It is intuitively clear from the inspection of these three terms that existing controlling shareholders will increase their effort if: (i) the project has high intrinsic value; (ii) monitoring effort is not too costly; (iii) the new productive idea is not too superior to that of existing CSs; (iv) the ex ante share of existing minority shareholders is high; (v) the contribution of the new CSs to output with their monitoring effort is not too high; and (vi) the value of the benefits from control is high.

Consider also that the new social welfare for the going public choice is:

$$\mathrm{SW_p} = (1 - \lambda(e))[(q + e - I) - \psi(e)] + \lambda(e)[(q + i + e_{\mathrm{N}} - I)] - \mathrm{CF} \qquad (29)$$

If, as it is reasonable to assume, even the highest effort of existing CSs does not eliminate the positive probability of a takeover, the social best is a situation in which the new productive idea is high and existing shareholders exert an effort level which is not inferior to that of a situation in which the possibility of takeover does not exist. This occurs if ex ante ownership is more dispersed, the value of the benefits from control are high, monitoring effort is not to costly for existing CSs and remaining in control in the new project with a diluted share does not reduce substantially the benefits from control. By comparing welfare functions under the two choices we find that:

$$\mathrm{SW_p} - \mathrm{SW_{BH}} = i + \lfloor e_{\mathrm{N}} - e_{\mathrm{p}} - \psi(e_{\mathrm{N}}) + \psi(e_{\mathrm{p}})\rfloor + \frac{e_{\mathrm{p}} - e_{\mathrm{BH}} - \psi(e_{\mathrm{N}}) - \psi(e_{\mathrm{BH}})}{\lambda(e_{\mathrm{p}})} > 0 \qquad (30)$$

where $\mathrm{SW_p}$ and $\mathrm{SW_{BH}}$ are social welfare functions under the going public and blockholder choices. Social preference of going public is increasing in i and in $\lambda(e_{\mathrm{p}})$ if $e_{\mathrm{p}} < e_{\mathrm{BH}}$ as it is the case. Hence decreasing in e_{p}.

In sum, dispersed ownership and the possibility of takeover in an entrepreneurial environment with creative productive ideas may make the going public choice socially preferred to the blockholder choice.

What is the effect on CSs choice?

To check how the existence of a market for corporate control affects CSs choice one must compare WCS and:

$$V_{\mathrm{BH}} = \frac{-[\beta(1 + r)M] + \beta(q + e - I)(1 - \alpha_{\mathrm{MS}}) + \alpha_1'(q + e_{\mathrm{p}}^*)}{(1 + \beta)(q + e)(1 - \alpha_{\mathrm{MS}})} \times \left(q + \frac{1 - \alpha_{\mathrm{MS}}}{2}\right) - \left(\frac{1 - \alpha_{\mathrm{MS}}}{2}\right) - \mathrm{CF}' \qquad (31)$$

If the going public option yields a relatively higher expected value from going public, the blockholder option will have a relatively higher equilibrium share for

New Venture Investment: Choices and Consequences
A. Ginsberg and I. Hasan (editors)

Chapter 6

Exit Decisions of Entrepreneurial Firms: IPOs Versus M&As

ILGAZ ARIKAN*

Fisher College of Business, Ohio State University, Columbus, OH 43210, USA

ABSTRACT

When an entrepreneur decides to sell his/her company, he is faced with a series of single decisions among two major alternatives: he can either sell the firm to the public in an initial public offering, or sell to interested public or private entities in a merger and/or acquisition. These alternatives are mutually exclusive, and unordered. Using a sample of 2820 IPOs and 34109 M&As consummated between 1975 and 1999 in the manufacturing sector, I test for the choices that entrepreneurial firms make given five conditions: bargaining power, resource value, risk propensity, market thickness, and search costs. I find that the given the propensity scores for each entrepreneurial firm, the model predictions hold. This chapter develops a new model for determining the exit decisions of entrepreneurs based on these factors.

1. INTRODUCTION

We had a very strong name and a great brand, but needed financing. We approached a big investment bank to IPO our company. They said "In the world of finance, you are just a rounding error". There were companies that wanted to acquire us but we did not want that then. Therefore we went ahead with an IPO. Following the IPO, for one year, if you put our stock price to a medical chart, you would only see a horizontal

* Tel.: +1-614-688-3321.
E-mail address: arikan.2@osu.edu (I. Arikan).

> line. We had the EKG of a potato. Only after these problems did we negotiate the sale of our company.
>
> Reed Foster, Co-founder of Ravenswood Winery, 2002

There are three main types of market exchanges and firms often have to make strategic decisions regarding which market mechanism to use when competing in factor markets to acquire or sell resources: auctions with bidding, negotiations with bargaining and spot markets with posted prices. For a market mechanism to be used during an exchange, it should be optimal for both the seller and the buyer, otherwise the exchange would occur using a different market mechanism, or would not occur at all. The choice of market mechanism affects both the buyers' and sellers' rent generation and appropriation potential. Hence, while the buyer is concerned with how he/she should buy a resource, the seller is concerned with how he/she should sell it. While it is well established that firms should acquire rare, inimitable and valuable resources from factor markets to gain competitive advantages, the mechanisms by which firms acquire these resources have received little attention. In this chapter, I study the factors that affect firms' decisions to prefer one market mechanism to another.

In factor or product markets, it is convenient to use a resource or a product as a tradable unit. What if the resource or the product is the whole firm itself? Firms have been characterized as a bundle of linked and idiosyncratic resources and resource conversion activities (Rumelt, 1987). When a firm is sold, its new owners have a right to its future cash flows. Therefore, the dynamics of competition on product and resource markets by firms (Wernerfelt, 1984), and the ability of firms to generate and appropriate rents is fully applicable to the market for firms, where firms are exchanged through various market mechanisms. I argue that for private firms an IPO is an auction, and an M&A is a negotiation; where, one unit of a firm's stock of equity represents a future cash flow opportunity, and a numeraire resource.

Two major markets exist for the sale of an entrepreneurial firm: initial public offering (IPO) vs. mergers and acquisitions (M&A) markets. Based on the theoretical work by Campbell and Levin (2001) and Arikan (2002), I argue that the discrete choice between choosing to auction off a company through an IPO or to negotiate its sale as a privately held target rests on five factors: bargaining power, resource value, market thickness, risk propensity and search costs.

The purpose of this chapter is to test a general model of market mechanisms in strategic factor markets (Arikan, 2002) and to determine the conditions under which sellers will prefer one to the other. In order to test the performance differences of market mechanisms for the sellers, I use the market for firms as a factor market and study the propensity to choose one mechanism over another by looking at the choice of IPOs vs. M&As for private firms. In the paragraphs that follow, I review the literature linking market mechanisms and entrepreneurial choices. Hypotheses are developed on the basis of these arguments. I then describe the IPO and M&A data used in this study and provide a description of measures. The section on statistical methods is followed by a discussion of results and implications.

2. MARKET MECHANISMS

What makes an entrepreneur decide to sell his/her company is beyond the scope of this study. Only recently two formal explanations have been offered for going public: life cycle theory and market-timing theory. According to the life-cycle theory, entrepreneurs who would rather just run their firms, due to cash considerations, will go public when they grow sufficiently large because it is more optimal (Chemmanur and Fulghieri, 1999; Zingales, 1995). The market-timing theory is based on an asymmetric information model where firms decide when to exercise the sale depending on the favorable pricing of the stock, and the existence of competitors in the market (Lucas and McDonald, 1990). Entrepreneurs can in fact use timing as a strategy to signal growth opportunities (Schultz, 2000).

For the purposes of this study, I assume that founders and other initial shareholders desire to raise equity capital for the firm, and convert some of their wealth into cash at a future date. Also, while some entrepreneurs might have non-financial reasons that are driven by semirational considerations, it is assumed here that they will be inclined to sell their shares after public market valuations have reached optimal levels (Ritter and Welch, 2002). Hence, I consider the cases when the entrepreneur or the venture capitalist gets rewarded for his/her initial efforts through selling his company to public or private investors either through auctions (IPO) or negotiations (M&A). Entrepreneurs use both IPOs and acquisitions to raise capital. The magnitude and scope of these activities are highly sensitive to time trends, and show great variation across countries.

Firms use both acquisitions and IPOs to sell their companies. Privately owned companies are also subject to takeovers, where small entrepreneurial firms choose to divest the entire firm. In this chapter, I focus only on privately held entrepreneurial firms because in these firms the agency conflicts between managers and nonmanagerial shareholders are negligible (Field and Karpoff, 2002), since the rents generated from the firm's sale accrue to a relatively small number of principals who internalize both costs and benefits. Specifically, I test the model predictions developed in Arikan (2002) controlling for the exogenous variables discussed in the literature. In the sections that follow, first I discuss the factors that affect the choice between market mechanisms. Then I examine IPOs as auctions and extend the analysis to acquisitions as negotiations.

3. DETERMINANTS OF MECHANISM CHOICE

Firms' choice of a market mechanism is determined by both endogenous and exogenous factors. These factors in turn affect rent generation and appropriation potential. I identify five factors: bargaining power of the parties, resource's value to each bidder, the market thickness for both the demand and the supply side, risk taking propensity of the parties, and the existence of search costs. The existence of scarcity in factor markets creates the need for heterogeneous expectations and private values, which signal higher marginal returns if acquired. I argue that, if a bidder is to generate rents, this is determined by the mechanisms by which the good is acquired, in addition to the productive capabilities of the good itself. Similarly, if a seller is to generate rents, the optimal mechanism choice might inhibit or enhance rent generation potential.

Bargaining power is the power to bind an opponent, and parties to an exchange try to gain advantages that their counterpart does not have access to (Schelling, 1956). Information structures play a major role in determining the magnitude and the direction of bargaining power. Bargaining under symmetric information results in efficient outcomes, whereas asymmetric information results in ex post trade inefficiencies (Coase, 1960). Since both search and bargaining are time consuming and discount future utilities, both parties have an incentive to avoid or limit these inefficiencies (Tirole, 1988), and the contracting dynamics depend on relative bargaining power of the parties involved. When parties with bargaining power choose to auction, they might be wasting a valuable opportunity to extract a larger portion of the surplus by entering into negotiations.

Hypothesis 1. Entrepreneurs with low bargaining power are more likely to choose auctioning their firms through an IPO rather than negotiating the sale through an M&A.

Hypothesis 2. Entrepreneurs with some degree of bargaining power are more likely to choose negotiating the sale of their firms though an M&A rather than auctioning off through an IPO.

Resource value is represented by two components: private and common values. The value of a pure common value item is the same to all bidders, but the bidders do not know the value at the time they bid. Instead, they receive signals related to the value of the item. In pure private value auctions, each bidder knows with certainty the value of the item to him/her but only has probabilistic information about the value of the object to other agents (Kagel et al., 1987). In the context of RBV, the private value component of resources leads to heterogeneous expectations about the productive uses of those resources. The heterogeneous expectations are going to be sources of economic rents for both the seller and the buyer depending on the exchange mechanism of auctions vs. negotiations.

The problem in private value bidding is 'strategic'. Depending on the mechanism design, the winning bid either pays the highest value, or the second, and if there is overbidding, it has a small effect on the expected payoffs to the bidders (Roth, 1995). On the other hand, in common value auctions, inexperienced bidders are subject to systematic failure to account for adverse selection. Although each bidder receives unbiased estimates of the item's value and have homogeneous bid functions; they assume they have the highest signal value (Kagel and Levin, 1986), and overbid their value. Most often this will result in winner's curse.

Hypothesis 3. As the private value component increases, the entrepreneurial firm's probability of being sold through an auction (IPO) increases.

Hypothesis 4. As the common value component decreases, the entrepreneurial firm's probability of being sold through an auction (IPO) increases.

Market thickness is determined by the number of buyers and sellers in a market, and the number of sellers and buyers party to an exchange have a very strong effect on the outcomes of auctions or negotiations. For example, if there is only one bidder, and the sale of a resource is negotiated, the seller is always better off inviting a second potential buyer and holding an auction without a reservation price. The restriction of qualified bidders to enter the bidding has a direct effect on the seller's expected revenues. Entry or restriction into markets can be induced by the use of reservation prices, or simply by endogenous choices.[1]

As discussed earlier, the opening price for an issue may be regarded as the reservation price, and is inhibitive in independent private value auctions, where unrestricted entry is optimal. In common value auctions on the other hand, sellers will want to discourage entry (Kagel, 1995). In first-price independent private value auctions, increasing the number of bidders results in more aggressive bidding. In situations

[1] When Ravenswood Winery decided to sell the firm two years after their initial public offering, the owners held an auction, and invited 5 potential bidders. After a period of due diligence, 3 bidders dropped out. Theoretically, we know that with 5 bidders the first price auction would generate revenues about the same as a multilateral negotiation (Thomas and Wilson, 2002). However, once the number of bidders dropped to two, negotiations would be more favorable to first-price auctions. Ravenswood at this time released the information about the competitors' bids and terms for the acquisition, creating a multilateral negotiation, hence maximizing its revenues from the sale.

where the number of potential bidders is unknown, auction or negotiation outcomes are strongly affected by bidders' risk neutrality or risk aversion.

Hypothesis 5. Entrepreneurial firms are more likely to choose auctions as market thickness and private value component increases.

Hypothesis 6. Entrepreneurial firms are more likely to choose negotiations as market thickness decreases, and the common value component increases.

Risk propensity assumptions, risk aversion and risk neutrality of sellers (as well as the bidders), are seminal for conclusions regarding the expected revenues. From the sellers' perspective, an entrepreneur when choosing a mechanism will probably consider whether alternatives will result in equivalent revenues. If the expected revenues were equivalent, he/she would be indifferent. However, the mechanisms will rarely have equivalent outcomes, which makes the mechanism choice more salient. The ownership and debt structure of the entrepreneurial firm provide important signals about the risk propensity of the owners/managers. Varying degrees of ownership, and the incentive mechanisms in place alter the risk level in the firm.

From the buyers perspective, if the bidders are assumed to be risk neutral, then we would expect them to maximize their expected profits, whereas, risk averse bidders would maximize their expected utilities. With risk neutral bidders, the expected revenues from different auction mechanisms will be equivalent (Riley and Samuelson, 1981; Vickrey, 1961). With risk averse bidders, they tend to bid above the risk neutral Nash equilibrium levels, and first-price and Dutch auctions generate more revenues than the English or second-price auctions. The revenue equivalence theorem also fails in private value auctions: first-price auctions generate higher revenues than Dutch, and second-price auctions generate higher revenues than English auctions.

Hypothesis 7. As the level of institutional ownership increases, risk averse owners will prefer auctions (IPOs) to negotiations (M&As).

Hypothesis 8. As the level of institutional ownership decreases, risk neutral owners will prefer negotiations (M&A) to auctions (IPOs).

Search costs exist due to information asymmetries among parties to an exchange. For each buyer, there is a cost associated with switching from one seller to another, therefore each search is costly. In spot markets, there is no reason to search for resources, since all that is available is priced the same, and the prices reflect all relevant information about a good. The market reaches a competitive equilibrium under perfect information where search costs are negligible, search is costless but delaying the bargaining is (Arrow and Debreu, 1954).

However, market for firms is far from this frictionless Walrasian space, and price formation for firms, and buyers' search strategies are interdependent

(Rothschild, 1973). Each seller and each product is different, and the buyers search for other outside options if one bargain fails. Within industry effects are instrumental in the search cost arguments. Although information is released to all potential investors, some information is valuable or less costly to industry insiders, while the same information may have higher information costs associated with for outsiders. Hence, although the rents are low, some buyers might stay in the market to negotiate or bid for a seller in order to avoid information proccssing and search costs (Salop and Stiglitz, 1977).

Hypothesis 9. The higher the search costs, an entrepreneurial firm will be more likely to be sold though an auction (IPO).

4. IPOs AS AUCTIONS

When an entrepreneur decides to go public through an equity offering, he/she is faced with two choices: bookbuilding vs. fixed price contracts. In the bookbuilding model, the entrepreneur picks its IPO team, consisting of an investment bank (underwriter), an accountant, and a law firm, in order to establish a detailed financial report history, going back at least two years before the offering. The team works on a prospectus, that includes all the financial data of the company for the past five years, information on the management team, and a description of the company's target market, competitors, growth strategy, etc. The prospectus is filed with the Securities and Exchange Commission (SEC) and the National Association of Securities Dealers (NASD) and reviewed for accuracy. The lead underwriter assembles a syndicate of other investment banks, allocates a certain number of shares to sell to their clients, and takes the entrepreneurial firm on a road show. This trip involves meetings with large institutional investors and banks. The underwriter tries to come up with a demand schedule for various price levels to determine the offering price.

If the entrepreneur chooses the fixed price method, the goal is to generate information cascades among early and late investors (Welch, 1992). This type of IPO mechanism allows sequential selling in the market and this enables early investors to be informative for the late investors. The early investors have greater market power and can generate information cascades through their investments (Benveniste and Busaba, 1997). The fixed price method has the potential to exploit the market by initially pricing the offering low enough to lure investors, hence causing a buying frenzy. When subscription decisions are made simultaneously, the winner's curse is avoided.

The fixed price offering is a common value auction, and it is less attractive for IPOs involving private values because the mechanism does not allow the seller to discover bidder's valuations. Historically the dominant approach in most of Europe, and especially in the U.K. (and the British colonies) has been the fixed price method (Benveniste and Busaba, 1997). The fixed price method tries to discover the price without first soliciting investor interest. Investors make decisions based on correlated pieces of information about the true value of the stock, which is revealed after the offering (Welch, 1992). Since the bidders are unclear about the value of the item at the time of bidding, but receive signal values that are related to (affiliated with) the value of the item, judgment failures result in winner's curse. An investor's purchasing decision is informative about his signal, and impacts the reservation prices of late-mover investors. When an investor ignores his reservation price, and purchases the stock, a cascade is formed. If the selling price is higher than first-mover's signal, a negative cascade develops, and all investors refrain from investing (Benveniste and Busaba, 1997). To avoid this, the issues are more underpriced initially when compared to bookbuilding pricing.

The bookbuilding method on the other hand aims to secure honest responses from investors on the offer price and allocations. The underwriter tries to set an offer price that better reflects the aggregate market valuation. This aggregate information is revealed to potential investors; therefore no one individual investor can have an effect on the overall market valuation. Since the underwriter solicits non-binding indications of interest from institutional investors and other investment banks, the IPOs conducted through bookbuilding mechanism can be referred to as Vickrey auctions, where the bidders' best strategy is to bid their true values.

The marketing method used to auction off entrepreneurial firms—whether the issuer uses bookbuilding or fixed price—will have an effect over the outcomes. The choice between these two forms is dependent on the resource value, risk propensity of the entrepreneurial firm, the issue size, and the marketing costs. Bookbuilding generates higher expected proceeds, and provides an opportunity to sell additional shares at full price after the IPO, but exposes the issuer to a greater risk. On the other hand, the fixed price offerings guarantee the issuer a certain level of proceeds, but at a lower level when compared to bookbuilding (Benveniste and Busaba, 1997). Neither strategy dominates, since the issuer acts on behalf of the entrepreneur, whose choice may be influenced in the first place by future financing concerns, considerable costs of the IPO, and the market trends which are highly cyclical.

Hypothesis 10. As the common value component increases, an entrepreneurial firm is less likely to be sold through the bookbuilding method.

Hypothesis 11. As the private value component increases, an entrepreneurial firm is more likely to be sold through the bookbuilding method.

5. M&As AS NEGOTIATIONS

Many management scholars treat mergers and acquisitions as auctions, where several bidders strategically bid to acquire another firm. Moreover, the U.S. competition authorities use auction models to evaluate the impact of proposed mergers and determine the antitrust implications (Baker, 1997), and yet technically M&A activities are either bilateral or multilateral negotiations, and parties bargain to buy-sell corporations. Furthermore, as stated before auctions are disadvantageous when there are bidders with interdependent valuations, which is the case for the majority of M&As considered as auctions. To avoid confusion and mislabels, I would like to define the terminology I will use in this context.

When an entrepreneurial firm puts itself on the M&A block, several interested potential buyers approach the firm, and bid to acquire the controlling rights of the target. The seller has to negotiate a deal with the potential buyers, and this is called a multilateral negotiation.[2] If there is only one buyer and one seller, this negotiation is called a bilateral negotiation (Thomas and Wilson, 2002), which might be the case when the firm is so specialized in its assets, resource base or functions such that only one buyer can appropriate the full value given its unique characteristics.

The experimental results show that multilateral negotiations closely resemble first-price auctions and are outcome-equivalent with a large number of bidders, whereas transaction prices are higher with only two bidders in negotiations than in auctions (Thomas and Wilson, 2002). In negotiations involving less than four buyers, multilateral negotiations yield higher revenues for the seller than first price auctions. However, when the number of bidders reaches four or more, there are no statistical differences with first price auction outcomes and negotiation outcomes. Two issues in M&A activities are critical for my analysis: uncertainty regarding the outcome of the acquisition, and uncertainty regarding the value of the target firm.

Negotiation outcomes are uncertain and bargaining is indeterminate (Roth, 1995). From a cooperative game theory perspective there are a number of Pareto optimal outcomes that can be achieved, and from a non-cooperative perspective, parties to an exchange can reach a number of feasible outcomes based on their expected utilities (Nash, 1950). In practice, firms may negotiate the sale of assets and/or equity stock, and the decision to sell or not may depend on a variety of factors. For example, Seth and Easterwood (1993) talk about managerial and shareholder interest alignment in the case of management buyouts.

---

[2] When two or more parties bargain over the division of a common surplus, this is called multilateral bargaining (Krishna and Serrano, 1996). For example, a merger or an acquisition as a result of a renegotiation between parties to a joint venture would in fact be an outcome of a bilateral (or multilateral) bargaining.

Acquisition outcomes are also altered in the case of takeovers. For example, in the case of initial partial equity ownership (toehold), overbidding is an optimal rational strategy, but can lead to an inefficient outcome (Burkart, 1995; Hirshleifer and Titman, 1990). A toehold makes a bidder more aggressive, and it increases the winner's curse for a non-toeholder. But, owning a toehold helps a bidder win cheaply only if there are no other toeholds. If there are other toeholds of equal or larger sizes, prices will be higher, and bidders sometimes exceed their ex ante expected profits that would occur in ex post acquisition (Bulow et al., 1999). The uncertainty surrounding the potential outcomes of bargaining behavior increases the winner's curse potential for some acquirers. Such uncertainty also discourages risk adverse owners/managers to choose negotiations as a mechanism.

Hypothesis 12. Entrepreneurial firms with partial equity owners are more likely to be sold through negotiations (M&As) than auctions (IPOs).

Empirical evidence suggests that markets for buying and selling companies are reasonably competitive, and an acquirer pays approximately the discounted present value of the target firm. If any, all above normal returns go to stockholders of the acquired firms (Porter, 1980). Firms can only generate above normal returns if the cost of resources to implement product market strategies is significantly less than their economic value (Barney, 1986). This requires firms to exploit competitive imperfections in factor markets. In other words, valuation of a target firm requires heterogeneous expectations about its future cash flow.

There is an important difference between bidding a premium and overbidding. If the bidder overestimates the value of the firm after the acquisition when incorporating the ex post synergistic gains, he/she will be subject to winner's curse. Two problems need to be addressed: valuation error and managerial hubris. Assuming decision makers are rational, those bidders that repeat acquisitions will learn from their past mistakes. It is also possible that "markets learn" as well as the individuals, as bankruptcies drive out aggressive bidders (Kagel, 1995) and individuals can learn by observing and adjusting their bids. However, in bilateral bargaining games, observational learning is not possible since the bidders can only see the outcome of their own choices (Garvin and Kagel, 1994).

The hubris arguments on the other hand are strongly related to the assertion that all markets (product, financial and labor) are strong-form efficient (Roll, 1986). This may be true, especially for the takeover markets of public targets, where the bidder receives a value from the market, and all he/she has to do is to determine whether this value represents the true value or not. However, the market for privately held targets is expected to have strong information asymmetries due to the existence of private information among targets' insiders. The privately held target firm will accept at least its reservation price, which is a lower bound. If the bidder expects that there will be potential synergies after the acquisition, he/she makes an offer. If this valuation is too high, and in error, the markets observe only the right hand side tail of the distribution; since the left side is never revealed (the error is always in the same direction). In the case of private targets, the hubris arguments do not apply in full effect because most of the bidders do not have full information about the target at the time of negotiations.

Related to the above discussion, one might ask what exactly do targets sell and bidders buy. In essence, ownership and control rights between the entrepreneur and the acquirers exchange hands. Control rights matter because it is impossible to write comprehensive contracts, and as new opportunities come along, it makes a difference who makes investment decisions. Furthermore, the control rights over assets and equity influence the ex post division of surplus (Hart, 1988). Equity sale for the control of cash flow rights places the responsibility for immediate decisions on the hands of the entrepreneur (or manager), where as ownership rights over assets shift the ex post rent appropriation to the acquirer. The capital structure for raising funds are strongly influenced by volatility of cash flows, and the net present value of the projects the firm is undertaking. If the assets that are currently held by an entrepreneurial firm have similar foreseeable economic rents among a number of potential bidders, then the entrepreneur is more likely to sell the assets. Unless these assets or resources generate heterogeneous expectations among the potential bidders, it is not possible to generate economic rents from the sale.

Hypothesis 13. As the common value component increases, an entrepreneurial firm is more likely to sell assets (ownership rights).

Hypothesis 14. As the private value component increases, an entrepreneurial firm is more likely to be sell equity (residual cash flow rights).

6. EMPIRICAL ANALYSIS

6.1. Empirical design

The empirical predictions for the market mechanism choice can be described as the choice tree depicted in Figure 1, and are summarized in Table 1. It is argued here that an entrepreneur will choose to auction off his/her company to public investors, or negotiate to private buyers if five conditions provide a suitable environment for the transaction. Once the decision is made, the entrepreneur follows the deal through, or he/she can back down, but all actions will have transaction costs associated with them. If the entrepreneur negotiates the sale, the deal is either realized or not; if he/she decides to auction off the firm, he can choose either the bookbuilding methods

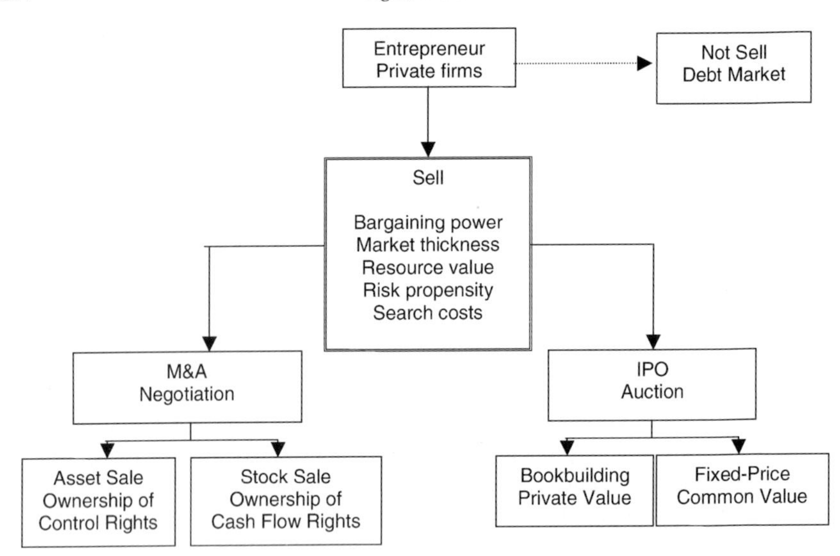

Figure 1. Empirical model of auction-negotiation mechanism choices.

or the fixed price contract method, but these choices are non-sequential. Whether an entrepreneur wants to sell his/her company using a specific market mechanism is an endogenous decision, but who buys the firm is not. In other words, for the seller deciding whether to auction the firm using an IPO, or to negotiate through an acquisition is a deliberate choice. The choice whether the sale is to a public or private buyer depends partially on the seller's disclosure preferences, as well as other exogenous factors. In the sections that follow, I will discuss the data, the variables, and the methodology.

6.2. Data and sample description

I constructed a panel dataset containing information on initial public offerings of private firms and mergers and acquisitions of private targets, across time, industry and country, using four different databases. For the initial public offerings, I used the Global New Issues Database maintained by the SDC Thompson Financial Securities. The data from this source covered private companies that issued common stocks between 1971 and 2000 in the U.S., 1991–2000 in Continental Europe both Euro and foreign common stock. Unit offers, ADRs, closed-end funds, REITs, bank and S&L IPOs are excluded from the sample.

Table 1. Summary of empirical predictions for the choice of auctions vs. negotiations

Hypotheses	Variables Prediction for . . .	Auction vs. Negotiation IPO vs. M&A Probability of Auction	Fixed vs. Bookbuilding Common vs. Private Probability of Fixed	Asset vs. Stock Control vs. Cash Flow Probability of Stock
Resource value	Patent originality	No prediction/Not included	–	No prediction/Not included
	Patent generality	No prediction/Not included	+	No prediction/Not included
	Tobin's q	+	–	–
	High-tech dummy (high-tech = 1; other = 0)	+	–	–
Bargaining power	Patent dummy	No prediction/Not included	–	No prediction/Not included
	Same industry dummy	No prediction/Not included	No prediction/Not included	–
	Prior relationships	–	No prediction/Not included	–
Market thickness	# of firms in target's industry	+	+	+
	# of offers sought	No prediction/Not included	No prediction/Not included	+
	# of tender offers extended	No prediction/Not included	No prediction/Not included	+
Risk propensity	Debt/total assets	–	–	+
	Cash/total assets	–	–	+
	Investor involvement	+	+	+
	VC involvement	+	–	+
Search costs	# of advisors of the target	+	–	+
	Investment bank reputation	+	–	+
	VC activity in focal industry	+	–	+
Interactions	# of firms × High-tech	+	–	No prediction/Not included
	# of firms × Low-tech	–	No prediction/Not included	–

From SDC, I obtained the M&A data from Mergers & Acquisitions Database for both U.S. and Foreign targets, with private sellers (targets). Recapitalizations, privatizations, self-restructurings, and leverage-buyouts are excluded from the sample. In practice, although some deals are announced, it is sometimes the case that these deals are not realized; hence I checked the accuracy of the SDC data using the DoneDeals Database from NVST, Inc. (Private Equity Network). The DoneDeals data is available from 1994 to 2000, so for the period 1991–1994, I used the LexisNexis search engine to randomly verify the announcements. The DoneDeals data reports mid-market M&As of consummated transactions with transaction values between $1 million and $250 million. The database provides both private and public sellers, but public sellers are excluded from the sample.

From the Global New Issues Database, for the following regions, initial public offering information is provided (Tables 2 and 3): The United States market-U.S. domestic public offerings of common stock for private companies, 1970–2000. The EuroMarket-common stocks sold in the EuroMarket, including foreign market issues, 1983–2000. The Continental European market-common stocks sold in most major Continental European nations, 1991–2000. The countries included are: Austria, Belgium, Bulgaria, The Czech Republic, Denmark, Finland, France, Germany, Greece, Hungary, Ireland, Italy, Luxemburg, Netherlands, Norway, Poland, Portugal, Slovakia, Spain, Sweden, and Switzerland. The International market (1983–2000), and the Domestic U.K. market for the public offerings of common stock in England, 1989–2000. Rest of the World—common stock sales of private companies from Australia and Asia Pacific, 1984–2000.

For the mergers and acquisitions of private companies of both U.S. targets and non-U.S. targets, I used the Worldwide M&A database (Tables 2 and 3). The domestic U.S. targets sample covers firms between 1979 and 2000, and for non-domestic targets, between 1985 and 2000. I separated the M&A dataset into two samples: stock sale and asset sale samples. A firm is considered to conduct a stock sale if: a combination of business takes place or 100% of the stock of a private company is acquired (merger), a deal in which 100% of a company is spun off or split off (acquisition), the acquirer holds less than 50% of the company's stock, and seeks to acquire 50% or more but less than 100% of the total number of stocks (acquisition of majority interest), the acquirer holds less than 50% of the company's stock, and seeks to acquire less than 50% or less than 100% of the total number of stocks (acquisition of partial interest), the acquirer holds over 50% of the stocks, and seeks to acquire 100% of the rest (acquisition of the remaining interest).

For the asset sale sample, a firm is considered to conduct an asset sale if: the assets of a company are acquired (acquisition of assets), a certain portion of the total assets is acquired (acquisition of certain assets). From both samples leveraged recapitalization such as one-time dividends, preferred stock or debt securities sales

Table 2. Manufacturing sector by (a) 2-digit SIC codes and (b) target nation

	IPO		M&A	
	N	Fraction (%)	*N*	Fraction (%)
(a) 2-digit SIC codes				
SIC				
20	169	4.04	3795	11.24
21	7	0.17	60	0.18
22	51	1.22	849	2.51
23	91	2.18	866	2.57
24	41	0.98	657	1.95
25	36	0.86	728	2.16
26	67	1.60	962	2.85
27	120	2.87	3403	10.08
28	681	16.30	3179	9.42
29	18	0.43	239	0.71
30	78	1.87	1424	4.22
31	27	0.65	231	0.68
32	48	1.15	1365	4.04
33	111	2.66	1256	3.72
34	100	2.39	1987	5.89
35	649	15.53	4200	12.44
36	859	20.56	3538	10.48
37	168	4.02	1635	4.84
38	718	17.19	2399	7.11
39	139	3.33	988	2.93
Total	4178	100	33761	100
(b) Target nation				
Country				
Canada	46	1.10	1090	3.23
China	3	0.07	357	1.06
France	13	0.31	2442	7.23
Germany	9	0.22	2528	7.49
Hong Kong	18	0.43	169	0.50
Israel	55	1.32	92	0.27
Italy	6	0.14	1335	3.95
Japan	4	0.10	396	1.17
Mexico	13	0.31	167	0.49
The Netherlands	19	0.45	683	2.02
Singapore	8	0.19	154	0.46
Sweden	9	0.22	414	1.23
Switzerland	5	0.12	432	1.28
United Kingdom	14	0.34	5207	15.42
United States	3898	93.30	11858	35.12
Other	58	1.39	6437	21.10
Total	4178	100	33761	100

Table 3. Logit model coefficient estimates for the choice of auctions (IPOs) vs. negotiations (M&As)

Probability modeled is mechanism = IPO

Variable	Panel A: Model 1			Panel B: Model 2			Panel C: Model 3		
	Estimate	Wald χ^2	Pr > χ^2	Estimate	Wald χ^2	Pr > χ^2	Estimate	Wald χ^2	Pr > χ^2
Intercept	236.780	3.540	0.0599	−206.963	113.650	<0.0001	239.000	3814.281	<0.0001
Bargaining power									
Patent dummy (patent = 1; no patent = 2)	−5.898	31.710	<0.0001	–	–	–	−5.540	30197.282	<0.0001
Resource value									
High-tech dummy (high-tech = 1; other = 0)	–	–	–	–	–	–	−0.438	310.760	<0.0001
Patent originality	–	–	–	1.047	52.0150	<0.0001	–	–	–
Patent generality	–	–	–	−0.490	11.545	0.0007	–	–	–
Exogenous factor									
Deal year	−0.114	3.240	0.0720	0.107	118.345	<0.0001	−0.115	3527.339	<0.0001
Wald test of model fit									
χ^2	46810.690			31142.780			34694.957		
Prob. > χ^2	<0.0001			<0.0001			<0.0001		
R^2	0.625			0.513			0.547		
Number of observations									
IPO	112664			61321			61740		
M&A	112317			2656			111230		
Log-likelihood	46810.692			−3537.845			136863.630		

are excluded. Also, if a company buys back its equity securities or converts its securities on an open market (either through private negotiations or through a tender offer), these transactions are excluded from the sample.

Finally, I obtained the patent data from the National Bureau of Economic Research (NBER) for the period 1963–1999. The dataset (PAT63_99) is complied by Hall et al. (2001). I matched the patent data with the Compustat data using the assignee (number). The patent dataset includes all utility patents in the USPTO's TAF database, and has 2,923,922 observations. Apart from the ten original variables issued by the USPTO, I also used ten other variables constructed by Hall and colleagues. The sample does not cover minor patent categories (design, reissue, and plant). In the data there are seven categories of assignees: unassigned (not yet granted, 18.4%), U.S. corporations (47.2%), non-U.S. corporations (31.2%), U.S. individuals (0.8%), non-U.S. individuals (0.3%), the U.S. government (1.7%), non-U.S. governments (0.4%). I excluded firms without cusip numbers, all U.S. government and non-U.S. government patents. In the final sample, I focused on manufacturing sector since approximately 32% of all IPOs are in manufacturing, followed by 25% in services.

7. VARIABLES

In examining the discrete choice of the entrepreneur between the market types and mechanisms, in this section I focus on explanatory and control variables. The explanatory variables are bargaining power, market thickness, resource value, risk propensity and search costs. The control variables are industry related characteristics, market-timing factors, other deal-specific factors, and financing factors. In the following subsections, I discuss the possible impact of these factors on the mechanism choice of the seller.

7.1. Explanatory variables

Bargaining power

Entrepreneurial firms gain bargaining power as they gain expertise and experience in the market, and as they accumulate assets. These firms also gain advantages in negotiations and signal uniqueness if they hold a patent. Especially in the biotech industry, firms that have formed alliances are more likely to negotiate a deal with their alliance partners. However, this does not mean that an entrepreneurial firm will always chose to be acquired. If the firm's goal is to generate capital, hence sell a minority equity stake, an IPO may be more feasible. In this case, firms that have formed alliances will have a higher propensity to IPO than firms that have not formed alliances (Stuart et al., 1999). Hence patents, prior related deals and the same industry dummy are used to establish bargaining power of the entrepreneur. I use the number of patents prior to the deal to identify entrepreneurial firms' bargaining power.[3] I expect a positive relation between the number of patents and the bargaining power of the seller. Second, the same industry dummy shows if the target and the acquirer are in different 2-digit industry codes. If the target's industry is different than the acquirer, I expect the target's bargaining power to be higher. Finally, prior relationships (toehold by the bidder, and related prior deal dummy) are used: if the acquirer has a toehold, then the target has higher bargaining power. The prior deal dummy (yes or no) will have the same positive relationship.

Market thickness

From the perspective of the entrepreneur, the number of bidders (buyers) is critical. In most of the auction literature, the number of bidders n is determined exogenously. If the seller prefers one institution and fixes n, the bidders have fewer incentives to enter the auction in the first place. When the entry varies stochastically (Harstad et al., 1990) or is endogenously determined, this has an impact on the coordination costs associated with incentives. Experimental results provide interesting comparative static predictions when the number of bidders is varied and is common knowledge, and when the number of bidders is unknown. With constant or decreasing absolute risk averse bidders, expected revenues increase if the number of bidders is concealed. In second price auctions, the number of bidders in the market has no effect on expected revenues, and in third-price auctions result in lower expected revenues for the seller. Also, even if bidders are bidding for a resource based on their private values, if they are within the same industry and they repeatedly bid for the same resources, their values are going to be correlated, hence the private value component would decline. I use the number of firms in target industry prior to the deal, the number of offers considered, the number of offers sought, and the number of tender offers extended to measure how thick the market was for the sale.

---

[3] Patents grant temporary monopoly rights in exchange for disclosure, and this clearly results in increased bargaining power for the owner of the patent. Also, The choice of selling for the firm hence will be strongly affected by the prior history of forming alliances, and its innovative performance. If an entrepreneurial firm has operating experience (Baum and Ingram, 1998), competitive experience (Barnett et al., 1994), or collaborative experience (Anand and Khanna, 2000), it will gain skills to coordinate and manage them (Sampson, 2002), and these will be a source of bargaining power.

Resource value

Empirically, it is very difficult to find pure common or private value goods. Instead, goods will have a combination of both common and private values. For the purposes of this paper, I will refer to a resource with more common values than private, a common value resource, and more private values than common, a private value resource. Patents provide a very unique opportunity to proxy for the value components of a resource.

I group firms in high-tech vs. not[4] and proxy the private component of the firms' innovative capability. Firms in high-tech industries such as biotech, computer equipment, electronics, communications, and general technology accumulate specific know-how and intellectual capital that are unique in their industries regardless of their patents. Most of these patents in high-tech industries will have private value components then common. Low-tech industries on the other hand, report R&D spending, which is not a good measure of the output of their R&D activities (Griliches, 1990). This grouping is consistent with the argument that firms that are in "complex product industries" (high-tech) use patents to force rivals into negotiations, whereas, firms in "discrete industries" (non-high-tech) use patents to block the development of substitutes by rivals (Cohen et al., 2000). By definition, common value components would be most easily replicated or replaced by competitors.

I use originality of the patent, generality of the patent, and the industry cluster to measure the resource value. In the sample, the "originality" vs. "generality" of patented innovation, provides basis for the resource's common and private value components. An original patent will have heterogeneous expectations associated with it, and the interested bidders most likely will have diverse uses, whereas with a general patent, the signals about the true value of the patent will be closely affiliated.[5] Hence I use the Trajtenberg et al. (1997) measures. According to this

---

[4] The industry type of the firms also affects market reaction. Especially with high-tech firms, it is shown that an IPO is more likely regardless of any existing positive earnings (Maksimovic and Pichler, 2001). Hence, firms are categorized as high-tech vs. low-tech using classifications provided by the SDC. High-tech industries would be biotechnology, chemicals, communications, computers, defense, electronics, medical, and pharmaceuticals.

[5] The generality component of the patent refers to forward citations as indicative of the impact of the patent. The originality component refers to citations made to, indicative of the depth and breadth of technologies it uses. This is similar to the weighted patent citation Sampson (2002) uses. She assigns a weight to each patent using citations made by later patents to identify the technological lineage of the invention. An alternative approach to this valuation problem is by studying the number of subclasses into which the patent office assigns the patent as in Lerner (1994). Lerner found that an increasing scope of patent is associated with higher valuations. In other words, a lot of firms found the patent valuable, and applicable to their existing resources and capabilities. Thus, if they were to bid for this resource, they would be bidding for a common value resource, and the highest bidder would most likely be subject to winner's curse.

measure, if a patent is generalizable, it will have a widespread impact by influencing subsequent innovations.[6]

The originality of the patent will render it only applicable to a narrow set of technology applications (Hall et al., 2001). Due to the count nature of the underlying data, generality and originality measures are downward biased when the citations for a patent are small (e.g. for patents granted later in the dataset between 1963 and 1999). I use the unbiased estimators for both measures correcting for the bias.

Risk propensity

In an entrepreneurial firm, if the owner's or managers' reputations or wealth are affected by firm performance, their actions will be influenced by their attitudes towards risk; and their propensity to take risks (or avoid them) signal other firms and investors information about the value of the firm, and its future prospects (Blazenko, 1987). When the entrepreneur retains the fractional ownership of a firm, this is a signal of firm quality. As the fraction increases, the value of the firm increases, and if the entrepreneurial ownership is sufficiently great, increases in ownership are associated with increased debt (Leland and Pyle, 1977). In other words, if managers are sufficiently risk averse, they signal low firm value with equity, and high firm value with debt. In the Modigliani–Miller paradigm where capital markets are frictionless, all individuals have homogeneous expectations and can borrow and lend at a risk-free rate, they do not pay taxes and bankruptcy costs do not exist; shareholders are indifferent to capital structure but managers prefer equity (Blazenko, 1987). Risk averse managers will try to avoid debt because increased debt would increase the total risk of share ownership. Owners (if they are not managing the firm), can tie the managerial performance to equity performance, and risk averse managers will prefer equity over debt.

To measure risk propensity of the firm, I use the debt/total assets and debt/cash ratios, employee involvement, management involvement, and venture capital involvement. If there is investor involvement, the number and the concentration of shareholders will also influence risk propensity of the owners/managers, thus the mechanism choice. If the manager is the owner, and a tender offer is made to takeover the company, an acquisition is more likely to occur (Buchholtz and

[6] Firm's innovative performance is commonly measured via its citation count (Sampson, 2002). The citations of previous patents identify technological lineage, and there is strong correlation between the value of the invention and its citation count. Generality measure is the percentage of citations received by patent i that belong to patent class j, out of n_i patent classes. The aggregate is the Herfindahl concentration index.

Ribbens, 1994). When the entrepreneur is also the manager, the principal and the agent are the same, and the nature of incentives change. Further, if the number of shareholders is small, the owner's incentive to accept an acquisition is not fully aligned with the small number of shareholders (Alchian and Demsetz, 1972). Small number of shareholders would prefer to offer the stocks through an auction.

I intentionally use entrepreneurial firms and the entrepreneur as the decision maker and study entrepreneurship as a function of actions.[7] Consistent with the agency perspective, an individual will shy away from risky modes, hence, in the case of selling his company, he/she will prefer auctions to negotiations since negotiation outcomes are difficult to determine ex ante. On the other hand, as the ownership claims are less, entrepreneurs will decrease their efforts, unless the owner disperses hierarchy (and the residual risk) to a small number of team entrepreneurs. Ownership and debt structure variables are used to test the variance in performance of individual vs. team entrepreneurs.

Search costs

In this chapter, IPO selling is less costly than the acquisition option. When an entrepreneurial firm is to be acquired both the search for the firm and the time to negotiate are cost items, whereas, the IPO process reveals more information about the firm to all interested potential buyers publicly. Related to both bargaining power and search costs, empirical evidence on biotech firms shows that biotech-pharma alliances are subject to both dynamics. While small entrepreneurial firms do come up with new innovations, it is more likely to pass the clinical trials if they form alliances with pharmaceutical firms with heavy discounts; and that these discounts are rational (Nicholson et al., 2002). Further, these biotech firms receive substantially higher valuations from venture capitalists, and the public equity market after the alliance. Hence, an alliance in the first period lowers the search costs of the sellers and the buyers, and increases rents of the pharmaceutical firm, which has a higher bargaining power. In the second period, the formed alliance increases the bargaining power of the entrepreneurial firms and they are more likely to send a positive signal to investors.

[7] This approach is consistent with Busenitz and Barney (1996). Further, a distinction between an "individual" versus "team entrepreneurs" needs to be made. It is argued that the team entrepreneurial forms will exhibit more variation in performance than will individual entrepreneurial forms, ceteris paribus because of risk aversion (Mosakowski, 1998). This argument is closely tied to risk aversion (Fama and Jensen, 1983) and shirking (Jensen and Meckling, 1976).

Another item that the entrepreneurs use to signal quality and value of their firm (thus lower the search costs of the buyers) is the investment bank they use as underwriters. IPOs underwritten by high prestige investment bankers yield smaller initial underpricing and less negative long run returns than the IPOs underwritten by lower reputation investment banks (Michaely and Shaw, 1994). Similarly, IPOs backed by venture capital outperform nonventure-backed offerings in the long-run (Brav and Gompers, 1996). To incorporate the fixed effect of the lead underwriter, I use the Carter-Manaster (1990) rankings for the investment banks. I proxy for search and coordination costs by using the number of advisors of the target, the lead underwriter, and whether the industry of the focal firm is one where venture capital firms are actively involved or not.

7.2. Control variables

Industry related factors

Herfindahl index for industry concentration is used to control for firm survival. Survival may be difficult in highly concentrated industries for private entrepreneurial firms (Sharma and Kesner, 1996); since small firms may be a target for acquisition (or takeover). The Herfindahl index is a measure of degree of competition within an industry, and is the sum of squared market shares of all members of a particular industry. I use the sales data to form an index value. A high index value shows higher industry concentration.

Another categorization is financial vs. service industries, where fragmentation and consolidation is known to affect competition, and firm values across industries (Berger et al., 1999). In highly fragmented industries such as the financial services industry, deregulation has increased consolidation and acquisitions are more likely to occur, and an IPO becomes less common (Holmström and Tirole, 1993).

In the risk propensity measures above, I use firm's debt level as a proxy for risk preferences of the owner/managers. To control for industry wide effects, leverage ratios of firms across the same industry are proxied to give a normal debt level (Harris and Raviv, 1990). This is important to distinguish the differences between high debt/asset firms from the rest across the industry. Some industries are more appealing for IPOs whereas others are targets for acquisitions. Similar to the leverage ratios, the industry market-to-book (M/B) ratio is calculated to control for industry characteristics as a baseline for each firm within the same group. In industries with high M/B, IPOs are more likely to occur (Pagano et al., 1998), whereas low M/B attracts takeovers.

Market-timing factors

In market timing theories of Lucas and McDonald (1990), and Choe et al. (1993), firms delay their initial public offerings, either because they are undervalued or because there are other competing IPOs being issued with favorable pricing. Apart from these rational models of market timing, Ritter and Welch (2002) provide an alternative explanation for semi-rational perspective without asymmetric information, incorporating the lag the entrepreneurs use to adjust their valuations. On the other hand, the cyclical nature of the public offerings results in a time trend, where entrepreneurs cluster the offers predominantly over the same time frame (Lowry and Schwert, 2002; Ritter, 1984). To proxy for the market cycle, I use the return on the stock market, and the volume of public offerings vs. acquisitions to establish the hot vs. cold markets.

Other deal-specific factors

Firm and deal size are two factors that need to be controlled for various reasons. First, firm size is argued as a clear indication whether a firm can compete after its initial offering (Pagano and Roell, 1998). Second, the process of initial public offering involves very high fixed costs (Ritter, 1987). If the firm is pursuing a bookbuilding method, it involves further costs for trips to potential institutional investors. Small private firms will prefer less costly alternatives to high cost alternatives (Holmström and Tirole, 1993). Third, insider ownership and changes in the ownership-control structure has a potential effect on the choice of entrepreneurial firm's sale type.

In an acquisition, the controlling stake changes hands, but in an IPO, the entrepreneur still controls the firm if the owners design the offering or acquisition as such. Owners will prefer public offerings for smaller liquidity, and acquisitions for maximum liquidity and completely cash out (Zingales, 1995). To proxy for firm and deal size, three measures are used to accommodate different industry norms. For example, in the biotechnology industry, most firms do not reach the marketing stage of their innovations; hence the number of employees is adequate to establish firm size (Shan, 1990). On the other hand, some industries invest heavily in their initial setup especially in manufacturing and transportation industries; hence proxies for scaled transaction values on the total assets are more appropriate measures for firm size (Pagano et al., 1998). For services firms, sales are used as another proxy for firm size. In the sample, firms are controlled for the total assets, number of employees and sales.

Financing factors

In a general equilibrium setting, it was argued that given uncertainty, risk neutral and risk seeking individuals would become entrepreneurs, and risk adverse individuals would become laborers (Kihlstrom and Laffont, 1979). This assertion assumes that all potential entrepreneurs are equally able and industrious, which is a highly restrictive assumption. Relaxing this assumption, it was argued that risk adverse entrepreneurs would sell to risk neutral investors if they thought they would get full value for their ideas (Amit et al., 1990). This raises the issue on how the entrepreneurial firm is financed, and what kind of signals this structure has on the expected revenues. Firms that signal quality and high future returns through venture capital funding and endorsements are more likely to conduct IPOs (Shane and Stuart, 2002).

8. STATISTICAL METHOD

When an entrepreneur decides to sell his/her company, he is faced with a series of single decisions among two or more alternatives. Each of these alternatives is mutually exclusive and is unordered. In other words, the focus of this empirical study is the decision after the entrepreneur decides to sell, and when he sells, he/she can either sell to another firm through negotiations, or he/she can auction off the firm to a number of public investors. Hence, I use a modified unordered conditional logit model. In a standard conditional logit model, it is assumed that the variability in scores for one variable is roughly the same at all values of the other variable, which is related to normality. When we relax the homoscedasticity assumption, we can group the alternatives into subgroups with their variance differing across and maintain the independence of irrelevant alternatives assumptions[8] (IIA) within the groups. This slight modification of the stochastic specification in the original conditional logit model defines a nested logit model (Greene, 1997).

Originally developed by McFadden (1979, 1981), the nested logit model maintains the dependence among alternatives between levels of the nest, with the equal pattern of dependence occurring within a level of the nest. This approach links different but interdependent decisions, and also decomposes a single decision so that

[8] The IIA assumptions imply that the odds-ratio between two alternatives does not change by the inclusion of any other alternative. In estimation of discrete choice models the major issues arise due to computational intractability for more than three alternatives (Maddala, 1983; McFadden, 1981). Nested structures solve these problems for single choices around many alternatives, and can be estimated either sequentially, or simultaneously. Please refer to Hensher (1986) for a detailed discussion on nested structure estimation.

the potentially restricted condition of cross-alternative substitution is minimized. The nested logit model is built around an inclusive value whose parameter provides a basis of identifying the behavioral relationship between choices at each level of the nest and also registers as a test of the consistency of the structure with utility maximization (Hensher, 1986). The alternative method to the nested model is an unordered choice model motivated by a random utility function, which works if and only if the choices are independently and identically distributed with Weibull distribution (McFadden, 1973). Although there are theoretical differences between the two methodologies and the distributions at the extreme ends of the two tails, the choice of one over the other makes no difference in practice.[9]

In order to construct the nested logit model, I start with L alternatives (auction vs. negotiation) divided into J subgroups (asset or stock sale, bookbuilding or fixed-price contracts). The entrepreneur first makes the L choice, and then makes specific choices within the set. The data consists of observations on the attributes of the choices $x_{j|l}$ (variables for explaining mechanisms in the second level) and attributes of the choice sets z_l (attributes of markets in the first level). First, using the unconditional probability, we define an inclusive value for the lth branch:

$$\text{Prob}[\text{mechanism}_j, \text{market type}_l] = P_{jl} = \frac{e^{\beta' x_{j|l} + \gamma' z l}}{\sum_{l=1}^{L} \sum_{j=1}^{j_i} e^{\beta' x_{j|l} + \gamma' z l}}$$

$$\text{Inclusive value for the } l\text{th market} = I_l = \ln \sum_{j=1}^{J_l} e^{\beta' x_{j|l}}$$

Nested logit models can be estimated using either limited information, two-step maximum likelihood approach (SML), or by full information maximum likelihood method (FIML). Since I am using a large sample that accommodates the parameters, the sequential approach was not used. Also, since the information matrix is not block diagonal in β and (γ, τ) the FIML estimation was more efficient than sequential approach (Greene, 1997). One critical drawback of the nested logit models is that when partitioning the choice set, the results might be dependent on the branches. Although this issue with the model can be resolved by relaxing the homoscedasticity assumption of equal variances, it is not a concern in this study. This is mainly because the partitions are very clear, and the choices are publicly announced. Even when the transaction involves private targets and acquirers, all choices are mutually exclusive, and recapitalizing firms are excluded from the sample.

[9] Both probit and logit use the same scores and the Hessian matrices. The estimator differences depend on how the probability functions differ, hence the tails are differentiated.

Since I am using a large sample that accommodates the parameters, I use the full information maximum likelihood method, and I solve for the objective function V_e (value maximization by the entrepreneur). *S* are selling alternatives, *N* are the number of firms in the sample, *V* is the value, firm (f), entrepreneur (e), $V_{e,s}$ is a function of firm characteristics, and an the error term which absorbs omitted firm characteristics and idiosyncrasies. *A* is the vector of variables for market type, and *B* is the vector of variables for market mechanism. The entrepreneur maximizes his/her value function:

$$V_{e,l,j} = \alpha' A_{e,l} + \beta' B_{e,l,j} + \varepsilon_{e,l,j}$$

The choices discussed in this paper are discrete and not continuous, the error terms are heteroscedastic, and the dependent variable (the choice) is not normally distributed. I assume that the decision makers choose the alternative from which they derive the highest utility; therefore I use the random utility maximization (RUM) model. The RUM outcome probabilities are based on utility difference only, and since the models normalize the utilities, the scales need not to be identified.

9. RESULTS

The choices discussed in this paper are discrete and not continuous, the error terms are heteroscedastic, and the dependent variable (the choice) is not normally distributed. I assume that the decision makers choose the alternative from which they derive the highest utility; therefore I use the random utility maximization (RUM) model. The RUM outcome probabilities are based on utility difference only, and since the models normalize the utilities, the scales need not to be identified (Heiss, 2002). First, I ran three separate regressions: a standard logit, a conditional logit, and a multinomial logit regression between possible decision combinations. The main results of these models are presented in Tables 3–7. Then, to run nested logit tests, I generated two categorical variables to identify first-level alternatives (IPO, M&A), and the second level alternatives (book, fixed, asset, stock). The results of the nested logit model with multiple specifications for robustness are presented in Table 8.

In the conditional logit model, the branches are assumed to be not correlated, and the main effects are due to the choice-specific attributes. In the multinomial logit model individual-specific set of probabilities for a set of choices are considered. The multinomial logit is testing single choices involving many alternatives, and this is resulting in a cross-alternative substitution (Hensher, 1986). As in the conditional logit model, the assumption of IIA should also hold for the multinomial logit model to have unbiased estimates. Furthermore, due to the globally concave likelihood function, the maximum likelihood maximization is straightforward, assuming

Table 4. Mean and median of explanatory variables

Panel A: Initial public offerings (IPOs)

Variables	All IPOs			Bookbuilding			Fixed-price		
	n	Mean	Median	n	Mean	Median	n	Mean	Median
Patent originality	372	0.36	0.36	95	0.36	0.42	277	0.36	0.37
Patent generality	382	0.43	0.41	103	0.40	0.40	279	0.44	0.35
Tobin's q	4394	1.04	1.00	92	2.32	1.00	741	1.89	1.00
High-tech dummy (high = 1, other = 0)	2814	0.34	0.00	337	0.44	0.00	2477	0.35	0.00
Patent dummy (patent = 1, no patents = 2)	4394	1.91	2.00	337	1.69	2.00	2477	1.89	2.00
# of firms in target's 2-digit SIC industry	2814	818.72	1012.00	337	752.75	1006.00	2477	793.81	1006.00
Debt/total assets (%)	4394	82.07	100.00	337	86.18	100.00	2482	83.26	100.00
Number of institutional owners	58	148	138.50	29	161.00	221.00	29	135.00	81.80
Institutional ownership (%)	58	51.66	54.35	29	57.40	73.00	29	45.92	40.20
VC backed dummy (VC backed = 1, not = 0)	2817	0.48	0.00	337	1.00	1.00	2477	0.41	0.00
Lead underwriter reputation	2569	3.98	6.10	220	2.58	6.00	2349	3.63	5.10
VC activity in focal industry	2814	0.76	1.00	337	0.74	1.00	2477	0.76	1.00
Debt/equity	964	16.82	3.60	80	13.94	4.80	537	22.59	4.30
Number of employees	44	964	164	23	564.00	947.00	21	1402.00	280.00
Return on assets (LTM)	1339	−18.28	−4.00	337	−22.42	1.25	736	−14.90	−1.40
Number of observations in the full sample		4394			337			2477	

Panel B: Mergers and acquisitions (M&As)

Variables	All M&As			Asset sale			Stock sale		
	n	Mean	Median	n	Mean	Median	n	Mean	Median
Tobin's q	34109	1.06	1.00	15243	1.87	1.00	18866	1.39	1.00
High-tech dummy (high = 1, other = 0)	34109	0.69	1.00	15243	0.72	1.00	18866	0.66	1.00
Same industry dummy	34109	0.35	0.00	15243	0.35	0.00	18866	0.36	0.00
Prior deals dummy (yes = 1, no = 2)	31326	1.95	2.00	14068	1.98	2.00	17258	1.93	2.00
# of firms in target's 2-digit SIC industry	34109	590.48	412.00	15243	552.40	343.00	18866	621.25	412.00
# of offers sought	30868	1.01	1.00	14059	1.00	1.00	16809	1.02	1.00
# of offers considered	30786	1.14	1.00	14049	1.11	1.00	16737	1.16	1.00
Debt/total assets (%)	793	19.03	11.19	512	16.60	10.67	281	23.45	12.54

Table 4. *(Continued)*

Panel B: Mergers and acquisitions (M&As)

Variables	All M&As			Asset sale			Stock sale		
	n	Mean	Median	*n*	Mean	Median	*n*	Mean	Median
# of advisors of the target	1973	1.04	1.00	1356	1.03	1.00	608	1.04	1.00
VC activity in focal industry	34109	0.55	1.00	15243	0.74	1.00	18866	0.40	0.00
Debt/equity (%)	688	1.17	0.36	459	1.15	0.34	229	1.22	0.37
Toehold of the buyer (%)	23026	0.81	0.00	11131	0.00	0.00	11895	1.58	0.00
Number of employees	2301	385.67	120	1309	259.81	100.00	992	551.74	165.00
Return on assets (LTM)	1279	−2.76	3.11	848	2.81	3.10	431	−13.71	3.25
Number of observations in the full sample		34109			15243			18866	

IPO Panel: All the firms in the sample are original IPOs of private firms. Their common stock has never traded publicly in any market and is offered in its initial public offering. Patent originality and generality are from Hall et al. (2001). Tobin's q is defined as the sum of the book value of assets and the market value of equity net of the book value of equity over the book value of assets. I use the middle price, in the original filing price range, at which the issuer expects securities to be offered. Firms' grouping under high-tech dummy is based on SDC definitions of the high-tech industry in which the issuer is involved as its primary line of business. Patent dummy is constructed by matching the NBER Pat63_99 dataset with the Compustat assignee number, excluding all firms without cusips. The # of firms in target's industry in the deal's year are determined using the Compustat given 2-digit SIC codes. Return on assets to control for profitability, (earnings from total operations divided by the total assets) is expressed as a percentage. The debt/total assets ratio is the fiscal year total debt divided by the total assets, expressed as a percentage. Institutional ownership is the percentage of common stock held by all reporting institutions on the corresponding institutional holdings date. VC involvement is total investment received by company to-date, and is measured by the sum of all rounds of financing a company has received throughout its lifetime. The # of advisors of the target are the number of book managers that maintain records of activity for the syndicate and underwrites the largest portion of the securities. For the lead underwriter reputation, Carter-Manaster (1990) rankings with Jay Ritter's modifications (2001) are used. Debt/equity ratio is the fiscal year long-term debt divided by the total common equity.

M&A Panel: All the firms in the sample are completed deals involving private target firms. For related prior deals, I use the dummy for the type of the deal that is related to the merger transaction, e.g. JV. Same industry dummy is used using targets' and acquirer's 2-digit SIC codes. The toehold in the firm is shares owned after the acquisition minus shares acquired. The number of offers sought is the number of offers made by the acquirer to the target to be considered. The number of offers extended is the number of times, which the offer was extended of common, or common equivalent, shares outstanding sought by the acquirer. Revenue per share equals the revenue divided by the most recent common shares outstanding.

Table 5. Conditional logit estimates

	Model A				Model B				Model C			
		Coef.	z	p-Value		Coef.	z	p-Value		Coef.	z	p-Value
Tobin's q ×	Book	10.964	7.27	0.000***	Fixed	−0.357	−2.88	0.004***	Fixed	−2.030938	−6.93	0.000***
	Asset	14.111	1.14	0.256	Asset	0.319	3.72	0.000***	Asset	0.3875427	3.66	0.000***
	Stock	14.022	0.92	0.358	Stock	0.327	3.81	0.000***	Stock	0.3960622	3.74	0.000***
VC industry ×	Book	0.185	0.50	0.619	Fixed	1.290	5.86	0.000***	Fixed	−0.0694672	−0.21	0.833
	Asset	−0.176	0.71	0.805	Asset	−0.793	−2.79	0.005***	Asset	−0.6983745	−1.73	0.083*
	Stock	−0.090	0.72	0.901	Stock	−0.716	−2.44	0.015**	Stock	−0.5459798	−1.31	0.190
High-tech ×	Book	0.241	0.61	0.544	Fixed	0.664	2.18	0.029**	Fixed	−0.0189906	−0.06	0.954
	Asset	3.523	4.78	0.000***	Asset	3.009	8.67	0.000**	Asset	3.918659	9.05	0.000***
	Stock	2.235	3.03	0.002***	Stock	1.787	5.04	0.000**	Stock	2.719611	6.19	0.000***
# of firms in industry ×	Book	0.000	1.11	0.266	Fixed	0.001	4.97	0.000***	Fixed	−0.0000731	−0.23	0.820
	Asset	−0.001	−1.25	0.211	Asset	−0.001	−1.83	0.067*	Asset	0.0001284	0.31	0.760
	Stock	−0.001	−1.86	0.063*	Stock	−0.001	−3.16	0.002**	Stock	−0.0001714	−0.39	0.696
Debt/assets ×	Book	−1.146	−0.98	0.328	Fixed	0.479	1.49	0.136	Fixed	−0.8228916	−2.59	0.009***
	Asset	−7.414	0.00	0.000***	Asset	−2.696	−5.36	0.000***	Asset	−2.148072	−4.21	0.000***
	Stock	−6.622	−3.42	0.001***	Stock	−1.829	−3.52	0.000***	Stock	−1.11567	−2.23	0.026**
Debt/equity ×	Book	−12.358	−6.71	0.000***	Fixed				Fixed	−0.0025366		
	Asset	−13.251	−1.07	0.286	Asset				Asset	−0.0079075		
	Stock	−13.157	−0.86	0.388	Stock				Stock	−0.0076475		
ROA ×	Book	0.013	2.48	0.013**	Fixed	−0.002	−1.06	0.289	Fixed	−0.002	−1.32	0.187
	Asset	0.013	0.89	0.374	Asset	−0.007	−2.86	0.004***	Asset	−0.007	−3.09	0.002***
	Stock	0.013	0.90	0.369	Stock	−0.007	−2.8	0.005***	Stock	−0.007	−3	0.003***
Constant ×	Book	−10.419	−5.41	0.000***	Fixed				Fixed	4.970665	7.87	0.000***
	Asset	−15.421	−1.23	0.219	Asset				Asset	−1.622371	−2.34	0.019**
	Stock	−15.018	−0.98	0.327	Stock				Stock	−1.454017	−2.17	0.03**
N			5800				6016				6016	
LR χ^2			2843.88				2179.89				2337.44	
Prob. > χ^2			0.0000				0.0000				0.0000	
Log-likelihood			−600.402				−995.04338				−916.2647	
Pseudo R^2			0.703				0.5607				0.5605	

*$p < 0.10$.
**$p < 0.05$.
***$p < 0.01$.

Table 6. Multinomial logistic regression

Choice		Model A			Model B			Model C		
		Coef.	z	p-Value	Coef.	z	p-Value	Coef.	z	p-Value
Fixed vs. Bookbuilding	Tobin's q	−0.3740	−3.10	0.002**	0.3020	2.42	0.015**	0.5394	3.57	0.000***
	ROA	0.0034	1.84	0.066*	0.0020	1.03	0.303	0.0025	1.30	0.193
	Debt/assets	−2.5352	−7.65	0.000***	−0.7700	−2.21	0.012**	−0.1873	−0.56	0.574
	High-tech				−0.7520	−2.52	0.027**	−0.7055	−2.35	0.019**
	Industry M/B				−0.6910	−2.59	0.01***	−0.2780	−2.86	0.004***
	Industry Tq				0.7240	1.34	0.181			
	# of firms				−0.0010	−2.66	0.008***	−0.0005	−1.82	0.068*
	VC industry							−1.2742	−5.73	0.000***
Fixed vs. Asset	Tobin's q	0.2520	9.38	0.000***	0.6590	6.29	0.000***	0.8754	6.16	0.000***
	ROA	−0.0007	−0.73	0.464	0.0040	−3.04	0.002***	−0.0043	−3.14	0.002***
	Debt/assets	−4.9070	−16.91	0.000***	−2.7920	−7.05	0.000***	−2.0800	−5.26	0.000***
	High-tech				1.9360	8.78	0.000***	2.1247	9.36	0.000***
	Industry M/B				−0.5060	−1.91	0.056*	−0.5734	−5.17	0.000***
	Industry Tq				−0.5730	−1.01	0.312			
	# of firms				−0.0010	−2.72	0.007***	−0.0007	−2.36	0.018**
	VC industry							−1.5406	−6.74	0.000***
Fixed vs. Stock	Tobin's q	0.2374	8.83	0.000***	0.6680	6.38	0.000***	0.8840	6.22	0.000***
	ROA	−0.0003	−0.29	0.777	−0.0400	−2.68	0.007***	−0.0037	−2.65	0.008***
	Debt/assets	−4.9024	−16.81	0.000***	−1.2310	−3.38	0.013**	−0.5983	−1.74	0.082*
	High-tech				0.5560	2.49	0.001***	0.6421	2.84	0.004***
	Industry M/B				−0.4670	−1.76	0.078*	−0.4546	−4.06	0.000***
	Industry Tq				−0.3340	−0.58	0.561			
	# of firms				−0.0020	−5.24	0.000***	−0.0016	−4.93	0.000***
	VC industry							−1.2960	−5.62	0.000***
	N		1597			1597			1597	
	Log-likelihood		−1390.419			−1116.364			−1086.956	
	Wald test		1646.990			2195.100			2253.910	
	Prob. > χ^2		0.000			0.000			0.000	
	Pseudo R^2		0.372			0.496			0.509	

Omitted	χ^2	z-Statistic	Omitted	χ^2	z-Statistic	Omitted	χ^2	z-Statistic
Small-Hsiao tests of IIA assumption[a]								
2	27.14	0.002**	2	29.622	0.128	2	24812.05	0.000***
3	10.62	0.388	3	21.42	0.495	3	2856.94	0.000***
4	13.70	0.187	4	43.15	0.005***	4	17121.90	0.000***
Hausman tests of IIA assumption[a]								
2	656.55	0.000***	2	8.71	0.795	2	28.23	0.008***
3	422.23	0.000***	3	97.06	0.000***	3	−89.61	0.000***
4	246.23	0.000***	4	178.18	0.000***	4	106.542	0.000***

[a] Ho: Odds (Outcome-J vs. Outcome-K) are independent of other alternatives.

Table 7.

Panel A: Logit modal for IPO vs. M&A comparison group is the firms that chooses IPO

	1	2	3	4	5	6	7	8
Tobin's q	−1.264	−0.077	0.895	0.833	0.591	0.703	0.58	0.938
	−2.58	−0.88	6.97***	5.88***	6.84***	5.19***	6.68***	5.81**
VC industry	−2.169	−0.688	−0.667	−0.376	−1.264	0.257	−1.64	−0.474
	5.26***	−1.38	−2.77***	−1.45	−6.48***	−0.6	−7.52***	−1.71
High-tech dummy	−3.920	−2.685		1.962	1.415	0.641	1.44	2.365
	−8.62***	−5.96***		8.94***	7.39***	2.17**	6.68***	8.98***
# of firms in industry	0.0004	0.002	−0.002	−0.001	−0.002	−0.001	−0.002	0
	1.01	3.60***	7.52***	3.78***	−9.15***	−4.14***	−9.41***	−1.46
Debt/asset			−0.633	−0.546	−1.563	0.79	−2.147	−0.586
			−1.92*	−1.74*	−5.09***	2.05**	−6.12***	−1.62
ROA			−0.003	−0.004	−0.004	−0.004	−0.005	−0.004
			−2.51**	−2.77**	−2.73***	−3.03***	−2.62***	−3.25***
Number or employees	0.0007	0.001						
	12.44***	12.65***						
2-SIC Ind. market/ book of equity								−0.229
Nation dummy (US)								−1.93
						0.127	0.299	
Nation dummy (Canada)						0.16	−0.36	
						−0.886		
Nation dummy (UK)						−1.92*		
						−5.639		
Nation dummy (China)						−0.29		
						5.491		
Year dummy (1990)						6.24***		
							0.532	
Year dummy (1991)							0.59	
							1.604	1.358
							3.11***	2.61**
							1.195	1.354
Year dummy (1992)							2.56**	3.19**
							2.338	2.849
Year dummy (1993)							7.37***	8.92**
							1.115	1.22

	1	2	3	4	5	6	7	8
Year dummy (1994)							3.01***	3.51**
							0.693	
Year dummy (1995)							1.71*	
							0.325	
Year dummy (1996)							0.63	
							0.051	
Year dummy (1997)							0.12	
							0.448	
Year dummy (1998)							1.03	
							0.691	
Year dummy (1999)							1.35	
							0.27	
Year dummy (2000)							0.36	
Constant		−4.141	−0.828	−2.428		−2.546		−3.108
		−5.62***	−2.28***	−5.65***		−3.41***		5.57**
N	3018	3018	1597	1597	1597	1597	1595	1595
Log-likelihood	−183.913	−167.075	−368.032	−325.245	−342.866	−208.26	−308.564	−277.31
LR χ^2	3816.03***	2975.20***	1475.67***	1561.24***	1528.18***	1795.21***	1594.01	1654.48*
McFadden's LRI	0.912	0.899	0.667	0.706	0.69	0.812	0.721	0.749

Panel B: Logit model estimates for the choice between fixed price vs. bookbuilding method comparison group is the firms that choose fixed price

	1	2	3	4	5	6	7	8	9	11	12	13
Tobin's q	3.947	0.818	3.945	0.976	0.314	2.981	0.802	0.008	0.964	2.375	2.082	−0.328
	8.93***	3.65***	8.91***	4.26***	1.09	6.38***	0.95	0.01	0.81	3.19***	2.90***	−0.33
ROA	0.011	0.002	0.011	0. 003	0.0001	0.006	0.022	0.017	0.022	0.01	0.017	0.027
	3.70***	1.34	3.70**	1. 86*	0.05	1.78*	1.18	1.08	1.00	0.71	1.16	1.37
Debt/asset	2.524	−1.304	2.527	−0.969	−1.08	2.024	1.1819	1.262	3.664	0.718	0.973	2.862
	4.30***	−3.44***	4.30**	−2.48**	2.23**	3.19***	1.64*	1.21	2.35**	0.78	1.06	2.18**
VC industry	0.121	−1.465	0.12	−1.457	−1.568	−0.127	−1.558		−2.546	−3.015	−3.2	
	0.35	−6.59***	0.34	−6.44***	−5.54***	−0.28	−1.66*		−2.04**	−3.20***	−3.43***	
# target firms in	0.0002	−0.001	0.0002	−0.002	−0.001	0.0003	−0.001	−0.001	−0.001	−0.002		
industry	0.72	−5.64***	0.66	−6.25***	−4.59**	0.7	−1.86	2.41**	−1.15	−2.53**		
High-tech			−0.025	−1.011			−0.689		−1.032	−0.282	−0.128	−0.166
			−0.07	−3.22***			−0.82		−0.97	−0.39	−0.18	−0.18
Industry market/book									−2.732	−0.066	−0.488	−2.892
of equity									−2.51**	−0.33	−0.87	−2.87***
Investment bank rep.					−0.001	0.033	0.003	−0.002	0.043			0.032
					−0.07	1.38	−0.09	−0.07	0.98			0.83
VC involvement									3.048	2.93	2.651	2.493
									2.87***	3.49***	3.38***	2.75***

Table 7. *(Continued)*

Panel B: Logit model estimates for the choice between fixed price vs. bookbuilding method comparison group is the firms that choose fixed price

	1	2	3	4	5	6	7	8	9	11	12	13
Patent dummy												
#of firms × high-tech												
Constant	−8.437		−8.422			−7.44						
	−9.12***		8.91***			−7.60***						
# of firms × low-tech												
Number of patents												
Patent generality									−5.462	−4.151	−4.461	−4.74
									−2.42**	−2.84***	−3.10***	−2.54**
Patent originality							−2.169	−2.628	0.506			
							−1.28	−1.6	0.22			
Year dummy (1992)												
Year dummy (1993)												
Year dummy (1994)												
Industry Tobin's q									−4.721		0.404	4.316
									2.31**		0.35	2.30**
N	828	828	828	828	729	729	113	113	112	150	150	114
Log-likelihood	−226.724	−297.013	−226.721	−291.038	−190.817	−148.265	−46.534	47.936	−35.318	−63.547	−66.98	−40.067
LR χ^2	124.22***	553.83***	124.22***	565.78***	628.97***	57.40***	63.58***	60.78***	84.63***	80.85***	73.98***	77.90***
McFadden's LRI	0.215	0.483	0.215	0.493	0.622	0.162	0.406	0.388	0.545	0.389	0.356	0.493

Panel C: Logit model estimates for the choice between asset sale vs. stock sale comparison group is the firms that choose asset sale

	1	2	3	4	5	6	7	8	9	10	11	12	13
Tobin's q	−0.007	−0.004	−0.004	0.01	0.003	−0.007	−0.008				−0.006	−0.006	−0.003
	−1.74	−1.05	−1.05	−0.98	−0.54	−1.58	(1.91)*				−1.35	−1.39	−0.59
ROA	0	0.001	0.001	−0.002	−0.006	0.001	−0.001	0.001	0	0	−0.001	−0.001	−0.001
	−0.33	−0.94	−0.94	0.33	−1.18	−0.95	−0.77	−1.22	−0.38	−0.42	−0.97	−0.74	−0.61
Debt/asset	−0.767	−0.63	−0.63	0.831	−1.297	−1.082	−0.989				−1.148	−1.044	−0.516
	(2.74)**	(2.15)**	(2.15)**	−0.6	(3.00)***	(3.05)***	(2.88)***				(3.15)***	(2.95)***	−1.62
VC industry	−0.367	−0.271	−0.271	0.123	−0.001	−0.366	−0.364	0.103	0.182	0.228	−0.379	−0.35	0.054
	(2.28)*	−1.57	−1.57	−0.29	−0.01	(1.90)*	(1.89)*	−0.68	−1	−1.25	(1.95)*	(1.80)*	−0.27
# of target	0.0000	0.001	0.001	0	0	−0.001	0.001				−0.001	0.001	
firms	(2.50)*	(2.71)***	(2.43)**	−0.58	−0.15	(1.81)*	(2.48)**				(1.81)*	(2.26)**	
High-tech	1.251	1.566	1.566				1.024	1.15	1.092	1.08		0.934	0.929
	(8.11)**	(5.91)***	(5.91)***				(4.62)***	(6.45)***	(5.40)***	(5.31)***		(4.13)***	(1.87)*

Same industry		−0.893	−0.893	−0.62		−0.675	−0.683	−0.576	−0.299	−0.257	−0.622	−0.583	−0.571
dummy		(4.86)***	(4.86)***	−1.14		(3.61)***	(3.68)***	(3.90)***	(1.76)*	−1.49	(3.30)***	(3.08)***	(2.91)***
Industry		−0.138	−0.138	0.308	0.096	−0.216	−0.26	−0.053	0.022	0.011	−0.197	−0.237	−0.077
market/book		−1.58	−1.58	−1.13	−0.8	(2.24)**	(2.68)***	−0.8	−0.27	−0.14	(2.02)**	(2.40)**	−0.73
# of firms ×			0			0.001					0.001		0.001
high-tech			−0.4			(5.10)***					(4.95)***		(1.87)*
Nation dummy												−1.442	−1.554
(US)												(2.87)***	(3.05)***
Industry Tobin's q													
Constant					0.268			−0.046					
					−0.69			−0.04					
Past-deal dummy				0.782		0.872	0.483	1.664	1.315	1.301	1.019	0.693	1.916
				−1.31		(5.68)***	(2.67)***	(6.23)***	(4.38)***	(4.45)***	(6.09)***	(3.56)***	(5.80)***
Same nation											−0.475	−0.568	−0.42
dummy											(2.41)**	(2.84)***	(1.97)**
# of offers sought				−2.49				−3.312	−3.004	−2.615			−3.072
				(2.05)**				(3.19)***	(5.10)***	(4.47)***			(4.51)***
# of offers										−0.218			−0.161
considered										(2.40)**			−1.48
# of firms ×		0											0
low-tech		−0.4											−0.55
Number of target													
advisors													
# of employees				−0.002									
				(2.43)**									
Toehold (%)					−32.651				−67.743	−70.315			
					−1.26				(2.28)**	(2.39)**			
N	769	769	769	143	466	679	679	1136	775	771	679	679	679

Notes: Coefficient estimates and z-statistics are reported.
Industry refers to the target firm's 2-digit SIC industry.
*Significant at 10%.
**Significant at 5%.
***Significant at 1%.

Table 8. Nested logit model coefficient estimates for the choice of sale method

Variable × Mechanism		Model						
		1	2	3	4	5	6	7
Tobin's q ×	Fixed			6.105				
				3.69^{***}				
	Asset	1.651	0.97	5.022				
		4.99^{***}	5.20^{***}	3.88^{***}				
	Stock		1.024	5.082				
			5.45^{***}	3.91^{***}				
VC industry ×	Fixed			−6.468				0.08
				-3.21^{***}				2.09^{**}
	Asset	−1.171	−0.701	−5.823				
		-2.12^{**}	−0.98	-3.25^{***}				
	Stock		−0.549	−5.796				
			−0.56	-3.05^{***}				
High-tech dummy ×	Fixed			−3.666				0.138
				-2.29^{**}				2.86^{***}
	Asset	6.44	9.767	6.379				
		11.19^{***}	5.74^{***}	292^{***}				
	Stock		−5.194	−8.386				
			-2.25^{**}	-2.85^{***}				
# of targets ×	Fixed			−0.009				
				-2.93^{***}				
	Asset	0.0002	0.002	−0.006				
		0.68	2.29^{**}	-2.29^{**}				
	Stock		−0.002	−0.01				
			-1.73^{*}	-3.36^{***}				
Debt/asset ×	Fixed			10.188				
				-2.11^{**}				

	Asset	−2.348	−9.943	−18.134				
		−2.90***	−5.23***	−4.34***				
	Stock		6.778	−2.501				
			3.77***	−0.5				
ROA ×	Fixed			0.015	0.0006			
				2.14**	0.61			
	Asset	−0.01	−0.014	0.001				
		−3.21***	−3.19***	0.1				
	Stock		0.004	0.015				
			0.85	2.01**				
# of targets ×	Fixed				−0.001		0.001	0.0002
Low-tech ×					−4.51***		1.99**	5.56***
# of targets ×	Fixed				−0.003	−0.0001	0.0005	
High-tech ×					−6.16***	−1.37	1.5	
VC involvement ×	Fixed					−0.284	−1.089	−0.36
						−4.20***	−2.46**	(–)
Number of patents ×	Fixed							
Patent originality ×	Fixed					0.527	0.969	−0.002
						1.73*	1.17	−0.02
Patent generality	Fixed					0.931	1.177	0.321
						(–)	(–)	2.87***
Investment bank rep. ×	Fixed					−0.001	0.003	
						−0.20	0.27	
Same industry ×	Stock							
Estimated probality of type choice from clogit ×	IPO						4.276	
							(–)	
Estimated probability of mechanism					8.474			
					18.46***			
Type dummy (Ipo-1, M&A-0)				−14.655				
				−4.30***				

Table 8. (*Continued*)

Variable × Mechanism		Model						
		1	2	3	4	5	6	7
Constant (IV for IPO)		102.876	13.845	13.479	−0.766	0.203	0.436	−52.856
		4.27***	3.19***	2.94***	−4.53***	4.37***	2.31**	−2.80***
Correlation within IPO					0.41	0.96	0.81	
Constant (IV for M&A)		97.297	19.416	−5.606	1.848	−26.374	−28.529	0.119
		4.09***	4.36***	−2.57**	−4.32***	−0.03	−0.01	6.61***
Correlation within M&A								0.99
LR test of homeskedaticity (IV-1)								
χ^2					128.40***	125.55***	116.36***	
Observations		6016	6016	6016	6016	1180	1180	1424
Number of groups		1504	2504	1504	2504	295	395	356
Log-likelihood					−1082.08	−118.79	−100.68	
LR χ^2					2005.82***	580.32***	616.56***	

Variable × Mechanism		Model						
		8	9	10	11	12	13	14
Tobin's q ×	Fixed		−0.182	−0.526	−0.399	−1.874	−0.904	
			1.68*	−1.01	−1.57	−2.40**	2.76***	
	Asset Stock							
VC Industry ×	Fixed	0.303	0.188	0.884	0.445	2.191	1.243	1.134
		2.10**	1.83*	2.44**	2.55**	1.73**	2.05**	2.92***

	Asset							
	Stock							
High-tech Dummy ×	Fixed	0.39	0.244	1.457	0.725	2.254		
	Asset	1.63	1.52	2.49**	2.60***	1.64		
	Stock							
# of targets ×	Fixed						0.0001	0.0001
							0.48	−0.28
	Asset							
	Stock							
Debt/asset ×	Fixed					0.674	0.052	0.104
						−1.1	0.13	−0.2
	Asset							
	Stock							
ROA ×	Fixed					−0.003	−0.002	−0.01
						−0.42	−0.39	−1.21
	Asset							
	Stock							
# of targets ×	Fixed	0.001	0.0001	0.003	0.001	0.002	0.001	0.001
Low-tech ×		5.61***	2.11***	5.29***	5.41***	1.61	1.09	1.79*
# of targets ×	Fixed	0.0001	0.0001	0.0001	0.0001	−0.002		
High-tech ×		0.84	0.7	0.72	0.28	−1.28		
VC involvement ×	Fixed	−1.362	−0.553	−2.622	−1.38	−1.828	−1.325	−2.108
		(–)	−2.50**	(–)	(–)	−1.67*	−2.24**	(–)
Number of patents ×	Fixed						0	0
							1.14	1.28
Patent originality ×	Fixed	−0.007	0.062	1.692				
		−0.02	0.28	2.09**				
Patent generality	Fixed	1.205	0.597		1.371	2.461	1.651	1.177
		2.86***	(–)		3.36***	(–)	(–)	1.69*
Investment bank rep. ×	Fixed							
Same industry ×	Stock							

Table 8. *(Continued)*

Variable × Mechanism		Model						
		8	9	10	11	12	13	14
Estimated probality of type choice from clogit × Estimated probality of mechanism	IPO							
Type dummy (Ipo-1, M&A-0)								
Constant (IV for IPO)		0.447	−54.584	0.997	0.483	0.606	0.42	0.582
		6.71***	2.50**	5.48***	3.43***	19.51***	2.51**	4.35***
Correlation within IPO		0.80		0.01	0.77	0.63	0.82	0.66
Constant (IV for M&A		−45.461	0.196	−103.443	−56.066	−68.33	−86.668	−54.27
		0	13.70***	−51.90***	−6.44***	−1.89*	−3.71***	−2.33**
Correlation within M&A			0.96					
LR test of homeskedaticity (IV-1)								
χ^2		217.41***		222.46***	238.45***	103.07***	104.32***	101.33***
Observations		1424	1424	1440	1480	584	576	576
Number of groups		350	356	360	370	146	244	144
Log-likelihood		−161.191		−166.27	−168.17	−56.35	−57.64	−54.27
LR χ^2		664.66***		665.59***	689.52***	292.10***	283.98***	274.29***

Coefficient estimates and z-statistics are reported.
*Significant at 10%.
**Significant at 5%.
***Significant at 1%.

independently distributed error terms. In other words, when the error terms for alternatives are correlated, the conditional logit estimates are biased. Hence by using a nested-choice approach, I decomposed a single decision and minimized cross-alternative substitution.

In the multinomial logit regression, the IIA is tested using Hausman and Small-Hsiao tests,[10] for the null hypothesis that the odds for outcome *j* vs. outcome *k* are independent of other alternatives. In both multinomial and conditional logit regressions, we cannot reject the null, and therefore cannot collapse the decision tree. For all the models of multinomial logit regressions (Table 6) additional Wald tests for combining outcome categories are conducted (not reported). The Wald tests for all pairs are conducted to test the null hypotheses of all coefficients except intercepts associated with given pair of outcomes are zero (i.e. categories can be collapsed) and all the hypotheses are rejected at the 1% significance level.

The results of the multinomial logit are presented in Table 6. For the model estimates to be unbiased, the IIA assumption should hold. Although the estimates and therefore the model have significant explanatory power, based on the theory discussed in this paper, we would expect the mechanisms to be strongly correlated into nests (IPO vs. M&A). Overall, the test results suggest that the null hypothesis of IIA should be rejected at the 1% significance level. This result also warrants the use of a more complicated nested model that does not restrict the choices to be uncorrelated.

Two problems need to be addressed when working with large panels of cross section data: missing data and self-selection (Griliches et al., 1978). In this dataset, a number of observations were missing from the manufacturing sector sample. Since the observations were missing randomly, using a subsample resulted in unbiased but inefficient estimates. To test for omitted variables and the inclusion of irrelevant variables, I conducted a likelihood ratio test. The test showed that the model has a strong fit. Furthermore, the Wald statistics showed a strong model fit as well. To test for multicollinearity I checked the correlation coefficients. In a nested logit, the presence of multicollinearity does not lead to biased coefficients, but the standard errors of the coefficients will be inflated. To interpret the coefficients of the continuous independent variables, I used the odds ratio and tested significance by using Wald statistic. Consistent with the nested logit method, I used McFadden's likelihood ratio index (LRI), which is the pseudo R^2.

[10] In the Small-Hsiao (1985) test for the IIA assumption, if the IIA condition holds, the maximized log-likelihood for the restricted choice set will not be too different from the log-likelihood computed over the restricted choice set using parameters obtained from the full choice set. The test statistics is asymptotically χ^2 distributed with degrees of freedom equal to the number of parameters. On the other hand, Hausman-McFadden (1984) test requires that the parameters of the restricted set model are approximately the same as those of the full choice set model.

The preliminary results of the logit models for the choice of auctions vs. negotiations provide support for the finding that firms that have some degree of bargaining power will choose to negotiate the sale of their firms through an M&A. Patent dummy variable is used to proxy the bargaining power of the entrepreneurial firm. The coefficient estimate of the variable patent dummy is statistically significant with the p-value < 0.001. Entrepreneurial firms' probability of being sold through an auction (IPO) is decreased if the firm has a patent.

The coefficient of patent originality as a proxy of private value component of the resource value is statistically significant with p-value < 0.0001 in explaining the probability of choosing auctions to negotiations. As the originality of the patent increases, the probability that the entrepreneurial firm will choose auctions to sell the company through an IPO increases. Therefore, Hypothesis 3 is supported. On the other hand, the coefficient of patent generality as a proxy for the common value component of the resource has a negative sign as expected, and is significant with p-value $= 0.0007$. As the generality of the patent increases, the probability that the entrepreneurial firm will choose auctions to sell the company through an IPO decreases. Therefore, Hypothesis 4 is supported. In both models, as expected, the time fixed effects is statistically significant. This finding supports the established literature on the cyclicality of the IPO and M&A activities.

Table 5-Panel C provides the better model with log-likelihood of −916.26. Explanatory variables associated with the resource value are Tobin's q, and high-tech dummy. In both models Tobin's q is significant for all mechanism types. However, increase in Tobin's q decreases the likelihood of choosing the fixed price mechanism over bookbuilding, whereas the same variable increases the likelihood of choosing either asset or stock sale over bookbuilding. High-tech dummy variable is positive and significant for all mechanisms in Model B, but insignificant as an explanatory variable for the likelihood of choosing fixed price over bookbuilding in Model C. Table 7 Panel A, presents the results of the logit model for the choice between IPO vs. M&A, in models 3 through 9 Tobin's q is positive and statistically significant. This result supports hypothesis 3 such that as the private value component increases, the entrepreneurial firms' probability of being sold though an auction increases. Also based on Panel B and C, increase in Tobin's q increases the likelihood of choosing fixed price over bookbuilding, as well as asset sale over stock sale. Same results hold for the high-tech dummy except that it is insignificant for the choice between fixed price vs. bookbuilding. In Table 8, the nested logit results are consistent with the above discussion.

Bargaining power as proxied by ROA for the whole sample is negative and statistically significantly associated with the likelihood of choosing asset or stock sales over bookbuilding as presented in the conditional logit model results in Table 5. ROA has no statistically significant explanatory power in explaining the likelihood of choosing fixed price vs. bookbuilding. Overall increase in ROA is associated with a decrease in the likelihood of choosing IPO over M&A as presented in the logit model exhibits in Table 7-Panel A. This result provides supporting evidence for Hypotheses 1 and 2. Entrepreneurs with low bargaining power are more likely to choose IPO rather than M&A. An increase in bargaining power is more likely to increase the likelihood of choosing M&A. The nested logit model results suggest that an increase in ROA increases the likelihood of choosing fixed price over bookbuilding, conditional on the firm choosing IPO over M&As.

Market thickness is proxied by the number of firms in the focal (target) firm's 2-digit SIC industry segment in the deal year. According to conditional logit results in Table 5, as the number of firms in the market increases the likelihood of choosing asset sale or stock sale over bookbuilding decreases. Also, according to the results in Table 7-Panel A, the logit model estimate for the number of firms in the industry is negatively associated with the likelihood of choosing IPOs over M&As: as the number of firms increases, the likelihood of choosing M&As increases. This result is consistent with the market power arguments in the literature. As the number of firms in the high-tech industries increases, the likelihood of choosing asset sale over stock sale increases. On the other hand, as the number of targets in low-tech industries increases, the likelihood of choosing fixed price over bookbuilding increases conditional on the choice of IPO over M&A (Table 8). Results provide supporting evidence for Hypothesis 5, which suggests that entrepreneurial firms are more likely to choose auctions as market thickness and private value components increase.

As summarized in Hypotheses 7 and 8, discussion on risk propensity suggests that increase in risk aversion corresponds to increase in the likelihood of choosing IPOs. Risk averse owners (managers) are more likely to have lower debt/asset ratios in their firms, which suggest that as the ratio of debt/assets increases the likelihood of choosing M&As increases. According to the conditional logit estimates, as the debt/asset ratio increases, the likelihood of choosing asset sale or stock over bookbuilding decreases. However, such an increase is negatively associated with the likelihood of choosing IPOs vs. M&As as presented in Table 7-Panel A. According to the results of the nested logit, increase in debt/asset ratio corresponds to the decrease in the likelihood of choosing fixed price over bookbuilding conditional on IPO being chosen over M&A. Moreover, an increase in debt to asset ratio increases the likelihood of choosing stock sale while decreases the likelihood of asset sale conditional on M&A being chosen. Venture capital involvement also proxies risk propensity for firms choosing IPOs. Although the logit model estimates presented in Table 7-Panel B for the choice between fixed prices vs. bookbuilding suggests that as firms with VC involvement are more likely to choose fixed price over bookbuilding, nested logit model suggest the opposite. When the probability of choosing IPO vs. M&A is accounted for, the firms with VC involvement are more likely to choose bookbuilding over fixed price. These results provide support for Hypotheses 7 and 8.

Hypothesis 9 suggests that as search costs increases, an entrepreneurial firm will be more likely to sell though an IPO. If the focal firm's industry is one where VC firms are active, then the search costs are expected to be lower for potential buyers and sellers as discussed previously. VC activity in the industry of the focal firm decreases the likelihood of choosing asset sale or stock sale over bookbuilding. Also VC activity in the focal firm's industry decreases the likelihood of choosing IPO vs. M&A (Table 7-Panel A). The results of the nested logit model are in accordance with the results of the conditional logit model.

Hypothesis 10 suggests that results based on the nested model suggests that as the common value component increases, an entrepreneurial firm is less likely to be sold through the bookbuilding method; and Hypothesis 11 suggests that as the private value component increases, an entrepreneurial firm is more likely to be sold through the bookbuilding method. Based on the nested logit results presented in Table 8, as patent generality increases, the likelihood of being sold through fixed price decreases, whereas patent originality is insignificant. These results provide support for Hypothesis 10 but not for 11.

As the initial toehold of the acquirer increases, the target firm's likelihood of selling assets vs. stocks decreases and the logit results provide support for Hypothesis 13. According to the nested logit results, for low-tech firms, the likelihood of stock sale decreases, conditional on being acquired. As the private value component proxied by Tobin's q increases the likelihood of choosing asset sale over stock sale decreases. However, the effect is reversed if the focal firm is U.S. based. According to the nested logit results, conditional on choosing M&A, Tobin's q is not a good variable in identifying a differential effect between asset sale vs. stock sale, since it is positive and significant for both. On the other hand, for a high-tech firm, the likelihood of being sold through a stock sale is increased conditional on being acquired. These results provide support for Hypothesis 14.

10. DISCUSSION

It was an implied assumption that when entrepreneurial firms grew large enough, they would need capital to finance more operations and they would go public. This propensity to offer stocks to public especially became visible in the mid 1980s, but the trends changed to neo-private companies when the leveraged buyout (LBO) trend picked up. This trend too changed when these neo-private companies became public again within seven years of LBOs (Kaplan, 1991). The reason why the trends have shifted in this fashion has to do with the rights to control an entrepreneurial firm (ownership), and the rights to have access to future cash flows (shareholding). Why firms want to go public, or private, is beyond the scope of this paper. Instead I ask, once the entrepreneur decides to go public, how should he sell his firm? I empirically study the entrepreneur's propensity to choose among different market mechanisms influenced by five conditions: bargaining power, market thickness, resource value, risk propensity, and search costs.

The findings show strong support for all first and second level predictions provided the entrepreneur chooses the conditional first level choice. By using a nested logit approach, I was able to decompose each discrete choice, and yield propensity scores for each individual choice. Except for Hypotheses 6, 11 and 14, the results provide strong support for all main and secondary level predictions. Hypothesis 6 is not supported, and Hypotheses 11 and 14 are partially supported.

The major shortcoming of this paper is that because I have used a large sample to test the arguments in an empirical setting, an in-depth analysis for these industries and companies could not be provided. By preferring breadth, I had to sacrifice details on several levels. Although implications of this study are applicable to firms in general, a closer study of entrepreneurial firms is needed. Also, since this study covers multiple industries, the measures for bargaining power and resource value are somewhat crude, mostly due to the well-known limitations of the patent data. Therefore, a natural extension of this research is a follow up study on how these firms chose a specific mechanism, using the survey method. This would enable us to develop better and more precise measures. As mentioned above, the question in this paper is "how" entrepreneurs should sell their firms, but "why" they sell is equally interesting. I believe that through a survey and interviews, we can construct indices for optimal market mechanisms. For example, if the full value of the entrepreneurial firm could be realized with combined resources of the acquirer, the target would choose to be acquired if the marginal resources were too costly to acquire or would require time to develop internally. Following this logic, an entrepreneur would choose auctions (IPO) if they had all the resources needed to realize the full value, or negotiate the sale (M&A) if they did not have the resources, and could not develop them.

For future research, I believe two areas should be emphasized: cooperative strategies, and new methods. The different choices among entrepreneurs in international markets may provide a lot of valuable information, if we hold the industry (ies) constant. For example, would the entrepreneur still sell his company to a foreign firm if the "fit" did not exist or could the firm form cooperative ventures to change its bargaining position? Future research on entrepreneurial firms should also incorporate experiments to address many of the theoretically important variables. For example, in this study, propensity to prefer among auctions and negotiations could also be tested in a controlled environment. Furthermore, we could also test our theories on whether entrepreneurial choices are subject to entrepreneurs' psychological biases and if these disrupt equilibrium predictions. Also, by constructing a market for firms in the lab, we could test our theories in the context of target and bidder firms in initial public offerings, bilateral or multilateral negotiations, and measure rent generation potential under various market conditions.

REFERENCES

Alchian, A., & Demsetz, H. (1972). Production, information costs, and economic organization, *American Economic Review*, *62*, 777–795.

Amit, R., Glosten, L., & Muller, E. (1990). Entrepreneurial ability, venture investments, and risk sharing, *Management Science*, *36*, 1232–1245.

Anand, B. N., & Khanna, T. (2000). Do firms learn to create value? The case of alliances, *Strategic Management Journal*, *21*, 295–315.

Arikan, I. (2002). Economics of strategic factor markets, Working Paper, Fisher College of Business, The Ohio State University, Columbus, OH (July).

Baker, J. (1997). Unilateral competitive effects theories in merger analysis, *Antitrust*, *11*, 21–26.

Barnett, W. P., Greve, H. R., & Park, D. Y. (1994). An evolutionary model of organizational performance, *Strategic Management Journal*, *15*, 11–28.

Barney, J. B. (1986). Strategic factor markets: Expectations, luck, and business strategy, *Management Science*, *32*, 1512–1514.

Baum, J. A., & Ingram, P. (1998). Survival-enhancing learning in the Manhattan hotel industry, 1898–1980, *Management Science*, *44*, 996–1016.

Benveniste, L. M., & Busaba, W. Y. (1997). Bookbuilding vs. fixed price: An analysis of competing strategies for marketing IPOs, *Journal of Financial and Quantitative Analysis*, *32*, 383–403.

Blazenko, G. W. (1987). Managerial preference, asymmetric information, and financial structure, *Journal of Finance*, *42*, 839–862.

Buchholtz, A. K., & Ribbens, B. A. (1994). Role of chief executive officers in takeover resistance: Effects of CEO incentives and individual characteristics, *Academy of Management Journal*, *37*, 554–579.

Burkart, M. (1995). Initial shareholdings and overbidding in takeover contests, *Journal of Finance*, *50*, 1491–1515.

Busenitz, L. N., & Barney, J. B. (1996). Differences between entrepreneurs and managers in large organizations: Biases and heuristics in strategic decision-making, *Journal of Business Venturing*, *8*, 9–30.

Campbell, C. M., & Levin, D. (2001). When and why not to auction, Working Paper, Department of Economics, Rutgers University (October).

Carter, R., & Manaster, S. (1990). Initial public offerings and underwriter reputation, *Journal of Finance*, *45*, 1045–1068.

Chemmanur, T. J., & Fulghieri, P. (1999). A theory of the going-public decision, *Review of Financial Studies*, *12*, 249–279.

Choe, H., Masulis, R., & Nanda, V. (1993). Common stock offerings across the business cycle: Theory and evidence, *Journal of Empirical Finance*, *1*, 3–31.

Coase, R. (1960). The problem of social cost, *Journal of Law and Economics*, *3*, 1–44.

Cohen, W. M., Nelson, R. R., & Walsh, J. P. (2000). Protecting their intellectual assets: Appropriability conditions and why U.S. manufacturing firms patent (or not), Working Paper No. 7552, NBER, Cambridge, MA (February).

Fama, E. F., & Jensen, M. C. (1983). Separation of ownership and control, *Journal of Law and Economics*, *26*, 301–325.

Field, L. C., & Karpoff, J. M. (2002). Takeover defenses of IPO firms, *Journal of Finance*, *57*, 1857–1889.

Garvin, S., & Kagel, J. H. (1994). Learning in common value auctions: Some initial observations, *Journal of Economic Behavior and Organization*, *25*, 351–372.

Greene, W. H. (1997). *Econometric analysis* (3rd ed.). New Jersey: Prentice-Hall.

Griliches, Z., Hall, B. H., & Hausman, J. A. (1978). Missing data and self-selection in large panels, *Annales De L'insee*, *30*, 137–176.

Hall, B. H., Jaffe, A. B., & Trajtenberg, M. (2001). The NBER patent citations data file: lessons, insights and methodological tools, Working Paper No. 8498, NBER, Cambridge, MA (October).

Harris, M., & Raviv, A. (1990). The theory of capital structure, *Journal of Finance*, *46*, 327–355.

Harstad, R. M., Kagel, J. H., & Levin, D. (1990). Equilibrium bid functions for auctions with an uncertain number of bidders, *Economics Letters*, *33*, 35–40.

Heiss, F. (2002). Structural choice analysis with nested logit models, *The Stata Journal*, *2*, 227–252.

Hensher, D. A. (1986). Sequential and full information maximum likelihood estimation of a nested model, *The Review of Economics and Statistics*, *68*, 657–667.

Holmström, B., & Tirole, J. (1987). The theory of the firm. In: R. Schmalensee, & R. Willig (Eds), *Handbook of Industrial Organization*, Amsterdam: North Holland.

Holmström, B., & Tirole, J. (1993). Market liquidity and performance monitoring, *Journal of Political Economy*, *101*, 678–709.

Jensen, M., & Meckling, W. (1976). Theory of the firm: Managerial behavior, agency costs and capital structure, *Journal of Financial Economics*, *3*, 11–25.

Kagel, J. H. (1995). Auctions: A survey of experimental research. In: J. H. Kagel, & A. E. Roth (Eds), *Handbook of Experimental Economics* (pp. 501–585). Princeton, NJ: Princeton University Press.

Kagel, J. H., & Levin, D. (1986). The winner's curse and public information in common value auctions, *American Economic Review*, *76*, 894–920.

Kagel, J. H., Harstad, R. M., & Levin, D. (1987). Information impact and allocation rules in auctions with affiliated private values: A laboratory study, *Econometrica*, *55*, 1275–1304.

New Venture Investment: Choices and Consequences
A. Ginsberg and I. Hasan (editors)

Chapter 7

Venture Capital in Financial Systems: Historical and Modern Perspectives

RICHARD SYLLA*

Stern School of Business, New York University, 44 West 4th Street, New York, NY 10023, USA

ABSTRACT

The prospects and promises of entrepreneurship are context-specific, with the relevant context being the nature of the financial systems available to entrepreneurs in historical and contemporary contexts. U.S. history and contemporary differences in conditions for new venture financing in the United States, Latin America, and East Asia illustrate the argument.

1. INTRODUCTION

Since new enterprises have started up throughout recorded history and probably even earlier, "venture" financing must be almost as old as the hills. But modern economic growth—sustained increases in income and product per person—is a phenomenon of the past two to three centuries and hence relatively new in human history. Much of our interest in new venture financing arises from the intimate connection we think it has with modern economic growth. Economic historians often trace sustained economic growth to continuing waves of innovation, especially technological changes, which were introduced first by entrepreneurs. The innovations raised productivity, enhanced the variety of goods and services available, and led to sustained economic growth.

* Tel.: +1-212-998-0869.
E-mail address: rsylla@stern.nyu.edu (R. Sylla).

Entrepreneurs and their innovative ventures had, of course, to be financed. In Schumpeter's (1934) celebrated theoretical analysis of economic development (a synonym for modern economic growth), the partner of the entrepreneur is the banker, and there is a strong implication that development will not occur unless both characters are present and working effectively together. Much of my interest as a financial historian in entrepreneurship and new venture financing arises from Schumpeter's insight. In pursuing that interest in my research, I have reached the conclusion that nature and effectiveness of venture financing at any time or place in history, including the present, depends on context. To be more specific, the nature and effectiveness of venture financing depends on the nature and effectiveness of the overall financial system of which venture financing is just a part, and often a minor part.

In this chapter I illustrate and amplify my contention that venture financing is financial-system-context specific. I do it in two ways, corresponding to the time-series and cross-section approaches used by econometricians in testing hypotheses and estimating parameters of their models with economic data. My "time series" consists of cases of new venture financing in several eras of U.S. history. My "cross section" consists of cases of new venture financing in different parts of the world today. Each type of evidence tends to support the argument that the nature and effectiveness of new venture financing depends on context.

Before turning to the case studies, I want to point out two policy implications of the argument, one negative and the other more positive. The negative implication is that because new venture financing is context-specific, one cannot at all easily transfer venture-financing practices from one context—say, a country that is a paragon of entrepreneurship and economic growth—where those practices function well to another context, namely, a context in which the overall financial system is different from and possibly less developed than the financial system of the paragon country.

The United States is arguably such a paragon country. For more than two centuries its institutions, economic and other, have created a climate hospitable to entrepreneurial innovation. Entrepreneurial innovation would be high on any list of the factors that have made Americans rich and their economy dynamic, as much now as in the past. People in other countries are curious about how Americans have been and continue to be so entrepreneurial. They observe, for example, the successes of American venture capital (VC) funds and the techniques they use to identify, nurture financially, mentor, and profit from new ventures. And they understandably develop the notion that they might emulate these in their own countries for personal and national enrichment.

Can they do this? In most cases, it is likely that they cannot. U.S. VC investors, individuals and firms, operate in small niches of what is probably the most complex, articulated, richly endowed, and highly developed financial system of any country in history. It is a financial system, as I and others have argued in recent work (Rousseau and Sylla, 1999; Sylla, 1998b, 1999, 2002; Wright, 2002), with a number of key institutional components: stable public finances and public debt management, a sound currency, a wide variety of banks and non-bank financial intermediaries, an effective central bank, securities markets (including money markets and futures, options, and other derivatives markets), and a variety of insurance companies and insurance markets.

Such an articulated financial system is highly conducive to the operations of VC investors. Individuals and institutions—financial and non-financial—can devote a portion of their funds to VC investments for possibly enhanced overall returns with minimal added risk because of all the opportunities the system provides for diversification and other forms of risk management. VC funds can be invested at relatively long horizons because of all the liquidity existing elsewhere in the financial system. When start-ups appear promising, bank and security-market financing becomes available. An initial public offering (IPO) mechanism orchestrated by specialized investment bankers with developed distribution channels to individual and institutional investors is in place to allow VC investors to exit from their VC investments in whole or in part, replenishing their funds for financing new ventures. New securities are attractive to investors because developed secondary (trading) markets give securities a high degree of liquidity. If would-be IPO investors want to invest more than their own available funds, the financial system offers numerous ways for raising outside funds to invest.

Such conditions conducive to VC investing in the United States do not exist in most other countries, even ones that are economically developed. In the mid-1990s I visited Italy in connection with the launching of an Italian edition of a book of mine. While there I talked with a banker whose bank had sponsored the translation, and we discussed financial systems. I mentioned to him that I had recently invested in an IPO of common stock in a new company formed to market wine through phone and mail orders. It quickly materialized that such a new venture would have found it impossible to raise equity funds to expand its operations in Italy, a wine-producing and wine-drinking country. As a new concept and venture in wine marketing it might even have had problems in obtaining bank financing, the common way of obtaining external finance for companies in Italy. More than likely, the Italian banker told me, such a firm would have to rely on the capital and credit possessed by the firm's entrepreneurs and maybe their families and friends, as many of the world's entrepreneurs past and present have done and still do, even in the contemporary United States despite its highly developed VC mechanisms. Banks in Italy would regard such a new venture as too risky to lend to, and Italian individuals and institutions are far less likely to invest in corporate securities than are Americans, a fact reflected in Italy's less extensive and active securities markets than America's. If such is the case in modern Italy, a developed country that centuries ago more or less invented modern banking and disseminated it to the rest of the world, how much more difficult must it be for lesser developed countries than Italy to emulate U.S. venture financing practices. The nature and effectiveness of new venture financing depends on financial-system context.

A more positive policy implication of the conclusion that new venture financing is context specific is that all of the efforts of national and international development agencies, the latter including prominently the World Bank and the IMF, focused on improving financial systems throughout the world are, as the British say, spot on. If these agencies achieve the hoped-for successes, then new venture financing possibilities should proceed apace with financial development. Of course, countries themselves need to take the lead in this; there is only so much the external development agencies can do. Both, however, can learn much about better and worse financial systems from studying financial history.

2. U.S. NEW VENTURE FINANCING FROM COLONIAL MERCHANTS TO DOT COMS

In this section of the chapter, I first present several case studies that illustrate changes in the ways important new ventures were financed in the early period of U.S. history when a modern financial system replaced a primitive one. Taken together, the cases demonstrate that new venture financing depends on financial-system context. Next, I discuss how the modern financial system, as it continued to develop, fostered a high level of entrepreneurship for virtually all of the two plus centuries of U.S. history.

2.1. Colonial merchants

In the American colonial period and for some decades after independence, general-purpose (that is, unspecialized) wholesale import-and-export merchants were, along with the great planters of the South, the dominant actors in the American economy. The colonial economy lacked anything remotely like the modern financial system described in the previous section. Public financial needs were minimal except in war periods. The colonies relied on foreign coins and their own fiat paper money issues. There were no commercial banks or a central bank, and both securities and insurance markets were primitive to non-existent. From their colonial relationship with Great Britain, however, Americans had access to the services of the British financial system with its established mechanisms—bills of exchange, for example—in financing foreign trade.

How did a colonial American get a start in merchanting, the primary path to wealth and influence in the urban, mercantile economy? Doerflinger (1986), in a thorough study of Philadelphia that examined the biographies and papers of approximately a hundred of that city's merchants, identifies four paths. One—easily understandable—was to have a rich father who perhaps also was a merchant. Such a fortunate young man could then enter and eventually take over his father's business—hardly an entrepreneurial achievement – or he could more entrepreneurially start his own firm with a capital stake from his parents.

A second path open to scions of less well endowed mercantile and professional families was to work in or with major mercantile firms until they had accumulated sufficient capital to start their own firms, or to utilize connections with such firms. Working in a firm was a form of apprenticeship. But one might also work with a firm, for example, by serving as a supercargo (business agent) on one of the firm's shipping voyages exporting American products to foreign markets and purchasing goods in foreign markets for sale in American markets. Or one might be lent money by an established merchant to whom one was known, in order to start a new mercantile firm. Or the established merchant, instead of lending his own funds to the newcomer, might recommend him for credit to supplying firms in England.

A third path to becoming a wholesale merchant was to start in lesser occupations that developed skills and experience useful to wholesale merchants. Doerflinger documents cases of retailers (dry goods sellers, grocers, and other shopkeepers), artisans, and mariners (e.g. ship captains). These lesser occupations inculcated skills in dealing with suppliers and customers, keeping account books, receiving and granting credit, and knowledge of both domestic and foreign markets. All of these were useful in becoming a wholesale merchant once one had accumulated enough capital and credit to embark on that more prestigious occupation.

Finally, the fourth path into wholesale merchanting was through foreign contacts and experience. A good number of successful merchants in Philadelphia were immigrants from England, Ireland, Scotland, France, and Holland. As such, they could exploit contacts and access to suppliers and credit in those countries. In some cases, the immigrants were sent by foreign firms to establish themselves in American markets.

A common theme of the various paths to entering colonial America's most prestigious and potentially lucrative enterprises is access to credit and capital. The VC providers in that period of primitive financial development tended to be successful merchants on both sides of the Atlantic. Established merchants, by providing capital and credit to newcomers to their business, advanced their own personal and business goals while allowing the newcomers to get started.

Much of the world continues to have a similar system of new venture finance. It is better than nothing, but not really adequate to calling forth the higher levels of entrepreneurship that we associate with modern economic growth, either in colonial America, which according to economic historians had little of the sustained increases in per capital product that define modern economic growth, or in modern less developed countries.

2.2. Textile manufacturing: Slater, and traditional new venture finance

The so-called industrial revolution is usually traced to late 18th century Britain, and it did not take long for key elements of it to make their way to America. Samuel Slater, a 21-year-old Englishman versed in the new mechanical textile spinning technology, migrated to America in 1789 (Chandler and Tedlow, 1985). He would later be termed the father of American manufacturing. But when he arrived on American shores, he was a typical would-be entrepreneur with skills and ideas, searching for financial support. New Yorkers put Slater in touch with the Browns, a wealthy family of merchants in Providence, Rhode Island, who already had been trying without much success to develop machine spinning. In April 1790, Slater and the Browns entered into a partnership, Almy, Brown, and Slater (Almy was a Brown son-in-law) calling for the Browns to supply the financing and Slater to build mechanical spinning machines and, by 1793, a factory to house them at Pawtucket near Providence, where the waters of the Blackstone River turned a waterwheel that powered the machines. Slater was to receive half the profits of the venture, the other half going to Almy and Brown.

The Slater example is a classic case of mercantile capital financing industrial entrepreneurship. Although three or four commercial banks had appeared in the United States by 1790 to ease the liquidity problems of merchants, there was still no financial system to speak of. All that would change in just a few years. Before the 1790s ended the United States would have the modern, articulated financial system described earlier in this essay. As we now will see, it changed the nature of entrepreneurial venturing.

2.3. Textile manufacturing: Lowell and modern new venture financing

Samuel Slater introduced mechanical spinning of cotton fibers into yarn to the United States in the early 1790s, but the other key step in textile manufacturing, mechanical weaving, was not perfected either in Britain or America for another two decades. Until that happened, and even for some years afterwards, the yarn was "put out" to handloom weavers to turn into cloth, which was then retrieved and marketed by the merchant-manufacturers.

Credit for introducing power looms and mechanical weaving in factories to the United States goes to Francis Cabot Lowell, a scion of well-to-do Boston merchants. Lowell, like Slater, relied to an extent on mercantile capital to finance his entrepreneurial achievement. But the manner in which he and his associates did so was totally different as a result of precocious U.S. financial development between 1793, when Slater and the Browns opened their Pawtucket mill, and 1814, when Lowell and his associates opened their first integrated (spinning and weaving under one roof) textile factory in Waltham, Massachusetts.

Lowell financed his venture by receiving a corporate charter for the Boston Manufacturing Company in 1813 from the State of Massachusetts. The charter authorized a capital of $400,000, of which $100,000 was initially raised to finance the development of Lowell's loom, and to secure a site and build a factory at Waltham on the Charles River, whose waters would turn a wheel to generate mechanical power (Chandler and Tedlow, 1985). Shares of stock in the enterprise were initially sold to a small group of wealthy individuals known to one another in the Boston mercantile community. By the time Lowell embarked on his venture, Americans—at least those who lived in major cities and towns—were quite familiar with banks, a central bank, corporations (banks and insurance companies were the largest of these), and stock and bond markets, a big change from 1790.

The venture proved a great success, in large part because the costs of manufacturing cloth in the Waltham factory were substantially lower than the prices the cloth commanded in the expanding American market. Economics teaches that in such a circumstance the proper thing to do is expand production in factories, and that is what Lowell's associates set out to do after his untimely death at age 42 in 1817. With the new textile technology perfected, the key need for expansion was power—waterpower at that time. The search for it led the associates to a site on the Merrimack River in the township of Chelmsford, Massachusetts, where the river fell about 30 feet in the space of one mile.

Earlier, the falls of the Merrimack at the site were considered a major inconvenience, and in 1792 a company called "the Proprietors of the Locks and Canals on Merrimack River" was incorporated by the State of Massachusetts in the early surge of U.S. incorporations to bypass the falls with a canal to facilitate navigation. That company sold 600 shares, raising $100 per share, to build the canal. But it was not a great success, earning only $3\frac{1}{2}\%$ on average, a low return on capital for that era. The Boston textile associates, realizing the potential of falling water to power their machinery, formed a new, again closely held company capitalized at $600 thousand. The company quietly through an agent acquired land along the river at the site and a controlling interest in the canal company at an average cost of $89–90 per share.

Having secured control of the site and its waterpower potential by means of land and stock purchases, the company widened and extended the canal, creating locations for a number of textile mills that were incorporated to take advantage of the new manufacturing technologies. The first of them, the Merrimack Company, started paying dividends in 1825, and according to one of the original Boston associates of F. C. Lowell, Nathan Appleton writing in 1858, it paid dividends averaging more than 12% per annum for the next 33 years (Chandler and Tedlow, 1985: 165). The village of East Chelmsford, with a dozen houses in 1821 when the Boston associates first visited it, quickly was renamed to become the major manufacturing center of Lowell, Massachusetts, with tens of thousands of denizens. Its success was the result of new manufacturing technologies, as has long been recognized, but the foregoing account indicates that a new and modern financial system with banks, corporations, shares, and securities markets made that success possible by offering entrepreneurs new ways of combining and expanding resources to exploit the new technologies.

2.4. Financial underpinnings of U.S. economic modernization

The contrast between the manner in which merchants and entrepreneurs such as Slater in 1790 and earlier were able to finance their start-ups, and those used by Lowell and his associates in 1813 and later to finance their innovative enterprises suggests that dramatic financial change took place during the intervening two decades. Most of that change in fact occurred in the early 1790s in what I like to call "the Federalist Financial Revolution" (Sylla, 2002). In 1789, when the new government under the Constitution first began to function, the United States was a bankrupt country paying neither interest nor principal on its domestic debts, and paying interest on some of its foreign debts only by issuing new debts. It then had three banks, one each in Philadelphia, New York, and Boston, but these were not linked together in anything resembling a banking system. It lacked a national currency and a central bank, and there were few business corporations. A few brokers in the largest cities dealt in evidences of debt left over from the War of Independence, most of which sold at 10–20 cents on the dollar because the national government lacked the means to service them, but there were no organized securities markets with regular trading.

By the mid-1790s, all that had changed. The federal government paid interest quarterly, in hard money or its equivalent, on most of a restructured and simplified national debt. The country had a central bank with branches in a number of states, more banks were being chartered by the states to compete with it, and all the new banks were forming a connected banking system. The United States had a new currency unit, the dollar defined in terms of specific weights of gold and silver, and a mint for coinage, although foreign coins continued to make up much of the monetary base. Active securities markets existed in major cities to trade the new debt and the shares of the new corporations that were chartered by the federal (just one, the central Bank of the United States) and state governments. The New York Stock Exchange, now far and away the world's largest stock market, traces its origins to an agreement forged by a number of the city's brokers in 1792, when the vast expansion of their business resulting from the federal debt restructuring and new equity issues created the need for organizational innovation. Foreign investors, attracted by higher returns than they enjoyed at home, were buying U.S. securities, in the process transferring capital to the new country.

The U.S. financial revolution of the early 1790s was the work of nationalist leaders, foremost among them Alexander Hamilton, the nation's first Secretary of the Treasury, who drew up the plans for it and executed them with the cooperation of President George Washington, the first Congresses, and business leaders from the merchant class. The modern financial system was a key development of U.S. history. It made the financial world of Francis Lowell in 1813 a very different one from that Samuel Slater encountered in 1790. Moreover, as much as any other early development, the financial system launched by the Federalists became the underpinning of the subsequent tremendous expansion of the U.S. economy and the rise of the country to great-power status in the world.

That the modern financial system put in place by the Federalists created a hospitable climate for entrepreneurship in the United States from the first years of the republic is abundantly evident in historical records. In a recent book based on study of hundreds of memoirs and autobiographies of what she terms the first generations of Americans, persons such as Francis Lowell whose lives encompassed the period from the Declaration of Independence in 1776 to the 1830s, historian Joyce Appleby (2000) found some telling commonalities in their experiences, and, although her economics are not all that sophisticated, she drew an appropriate conclusion:

> In personal reminiscences banks figure as the principal source for currency—welcome, unregulated, and amazingly good at flooding the nation with notes . . ..
>
> Debt acted as a mighty leveler in this ebullient society, and it figured in almost every autobiography . . .. The elaboration of a national market depended upon many, many young men leaving the place of their birth and trying their hands at new careers. The range and sweep of their entrepreneurial talents, defined best as the ability to take on novel undertakings as personal ventures, suggests the widespread willingness to be uprooted, to embark on an uncharted course of action . . .. To fail to mark this feature of the early republic is to obscure a very important element in American history: the creation of a popular entrepreneurial culture that permeated all aspects of American society (Appleby, 2000: 86–89).

Three recent monographs on U.S. financial history confirm with detailed and sophisticated economic analyses the ways in which banks, often working in tandem with money and securities markets, financed a great variety of new ventures in the period up to 1860, by which time the United States was clearly one of the largest and most dynamic of the world's national economies (Bodenhorn, 2000; Lamoreaux, 1994; Wright, 2002). In short, it seems that from the first decades of the country's history, the modern U.S. financial system launched in the 1790s did not merely cater to entrepreneurial instincts. It unlocked them.

The financial system would continue to perform that function from that early era to our own. Given the country's initially great size which then tripled through territorial expansion in the period before 1860, economic historians of the United States attach great importance to a "transportation revolution" that extended and integrated the U.S. markets during the early and middle decades of the 19th century (Taylor, 1958). Its main features were the construction of great canals, the proliferation of steamboats on the country's internal river systems, and the advent of railroads. In each case, the transportation innovations—the largest entrepreneurial start-ups of that era—were funded by a modern financial system that preceded them on the scene. When a public entrepreneur, New York State governor DeWitt Clinton, had a vision of a great canal traversing the state from the Hudson River to Lake Erie, the Erie Canal that resulted was financed by selling state bonds to domestic and European investors during its construction between 1817 and 1825. Other states attempted during the next two decades to emulate New York's success with canals of their own, issuing bonds to banking houses that resold them to American and European investors. Even the ill-fated Second Bank of the United States became heavily involved in this form of intermediation after losing its federal charter and central banking functions in 1836.

Steamboats individually were not such large investments; they could be financed by their owners or by loans from banks. And there were plenty of banks. The banking system grew spectacularly, increasing the number of incorporated banks from 100 in 1810 to 1,600 by 1860, and to eventually to more than 20,000 separate banking institutions at the peak of the 1920s. Such large numbers of competitive banking corporations chartered by legislatures, by separate acts and under general incorporation laws, were a distinctive feature of U.S. development, one that other nations later emulated. But there were also large numbers of "private" or unincorporated banks that were partnerships under the common law.

Issues of stock commonly financed early railroads, with some aid coming also from governments, both financially and with charters of incorporation and grants of eminent domain. Later, larger railroads and railway systems relied more on bond financing intermediated by specialized investment bankers with connections to domestic and foreign investors. The great banker J. P. Morgan earned his spurs and made most of his fortune in railroad finance during the late 19th century, while at times also financing the U.S. government and, around the turn of the century, great industrial corporations.

The 20th century witnessed many changes to the U.S financial system. Among them were the return of a central bank in 1913 after a seven-decade absence, and increased federal regulation of banks and securities markets dating from the 1930s. The 1950s brought the rise of new institutional investors such as pension funds and the expanded use by Americans of mutual funds. Exchange-traded financial futures and options contracts came in the 1970s, while so-called junk bond financing and leveraged buy-outs were innovations widely exploited in the 1980s. But the financial system's function of financing new ventures, always a niche business, continued throughout the century, even as new forms such as private VC funds and public aids to small business start-ups appeared.

A most recent and vivid example of new venture financing occurred in the late 1990s. To many observers, the dot com, internet, and other technology-related IPO frenzies of that period seemed to be unprecedented. In fact, they differed little from the earlier financial frenzies dating back to the financial system's inception. Consider the IPO of the first Bank of the United States in 1791. In July of that year, the initial subscribers to shares of the Bank, which was an integral component of Alexander Hamilton's new financial architecture for the country, paid $25 down for a right to buy a full share of Bank stock with a series of additional payments that would total $400, the par value of a share, over the subsequent three years. A month after the offering, the price of a right advanced from $25 to $300, before dropping back to a range of $130–$160 toward the end of the year (Sylla, 1998a). Investors who bought rights to Bank shares, or "scripts" as they were called at the time, at the peak in the aftermarket of 1791 were severely punished. More than two centuries later, American investors would be similarly punished after they bought dot com shares at aftermarket peaks at the end of the 1990s. In each case, however, new entrepreneurial ventures were financed. During the intervening two centuries, there were many similar occurrences—the canal and railway manias of the 1830s and 1840s, still larger railway manias in the 1860s and 1870s, the industrial-stock and merger manias at the turn of the 20th century, the radio and electrical utility manias of the 1920s, and the conglomerate (mergers of companies in unrelated businesses) and "nifty fifty" (growth stocks so great that they should be purchased at any price) manias of the 1960s and 1970s, to name several. From first to last in U.S. history, a modern financial system promoted new ventures, often in a frenzied manner.

In all of these manias, overly enthusiastic investors got burned as asset-price bubbles collapsed. Other investors were rewarded, provided the companies in which they invested survived and eventually prospered. One might well argue that when entrepreneurs can find investors willing to bear such risks and take such lumps, economic development can advance more rapidly than it would in a more risk-averse culture. The American experience is a prime example supporting the argument. For more than two centuries since the United States was blessed—and, some would say, now and then cursed—with a modern, cutting-edge financial system that included among its achievements a continual pushing out of the envelope of new venture financing.

3. NEW VENTURE FINANCING ACROSS THE WORLD TODAY

Differences in new venture financing across the world today are analogous to the differences one can identify at various times in U.S. history. A primary reason for both types of differences—the historical and the contemporary—is that financial systems differ. Before the 1790s, the U.S. financial system was primitive; from the 1790s to the present the U.S. financial system has been modern. One could say the same of U.S. entrepreneurship: before the 1790s it was primitive, and after that time it was modern. The nature of entrepreneurship depends, of course, on more than the nature of the prevailing financial system. In particular, new technologies create entrepreneurial opportunities in ways that are independent of prevailing financial arrangements. Slater and Lowell latched onto and later perfected in America new technologies that arose in Britain's technological environment, and these were only distantly related to Britain's financial environment around the turn of the 19th century.

Its does seem, however, that whether entrepreneurial opportunities are exploited, as well as the extent to which they are exploited, depends very much on financial arrangements. If that were not true, then we would not expect to observe such wide disparities in level of economic development and rates of economic growth as we do in fact observe in the contemporary world. Technological knowledge is pretty much available to all; students from most nations of the world, for example, are enrolled in the universities and technological institutes of developed countries. That has been true for decades, perhaps even more than a century. Yet discrepancies in levels of economic development and per capita incomes across countries are at present as wide as they ever have been in modern history. Although modern economic growth is spreading to more and more of the world's nations, gaps between the richest of them and the poorest of them continue to grow. Factors other than technology account for this. One of the most important of them is differences in financial systems and their ability to foster entrepreneurship across countries.

In support of this argument, I draw here on a recent study by a non-governmental organization, the Inter-American Development Bank (IADB), *Entrepreneurship in*

Emerging Economies: The Creation and Development of New Firms in Latin America and East Asia (Inter-American Development Bank, 2002). It is a detailed study, based on extensive surveys of more and less successful new ventures, of conditions fostering and hampering entrepreneurship in the two world regions. The study finds that conditions are substantially more favorable to entrepreneurship in East Asia than in Latin America, and that one of the main reasons for the observed differences had to do with access to financing. In an appendix to the study made up of articles and comments on the report, a Japanese expert (Hosono, 2002) summarizes the findings regarding access to finance as well as I can, so I quote from his summary at some length:

> The Study found that access to external sources for financing is clearly higher in East Asia than Latin America, both at the start-up stage and at the early development stage. At the start-up stage, the role of venture capital, especially of business angels, in East Asia is double that of Latin America . . .. Bank loans are also more frequently used in the former than in the latter . . .. Even loans from public institutions . . . are used much more commonly in East Asia than Latin America . . .. In addition to this basic difference, the Study found that the more dynamic firms use these sources to a much greater extent. Furthermore, compared to their East Asian counterparts, more Latin American entrepreneurs mentioned the limited or total lack of access to external financing as a significant factor.
>
> The Study also made it clear that sources of external finance in Latin America do not necessarily meet the needs of the region's entrepreneurs. When consulted about why they did not use external financing sources, two-thirds of the Latin American entrepreneurs stressed that those sources did not meet their needs because of such features as high interest rates or requirement of collateral. However, half of them preferred to avoid them because they wanted to maintain control of their business, did not want to take on more debt, and/or they lacked faith in [financial] institutions . . .. As regards East Asian entrepreneurs, on the other hand, less than 1 in 5 said they preferred to avoid the use of financial sources or they considered them inappropriate . . ..
>
> The number of firms that use bank loans and public institutions at the early development stage is almost two times greater than during the previous stage (start-up stage) in East Asia, while in Latin America only a small increase is observed. This impressive difference at the advanced stage reflects the much better access of East Asian entrepreneurs to . . . external financing as compared to their Latin American counterparts. The two most important sources of financing at this stage for dynamic enterprises in East Asia are loans from commercial banks and national public institutions. The percentage of dynamic firms that use these loans . . . is substantially higher than the case of Latin American firms . . ..
>
> An important contribution of the Study is that it made clear how the weakness in the financial system affects firms at the start-up and the early development stages . . ..
>
> An even more important and original contribution of the Study . . . is that it clarified the way Latin American entrepreneurs respond to the problem of external financing. Frequently they look for alternative finance sources, and if they cannot get enough

Lamoreaux, N. R. (1994). *Insider Lending: Banks, Personal Connections, and Economic Development in Industrial New England*, Cambridge: Cambridge University Press.

Rousseau, P. L., & Sylla, R. (1999). Emerging Financial Markets and Early U.S. Growth. National Bureau of Economic Research Working Paper No. 7448.

Schumpeter, J. A. (1934). *The Theory of Economic Development: An Inquiry into Profit, Capital, Credit, Interest, and the Business Cycle*, Cambridge, MA: Harvard University Press.

Sylla, R. (1998a). The first great IPO, *Financial History*, *64*, 4–5.

Sylla, R. (1998b). U.S. securities markets and the banking system, *Federal Reserve Bank of St. Louis Review*, *80*, 83–103.

Sylla, R. (1999). Emerging markets in history: The United States, Japan, and Argentina. In: R. Sato, R. V. Ramachandran, & K. Mino (Eds), *Global Competition and Integration* (pp. 427–446). Boston: Kluwer Academic Publishers.

Sylla, R. (2002). Financial systems and economic modernization, *Journal of Economic History*, *62*, 279–292.

Taylor, G. R. (1958). *The Transportation Revolution, 1815–1860*, New York: Holt Rinehart.

Wright, R. E. (2002). *The Wealth of Nations Rediscovered: Integration and Expansion in American Financial Markets, 1780–1850*, Cambridge: Cambridge University Press.

New Venture Investment: Choices and Consequences
A. Ginsberg and I. Hasan (editors)

Chapter 8

Canadian Labour-Sponsored Venture Capital Corporations: Bane or Boon?

DOUGLAS J. CUMMING[a,*] and JEFFREY G. MACINTOSH[b,1]

[a] *School of Banking and Finance, Faculty of Commerce and Economics, University of New South Wales, UNSW, Sydney, NSW 2052, Australia and School of Business, University of Alberta, Edmonton, Alta., Canada T6G 2R6*
[b] *Faculty of Law, University of Toronto, 78 Queen's Park, Toronto, Ont., Canada M5S 2C5*

1. INTRODUCTION

In the past 20 years, there has been a growing awareness that the comparative advantage of the Western industrialized economies increasingly lies in the "knowledge-based industries" (or KBI). A significant part of the KBI consists of various high technology sectors, including information technology (which includes the internet, as well as computer hardware and software), telecommunications, biotechnology, medicine, and the like. That the comparative advantage of developed economies should be shifting toward the KBI/technology sectors is fully consistent with the picture painted by development economists. Prior to significant capital formation and development of infrastructure, investment in education is thought to yield relatively modest returns. In highly developed economies, however, the return to education, and particularly technical education, shows a significant and rising premium. In the developed economies, this has led governmental policy makers to closely scrutinize KBI with a view to enhancing the development of these sectors.

* Corresponding author. Tel.: +1-780-492-0678; fax: +1-780-492-3325.
E-mail addresses: douglas.cumming@ualberta.ca (D. J. Cumming); j.macintosh@utoronto.ca (J. G. MacIntosh).
[1] Tel.: +1-416-978-5785; fax: +1-416-978-6020.

This in turn has led governments to an examination of venture capital (VC). Venture capitalists (VCs) participate in both the funding and development of fledgling enterprises in a variety of KBI sectors. For example, in both the United States and Canada, approximately 90% of all venture capital investments are made in technology investments. In some cases, the perceived importance of developing the KBI/technology sectors has led to various governmental programs designed to give a boost, either directly or indirectly to these sectors. In some cases, this consists of direct assistance for technology companies, whether through tax mechanisms, subsidies for research and development, or other means. In other cases, the assistance is indirect, as is the case, for example, with governmental subsidization of venture capital activity. Whether these government dollars are well or poorly spent is a question that is difficult to answer, sine the social return to innovative activity is generally thought to be perhaps double the private return, but with fairly large confidence intervals.

This chapter examines a governmental assistance program run by various Canadian governments, both provincial and federal, that takes the form of indirect subsidies to technology enterprises via tax subsidization of the *investors* in venture capital funds called "labour-sponsored venture capital corporations", or LSVCCs. The LSVCC differs in many important ways from a typical private fund. It may be incorporated pursuant to enabling legislation either federally or in a variety of provinces, and is subject to restrictions set out in the legislation. In particular, it must have a labour union sponsor, which controls the fund but has no ownership interest, is open only to individual investors (who need not have a high net worth), and must invest in firms based within the sponsoring jurisdiction. The primary motivation for the LSVCC is to expand the pool of venture capital under management in Canada, although such funds often operate under multiple statutory mandates that extend beyond profit maximization. The mechanism for inducing investment in LSVCCs has been the provision of generous tax subsidies to investors, consisting of a combination of tax credits and deductibility of the investment from income.

The primary motivation of this chapter is to examine whether the tax expenditures that underlie the LSVCC programs represent a useful expenditure of public monies. We suggest that they do not, and that the various government sponsors should seriously consider abandoning the LSVCC programs. We begin our examination with a sketch of the Canadian venture capital industry. This is followed by a description of the organizational structure of the LSVCCs and of the various statutory constraints that they operate under. We briefly compare this structure and these constraints with those applicable to the LSVCCs' private sector counterparts. At a theoretical level, we suggest that the LSVCC structure is highly inefficient and likely to lead to a high level of agency costs vis-à-vis funds investors.

We then turn to an examination of the profitability of the LSVCC funds. We adduce evidence suggesting that LSVCC performance has been extremely poor, and we summarize other research that points in the same direction. While the fixed and variable components of fund manager compensation (the management expense ratio, or MER, and the carried interest, respectively) are comparable to those of private funds, LSVCCs have performed extremely poorly compared to both Canadian and U.S. private funds, various Canadian and U.S. market indices, and even 30-day treasury bills.

While this suggests that the tax expenditures that underlie the LSVCC programs have not been wisely spent, many have claimed that the LSVCC programs are justifiable even in the face of poor returns, on the basis that they have significantly augmented the pool of Canadian venture capital investments. We summarize evidence in related research, however, that strongly suggests that LSVCCs have so energetically crowded out other types of venture capital as to lead to an overall *reduction* in the aggregate pool of Canadian venture capital. We conclude by suggesting that the Canadian LSVCC program has been a very costly failure for its government sponsors.

2. THE CANADIAN VENTURE CAPITAL INDUSTRY: TYPES OF FUNDS

There are five different types of venture capital funds in Canada (Amit et al., 1997, 1998; Macdonald, 1992; MacIntosh, 1994, 1997): private, corporate, government, hybrid, and the LSVCCs. As discussed further below, throughout the 1990s and up to the present, the LSVCC has been the dominant form of venture capital organization in Canada. Because of the statutory limitations placed upon LSVCC funds, LSVCCs are significantly different from private venture capital funds in organizational structure. Private funds, the second most important form of venture capital organization in Canada, are similar to U.S. venture capital limited partnerships, although historically they have tended to be less specialized than their U.S. counterparts, which often invest only in particular areas of the high tech spectrum (Gompers and Lerner, 1996, 1999; Halpern, 1997; MacIntosh, 1994). Canadian corporate VCs are analogous to U.S. corporate VCs (Gompers and Lerner, 1999), but tend to finance a somewhat more heterogeneous group of entrepreneurial firms (Cumming, 2000). Government venture capital funds in Canada (which comprised 5% of the overall pool of capital in 2001) are managed either by in-house managers or by independent professional managers and finance a wide variety of different entrepreneurial firms. Canadian hybrid venture capital funds, which constituted 6%

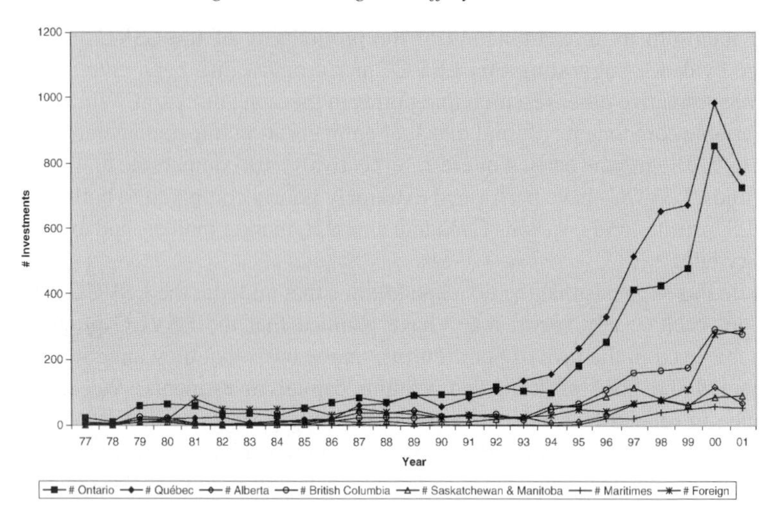

Figure 1. Geographic distribution of venture capital in Canada: 1977–2001.

of the pool of capital in 2001,[2] receive both government and private support and invest in all types of entrepreneurial firms (Cumming, 2000; MacIntosh, 1994).

3. THE CANADIAN VENTURE CAPITAL INDUSTRY: OVERVIEW STATISTICS

In this section, we provide a perspective on venture capital in Canada by presenting a variety of descriptive statistics.[3] Most of the data encompass the full 1977–2001 period, but some of the statistics presented below cover a shorter time frame where the full data are not available.[4]

The geographic distribution of venture capital investments in Canada is presented in Figures 1 and 2. Figure 1 shows the regional distribution by number of investments, while Figure 2 shows the distribution by dollars of investment. These figures show

[2] The category "Hybrid" was used by the Canadian Venture Capital Association (CVCA) until 2001. The new term used by the CVCA for these funds is "Institutional".

[3] These statistics are culled from various reports by Macdonald & Associates, Ltd. for the Canadian Venture Capital Association over the period 1977–2001. See Gompers (1998) and Gompers and Lerner (1999, 2001) for similar U.S. venture capital statistics.

[4] See also the Canadian Venture Capital Association Annual Reports, Amit et al. (1997, 1998) and Cumming (2000), for aggregate Canadian venture capital industry statistics.

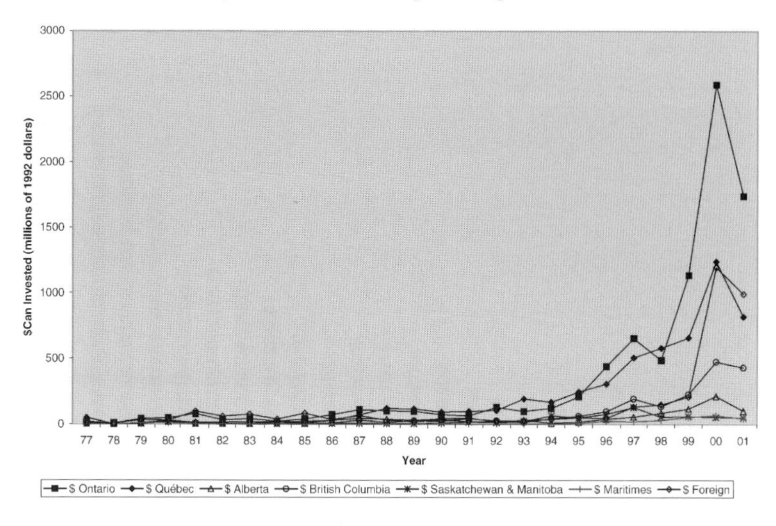

Figure 2. Geographic distribution of venture capital in Canada: 1977–2001.

that there has been a significant increase in the number and dollar value of venture capital investments in Canada since 1990, particularly in Ontario and Quebec, in which the vast majority of investments are made. The one noteworthy difference between Figures 1 and 2 is that since about 1995, the average size of investment in Quebec has been noticeably smaller than that in Ontario. The Quebec venture capital industry is dominated by one very large LSVCC fund, while Ontario's industry has a greater admixture of private and other types of funds.

The increase in the total number of venture capital investments and venture capital firms (that are full members of the Canadian Venture Capital Association; the total number of firms is larger, but unknown) in Canada is shown in Figure 3. Figures 1–3 clearly demonstrate the dramatic growth in the Canadian venture capital industry in Canada over the 1977–2001 period.[5]

Figure 4 presents data for capital under management, capital available for investment and new venture funds for the 1988–2001 period. The capital available for investment reflects the extent to which contributions to venture capital funds have outstripped the funds' ability to invest these contributions. It can be seen from Figure 4 that, historically, there has been a large "overhang" of uninvested capital in

[5] Similar trends in venture capital are documented elsewhere. Previous research on the distribution of venture capital investments within the U.S. includes Gompers and Lerner (1998) and Sorenson and Stuart (2001); research across countries appears in Black and Gilson (1998) and Jeng and Wells (2000).

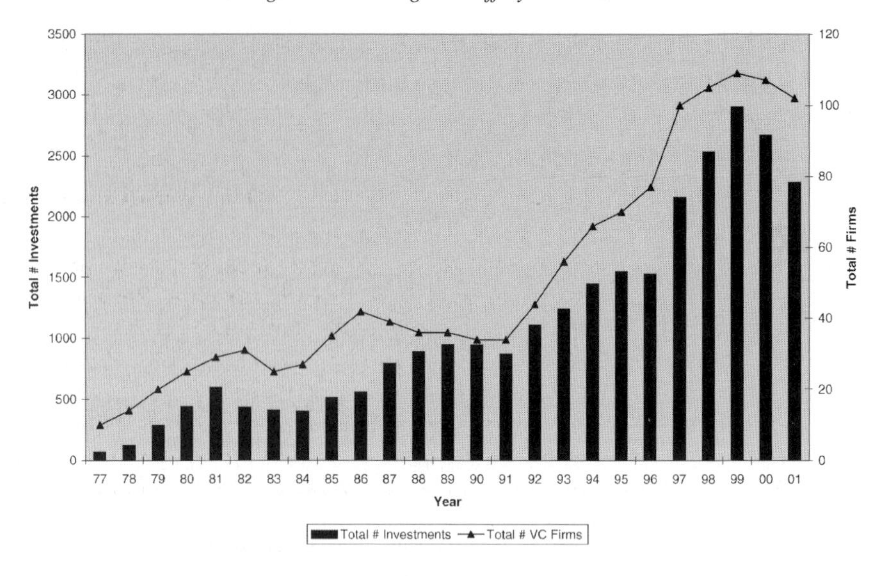

Figure 3. Venture capital firms in Canada (full members of the Canadian Venture Capital Association): 1977–2001.

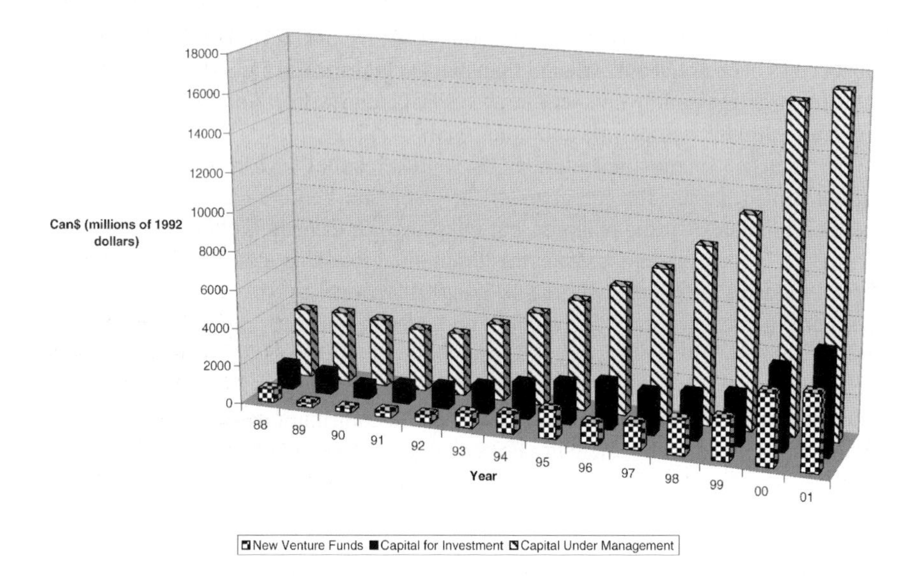

Figure 4. Venture capital funds in Canada: 1988–2001.

Canada. Much of this overhang is attributable to the LSVCCs. By the end of 1996, the overhang amounted to approximately three years of venture capital investments (Department of Finance, 1996). The problem of overhang forced Canada's second largest LSVCC (Working Ventures) to suspend new capital raising for two and a half years (from mid-1996 to the end of 1998). At the time of suspension, Working Ventures had only 19% of its contributed capital invested in eligible businesses.[6]

There appear to be a number of reasons for the LSVCCs' inability to invest all of their contributed capital. As discussed further below, the LSVCCs raise most of their money through contributions to individual registered retirement savings plans (RRSPs), which roughly correspond to 401K plans in the United States. Most of the fund raising of LSVCCs takes place in the last three months before the date after which a contribution cannot be used to reduce the individual's tax payable in for the previous tax year. This makes LSVCC fund raising "lumpy", concentrating contributions at one time of the year, raising the likelihood of a mismatch between funds flow and available investment opportunities. In addition, LSVCC investors were, until 1996, locked into their investments for only five years, following which they could demand redemption at net asset value. While the lock-in period has been increased to 8 years in most jurisdictions (although in Quebec, where shareholders have always had to hold until retirement), this is still a shorter lock-in period than that for private funds (ten years). This has prompted the LSVCCs to retain a certain proportion of capital in liquid investments such as treasury bills and bank deposits to satisfy demand redemptions. We also believe that the overhang problem is a function of the comparative lack of skill of the LSVCC managers, who have had more difficulty than their private fund counterparts in finding promising investments. Evidence consistent with lower skill levels is presented in Brander et al. (2002) and Cumming and MacIntosh (2001a, 2003a, b, c).

As can be seen from Figure 5, in the 1990s much of the new capital in the industry resulted from the growth of the LSVCCs. By contrast, there were relatively modest increases in the growth of private funds. While significant percentage increases are observed among corporate, hybrid and government funds, these increases nonetheless represent a relatively modest increase in aggregate dollar value. As discussed further below, the increase in LSVCC capital has taken place despite extremely poor returns, and is attributable to the strong tax incentives that have induced investors to contribute capital to these funds. Elsewhere (Cumming and MacIntosh, 2001b), we produce evidence consistent with the view that these

[6] See "Working Ventures Puts Capital Raising on Hold" at http://www.newswire.ca...June996/05/c0564.html

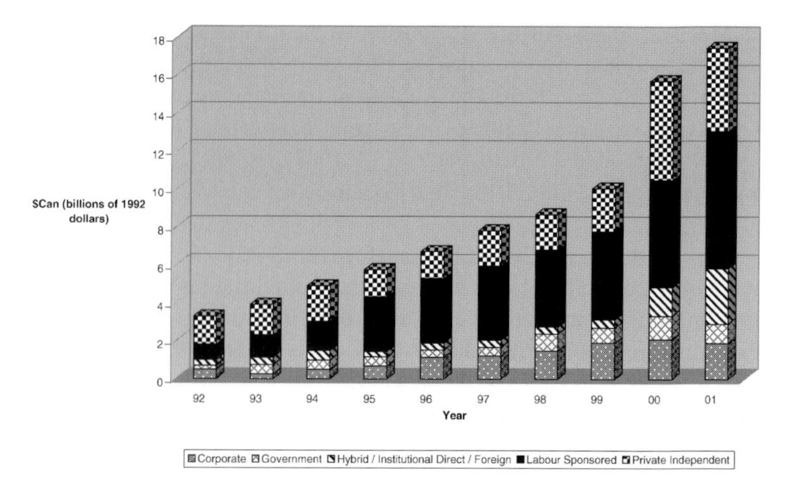

Figure 5. Venture capital under management by investor type in Canada: 1992–2001.

tax benefits have lowered the LSVCCs' cost of capital relative to other fund types, resulting in a crowding out of these other funds (discussed further below).

Figure 6 presents book value (cost) and market value estimates of venture capital in Canada over the 1981–1999 period. Market values have been declining relative to book values since 1995 (with the exception of 1999). This is consistent with increases in the frequency of less profitable exit vehicles such as buybacks over this period of time (see Figures 7 and 8).[7] The high frequency in the use of buyback and secondary sale exits exhibited in Figure 7 constitutes a marked departure from the U.S. experience, in which IPOs and trade sale exits are the two most important forms of exit (Cumming and MacIntosh, 2003a, b, c; MacIntosh, 1997). However, the profitability (gross returns)[8] of different forms of exit indicated in Figure 8 is roughly similar to that seen in the U.S.

Evidence of further differences between the U.S. and Canadian venture capital industries is presented in Cumming and MacIntosh (2001a) (re choice of investment duration), and Cumming and MacIntosh (2003b, c) (re choice of the extent of exit). Cumming and MacIntosh (2003b, c) also find that the risk and return to venture

[7] Cumming and MacIntosh (2003b) also find that the variance of the returns to venture investing in Canada is significantly lower than that in the U.S. for most exit vehicles over the 1992–1995 period.

[8] Annualized returns for each form of exit are not available on an industry-wide basis in Canada. See Cumming and MacIntosh (2003a, b, c) for evidence on each exit vehicle from a hand-collected sample. See Gompers and Lerner (1999) for statistics on U.S. VC-backed IPO performance; see also Francis and Hasan (2001) for recent evidence on the performance of VC-backed IPOs.

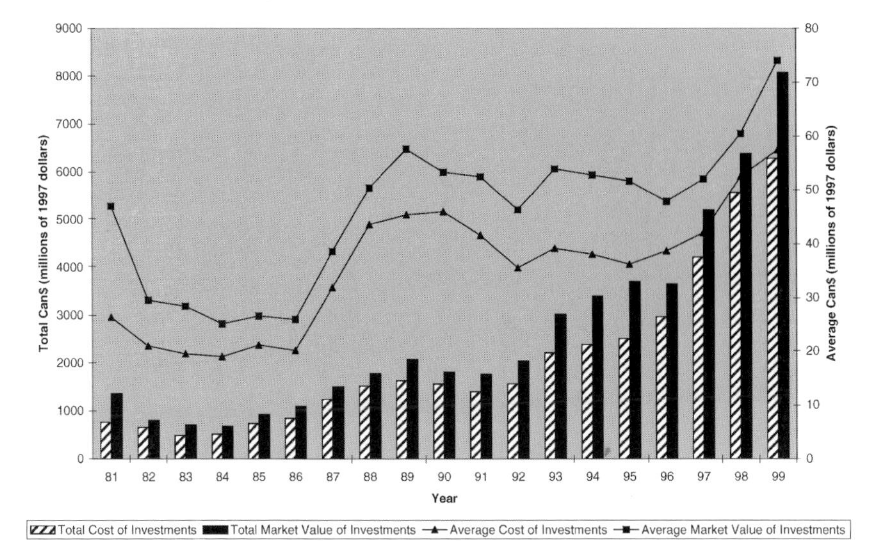

Figure 6. Venture capital market value estimates in Canada: 1981–1999.

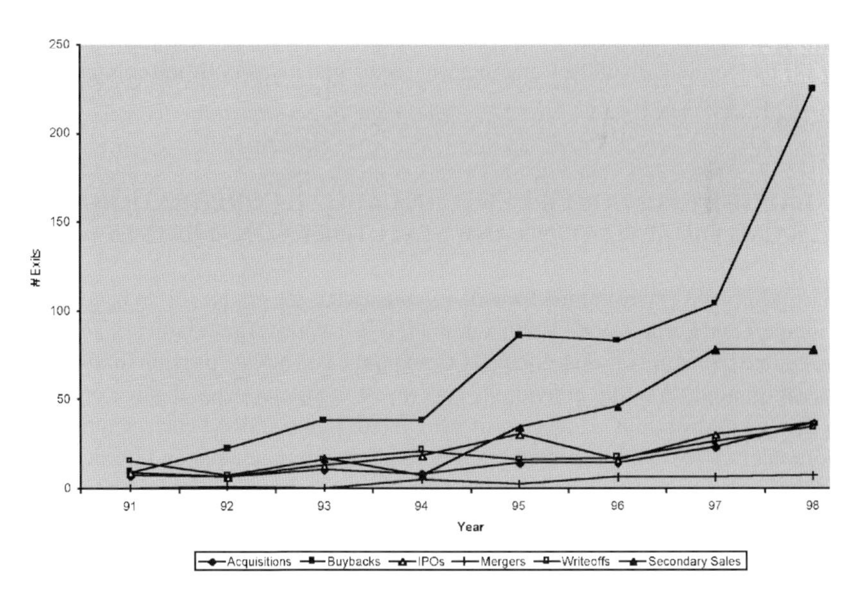

Figure 7. Venture capital exits in Canada: 1991–1998.

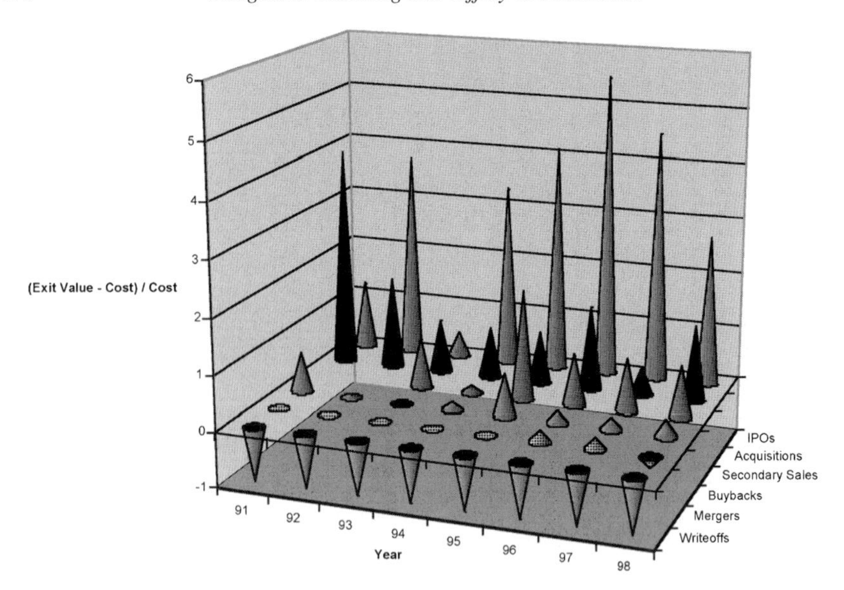

Figure 8. Venture capital exits in Canada: 1991–1998.

capital in Canada are lower in the United States. We attribute these differences in part to the domination of the Canadian venture capital industry by the comparatively inefficient LSVCCs.

4. LABOUR-SPONSORED VENTURE CAPITAL CORPORATION IN CANADA: ORGANIZATION AND STATUTORY CONSTRAINTS

The traditional venture capital form of organization is the private limited partnership (Gompers and Lerner, 1996, 1999; Sahlman, 1990). Private funds both in Canada and the United States are raised principally from public and private pension funds as well as corporations. Wealthy individuals may invest in private funds that are organized as limited partnerships. However, in the United States, individuals account for no more than 10–20% of all venture capital fund raising (Gompers and Lerner, 1999, 2001). The Canadian LSVCC (similar to the Venture Capital Trust in the U.K.) is a substantial departure from this traditional model. In an LSVCC, *only* individuals may invest. Moreover, *any* individual, regardless of his or her net worth, may invest, and there is generally no minimum (or a small minimum) investment required. Because the tax advantages of investing in an LSVCC are exhausted on investments in excess of CAN$5,000 (some jurisdictions have reduced this to CAN$3,500), the great bulk of contributions to LSVCC funds take the form of contributions through individual registered retirement savings plans (RRSPs) of that amount or less (Vaillancourt, 1997). LSVCCs are, in essence, a type of highly specialized mutual fund that invests mainly in private, and hence highly illiquid high growth companies (usually in the technology sectors) in the jurisdiction in which the LSVCC is based.

The first LSVCC was created in the province of Quebec in 1983. The legislative mandate of that fund (which is similar to that adopted by the other provinces) is three-fold in nature; to generate value for the unit holders, to create jobs within the province of Quebec, and to create economic development by fostering the growth of small and medium-sized enterprises. However, Vaillancourt suggests that an additional motive for the introduction of LSVCCs was to achieve labour peace in Quebec by using the LSVCC vehicle to divert economic benefits to unions (Vaillancourt, 1997). This is done in at least three ways. First, while a union must sponsor a LSVCC, the union will not typically run the fund (and will hire outside managers to do so), nor will it have an ownership interest in the fund. The union is thus able to charge a fee for "renting" its name to the fund (typically either a fixed fee or a percentage of net assets). Second, LSVCCs incorporated in Quebec (and in some other provinces) give priority to investments in unionized businesses. Third, the LSVCCs subsidize job creation, which will often redound to the benefit of unions.

In order to attract investment, the various jurisdictions allowing for the creation of LSVCCs offer individual investors generous tax credits. These tax credits are matched by the federal government, which also allows investors to deduct the amount of the investment (up to a stated ceiling) from their income for the year in which the contribution is made (but only so long as the contribution is placed in an RRSP). Until the mid-1990s, for example, on an investment of up to CAN$5,000, in most jurisdictions individual investors received a combined federal and provincial tax credit of 40% and could simultaneously use the investment as a tax deduction, for a total after-tax cost of less than CAN$500 on a CAN$5,000 investment (with the governmental sponsors effectively paying the rest). An individual investor holding for the (then) required five year period would reap a return on investment in excess of 100% even if the fund earned no more than 2% per year (Osborne and Sandler, 1998). Currently, in most jurisdictions, individuals contributing up to CAN$3500 will receive a combined tax reduction of up to CAN$1050, providing an immediate investment return of 30%.[9] The tax benefits in each of the provinces are

[9] Note however that there is a minimum holding period in each jurisdiction (typically 8 years). Early withdrawal of contributed funds results in a penalty fee. Note that all dollar figures discussed herein are in Canadian dollars.

indicated in Table 1, item #10. These incentives have made LSVCCs an attractive asset class for individual investors in a way that is at least partially decoupled from the fundamental quality of the investment. These tax incentives are the main reason for the growth of LSVCCs (Vaillancourt, 1997).

In what follows, we contrast some key features of the contractual and governance structure of LSVCCs with that of private funds, the LSVCCs' main competitor. In the case of private funds, both in Canada and the United States, limited partnership agreements govern the relationship between the limited partners (the investors) and the general partner (the venture capital management company). According to Gompers and Lerner, such agreements contain three types of restrictive covenants: those relating to the management of the fund (e.g. the size of investment in any one firm, the use of debt, coinvestment, reinvestment of capital gains); those relating to the activities of the general partners (e.g. coinvestment by general partners, sale of partnership interests, fundraising, the addition of other general partners); and covenants restricting particular forms of investment (e.g. investments in other venture funds, public securities, leveraged buyouts, foreign securities and other asset classes) (Gompers and Lerner, 1996). Importantly, the 'technology' of restrictive covenants has changed over time as experience with venture capital partnerships accumulates. Further, the relative frequency with which different types of restrictions are used changes over time with changes in economic conditions. Gompers and Lerner (1996) suggest that this adaptability is one of the more valuable attributes of the contractually-based limited partnership vehicle.

By contrast, LSVCCs are set up as corporations, which then enter into a contract with the venture capital manager to supply management services. From an agency cost perspective, one particularly startling feature of LSVCCs is that, while the sponsoring union will typically have no effective ownership interest, it will invariably have control of the board of directors of the fund (items 23–24, Table 1). The lack of an ownership interest obviously attenuates the incentives of the union sponsor to contract efficiently with the management company, to exercise its control in the interest of shareholders, and to monitor the fund's board of directors and the management company. In short, this structure is a receipe for a high level of agency costs.

There are three sources of restrictions on managerial activity in LSVCCs. Covenants that generally mimic those reported by Gompers and Lerner (1996, 1999) for private funds are often found in the contract between the LSVCC fund and the management company. In other cases, the board of directors of the fund will adopt policies which may vary from time to time, and which are imposed (by prior contractual agreement) on the management company. While no systematic analysis has yet been done of the extent to which these covenants and policies duplicate those of private funds, our preliminary investigations suggest that the covenants binding LSVCC managers to their investors are very similar to those discussed by Gompers

Table 1. Legislation governing labour-sponsored investment funds in Canada: An overview, by Jurisdiction (current as of March 1997)

Québec	Federal Government	British Columbia	Manitoba	Ontario	New Brunswick
I. THE STATUTE AND RELATED DETAILS					
1. What is the legislation called?					
Act to Create Fonds de solditarité du Québec (FTQ); and Act to Create the Fonds de development de la Confederation des syndicats nationaux pour la cooperation et l'emploi (Fondaction CSN)	Part X.3 of the Federal Income Tax Act.	The Employee Investment Act.	The Manitoba Employee Ownership Fund Corporation Act.	Labour Sponsored Venture Capital Corporations Act.	New Brunswick Income Tax Act and An Act Respecting the Workers Investment Funds.
2. When was it introduced?					
1983 (Fonds de solditarité FTQ) 1995 (Fondaction CSN)	1988	1989	1991	1992	1993/1994
3. What government department is responsible for it?					
Ministry of Finance, Quebec	Finance Canada	Ministry of Small Business, Tourism and Culture	Department of Industry, Trade and Tourism	Ontario Ministry of Finance	New Brunswick Department of Finance
4. What is the rationale for this statute?					
To permit establishment of a labour-sponsored investment fund directed by the FTQ that invests in Quebec enterprises with the goal of creating, maintaining or preserving jobs; facilitates training of workers in economic matters, stimulates the economy through strategic investments; and invites workers to participate in economic development through subscription to Fund shares.	To allow for establishment of national labour-sponsored investment funds that will supply risk capital to small and medium sized enterprises and thereby contribute to Canadian economic development, job creation and protection.	To permit establishment of a labour-sponsored investment fund that promotes job creation and protection in all parts of British Columbia through risk capital supply to value-added small- and medium-sized firms and that facilitates economic and financial education for workers.	To permit establishment of a labour-sponsored investment fund that promotes capital retention and a stable economy, worker ownership, employment and continued resident ownership of firms in Manitoba and that contributes to other goals, such as corporate social responsibility and worker economic education.	To allow for the establishment of labour-sponsored investment funds that supply risk capital to small and medium-sized enterprises and thereby contribute to economic development, job creation and protection in Ontario.	To permit establishment of labour-sponsored investment funds that promote capital retention, a stable economy, and job creation and protection in New Brunswick and, especially in relation to the Workers Investment Fund, that contribute to other goals, such as worker participation in economic matters.

Table 1. *(Continued)*

Québec	Federal Government	British Columbia	Manitoba	Ontario	New Brunswick
5. How many funds can be created?					
One Fund is established by each Act; i.e., an Act for the Fonds de solditarité (FTQ); an Act for Fondaction (CSN)	An indefinite number.	An indefinite number, though only one has been authorized by the provincial government to date.	Originally one, Crocus Investment Fund. Amendments to the Act are being considered to allow for more than one fund.	An indefinite number.	An indefinite number of national funds and one provincial fund.
6. Who can create a fund?					
The respective Acts created the Fonds solditarité; and Fondaction	A union, as defined by federal law, that represents workers in more than one province or that is composed of two or more affiliates.	A labour body or other work-related organization (with more than 150,000 members in British Columbia), as defined by provincial law.	The Manitoba Federation of Labour (MFL) is specified as the Crocus Fund sponsor.	A provincial labour body; an organization of worker co-operatives; or an entity registered under Part X.3 of the Federal Income Tax Act.	A union, as defined under the Federal Income Tax Act; in the case of Workers Investment Fund, the New Brunswick Federation of Labour.
7. How many funds have been established under this statute so far (March 1997)?					
Two; the Fonds de solditartié (FTQ) and Fondication (CSN)	Several are registered; however, only two – Working Ventures Canadian Fund, Inc., and Canadian Medical Discoveries Fund, Inc. – currently operate fully (i.e. they both raise capital and invest) as national funds in up to five provinces.	One. The Working Opportunity fund.	One. The Crocus Investment Fund; legislative changes are under consideration to allow for more Funds at the discretion of the Minister.	Twenty, including the First Ontario Investment Fund (and national funds, such as the Working Ventures Canadian Fund). One Fund's registration has subsequently been withdrawn.	One provincial fund; the Workers Investment Fund, Inc. So far, only the Working Ventures Canadian Fund, Inc., and the Canadian Medical Discoveries Funds, Inc., are fully operative (i.e. they both raise capital and invest) as national funds in New Brunswick.
8. What kinds of shares can a fund issue?					
Class A (common) shares issued to individuals; Class G shares without voting rights have been issued to the FTQ and the government of Quebec. The Fund administrators may issue other categories of shares which do not confer voting rights at the shareholders meeting.	Class A (common) shares issued to individuals; Class B shares issued to the labour sponsor, others determined as necessary by the fund and as approved by the Minister of Finance.	Class A (common) shares issued to individuals.	Class A (common) shares issued to individuals; Class G shares issued to Manitoba's Minister of Finance; Class I shares issued to institutional investors (e.g., pension funds); and Class L shares issued to the labour sponsor.	Class A (common) shares issued to individuals; Class B shares issued to the labour sponsor, others determined as necessary by the fund.	Class A (common) shares issued to individuals; Class B shares issued to the labour sponsor, others determined as necessary by the fund.

9. Which receive a tax benefit?					
Class A shares only.	Class A shares only.	Class A shares only.	Class A shares only.	Class A shares only.	Class A shares only.
10. What is the tax benefit?					
15% provincial credit (along with matching federal credit). This applies to a maximum of $3,500 in annual share purchases per taxpayer.	15% federal credit with or without a matching credit in every province except Alberta and Newfoundland (national funds obtain the second credit only by satisfying government needs on a province-by-province basis). This applies to a maximum of $3,500 in annual share purchases per taxpayer.	15% provincial credit (along with matching federal credit). This applies to a maximum of $3,500 in annual share purchases per taxpayer.	15% provincial credit (along with matching federal credit). This applies to a maximum of $3,500 in annual share purchases per taxpayer.	15% provincial credit (along with matching federal credit). This applies to a maximum of $3,500 in annual share purchases per taxpayer.	15% provincial credit (along with matching federal credit). This applies to a maximum of $3,500 in annual share purchases per taxpayer.
11. Who can be a (common) shareholder?					
Any person. Quebec residency is one of the factors determining if an individual is eligible for tax credits.	Any individual resident of Canada at the time of buying shares.	Any individual resident of British Columbia (defined as being employed on a continuing basis for at least 20 hours per week).	Any resident of Manitoba at the time of buying shares.	Any resident of Ontario at the time of buying shares.	Any resident of New Brunswick at the time of buying shares.
12. How long must shares be held?					
Until shareholder's retirement (age 60–65, or 55, if the shareholder avails himself of his right of retirement or early retirement).	Eight years (previously, it was five years).	Eight years.	Seven years.	Eight years (previously, it was five years).	Eight years (previously, it was five years).
13. Are there any exceptions?					
Yes. Shares can be redeemed earlier under special circumstances, e.g., planned retirement, a return to school, terminal illness, investment in one's company, emigration, an urgent need for liquidity, and a serious reduction in income.	Yes. Shares can be redeemed earlier in the event of the holder's death, severe illness/disability, change of nationality or in the event of sales/transfers (per set conditions).	Yes. Shares can be redeemed earlier in the event of the holder's death, severe illness/disability, bankruptcy, job loss, (persisting for at least six months) or in the event of sales/transfer (per set conditions).	Yes. Shares can be redeemed earlier in the event of the holder's death, severe illness/disability, retirement or financial hardship or in the event of sales/transfer (per set conditions).	Yes. Shares can be redeemed earlier in the event of the holder's death, severe illness/disability, or in the event of sales/transfers (per set conditions).	Yes. Shares can be redeemed earlier in the event of the holder's death, severe illness/disability, or in the event of sales/transfers (per set conditions).

Table 1. *(Continued)*

Québec	Federal Government	British Columbia	Manitoba	Ontario	New Brunswick
14. Are any payroll deductions encouraged?					
Yes. Quebec employers must remit deductions to the fund if the lesser of fifty employees or 20% of the total workforce so request.	No	No	Yes. Manitoba employers must remit deductions to the fund if the lesser of fifty employees or 20% of the total workforce so request.	No	Yes, but only for the Workers Investment Fund. NB employers must remit deductions to this fund if the lesser of 50 employees or 20% of the total workforce so request.
II. RULES GOVERNING SHARE DISTRIBUTIONS					
15. Is there a limit on how much capital can be raised per year through share sales?					
Not presently. There was a temporary ceiling imposed by provincial authorities in the period 1993–1994.	No	Yes. No more than a total of $40 million can be raised annually.	Yes. No more than a total of $30 million (or as determined by the provincial government) can be raised annually.	No	No
16. Does the act allow for the sale of shares by representatives trained by the fund, including employees and/or fund representatives?					
Yes	No	No	Yes	Yes (in the case of First Ontario Fund)	No
17. What public authority monitors a fund's sales activity?					
The Commission des valeurs mobilières du Québec	The securities commission or the appropriate authority in each province where sales occur.	The British Columbia Securities Commission	The Manitoba Securities Commission	The Ontario Securities Commission	The New Brunswick Department of Justice
18. What is the role of regulatory authorities?					
Protecting the public in share sales transactions, information disclosure requirements, etc.	Protecting the public in share sales transactions, information disclosure requirements, etc.	Protecting the public in share sales transactions, information disclosure requirements, etc.	Protecting the public in share sales transactions, information disclosure requirements, etc.	Protecting the public in share sales transactions, information disclosure requirements, etc.	Protecting the public in share sales transactions, information disclosure requirements, etc.
19. What provinces are currently open to national funds?					
No.	Not applicable.	No.	No. (But Saskatchewan is open.)	Yes.	Yes. (Nova Scotia and Prince Edward Island are also open.
20. What is the required period of fund shareholding?					
Until shareholder's age of retirement.	Eight years.	Eight years.	Seven years.	Eight years.	Eight years.

21. Does the jurisdiction allow for union-directed share distributions?					
Yes.	No.	No.	Yes.	Yes (First-Ontario LSVCC).	No.
22. What are the investment level enforcement measures (see also #31)?					
Restrictions of subsequent capital-raising.	Deficiency taxes.	Temporary suspension or revocation of fund registration.	Temporary suspension or revocation of fund registration.	Deficiency taxes.	Deficiency taxes.
III. FUND DECISION MAKING					
23. Who directs a fund?					
A Board of Directors, a majority of whom are nominated by the labour sponsor.	A Board of Directors, at least one-half of whom are nominated by the labour sponsor.	A Board of Directors, at least one-half of whom are nominated by the labour sponsor.	A Board of Directors, a majority of whom are nominated by the Manitoba Federation of Labour.	A Board of Directors, at least one-half of whom are nominated by the labour sponsor.	A Board of Directors, at least one-half of whom are nominated by the labour sponsor (in the case of the workers investment fund, the New Brunswick Federation of Labour).
24. Who else sits of a Board of Directors?					
– 2 members elected by shareholders – 4 members representing: individual enterprises, financial institutions, social-economic interests, and a fourth – the 17th member is the President/CEC of the Fund.	Shareholder representatives elected at an annual general meeting and others as determined by the labour sponsor.	Shareholder representatives elected at an annual general meeting and others as determined by the labour sponsor.	Elected or appointed representatives of Class A, Class G and Class I shareholders.	Shareholder representatives elected at an annual general meeting and others as determined by the labour sponsor.	Shareholder representatives elected at an annual general meeting and others as determined by the labour sponsor.
IV. REQUIREMENTS OF INVESTMENT					
25. In what kinds of business must a fund invest?					
A small- or medium-sized company/partnership (defined as having no more than $50 million in assets; or the net value of which is a maximum $20 million).	A small- or medium-sized company/partnership (defined as having no more than 500 employees and $50 million in assets).	A small- or medium-sized company/partnership in a new and/or value-added sector (e.g., manufacturing and processing industries, high technology, tourism, aquaculture).	A small- or medium-sized company/partnership (defined as having a maximum of $50 million in assets). One-quarter of newly-raised capital must go toward deal sizes of less than $1 million.	A small- or medium-sized company/partnership (defined as having no more than 500 employees and $50 million in assets). At least 10% of total investments must go to very small companies (defined as having no more than 50 employees and $5 million in assets).	A small- or medium-sized company/partnership (defined as having no more than 500 employees and $50 million in assets).

Table 1. *(Continued)*

Québec	Federal Government	British Columbia	Manitoba	Ontario	New Brunswick
26. Where can a business be located?					
Anywhere, as long as the majority of employees reside in Québec.	At least one-half of company activity (defined as 50% of salaries and wages paid) must take place in Canada.	At least one-half of company activity (defined as 50% of salaries and wages paid) and most assets must reside in B.C.	The majority of a company's assets and work-force must reside in Manitoba.	At least one-half of company activity (defined as 50% of salaries and wages paid) must take place in Ontario.	At least one-half of company activity (defined as 50% of salaries and wages paid) must take place in New Brunswick.
27. What is the nature of the investment?					
Any financial assistance in the form of loan, underwriting, equity, shares, etc.	New equity in a company, et al.	New equity in a company, et al.	New equity in a company, et al.	New equity in a company, et al.	New equity in a company, et al.
28. What is the required level of fund capital in equity (i.e., no debt securities) investments?					
60% of previous year's average.	60% within one year.	80% within three years of capital raising.	60% of previous year's average.	70% within two years.	60% within one year.
29. Are there limits as to how a business can use a fund's investment?					
No.	No.	Yes. For instance, a company cannot re-lend the money or invest in activity unrelated to the firm.	Yes. For instance, the money cannot be used to unionize workers.	Yes. For instance, a company cannot invest the money in land unrelated to the firm or outside Canada.	No.
30. What level of total capital must be invested in business projects?					
At least 60% of the previous year's average net assets.	60% of capital accumulated by each year's end must be placed in projects by the following year. (Special provisions apply for investments in very small companies, i.e. with up to $10 million in assets.)	80% of capital must be placed in eligible projects within three years of receipt.	At least 60% of capital of the previous year's average net capital. For the period 1996–1997, the requirement was 75%. A majority of assets should support worker ownership and participation in some form.	50% of capital must be placed in projects within one year of receipt and 70% within two years.	60% of capital accumulated by each year's end must be placed in projects by the following year. In the case of national funds, the provincial government determines individual agreements for re-investment of sales proceeds in N.B.

31. What happens if this level is not met?					
The fund is restricted in subsequent capital raising.	The fund pays a 20% deficiency tax and additional penalties (including possible revocation of the funds registration) depending upon the case.	A fund's registration may be temporarily suspended or revoked, depending upon the circumstances.	The fund's registration may be permanently revoked.	The fund pays a 20% deficiency tax. A rebate on this tax is available if appropriate action is taken by the fund.	The fund pay a 20% deficiency tax and additional penalties (including possible revocation of the fund's registration) depending upon the circumstances.
32. How are the rest of the assets to be invested?					
In reserves of liquid securities (e.g. cash, government bonds) or in other vehicles according to the investment policy approved by the Board of Dir.	Primarily in reserves of liquid securities (e.g. cash, government bonds) in the start-up period. The reafter, as determined by a fund.	Primarily in reserves of liquid securities (e.g. cash, government bonds). Generally, assets must be invested domestically.	Primarily in reserves of liquid securities (e.g. cash, government bonds) or as determined by the fund.	Primarily in reserves of liquid securities (e.g. cash, government bonds). Generally, assets must be invested domestically.	Primarily in reserves of liquid securities (e.g. cash, government bonds) in the start-up period. Thereafter, as determined by a fund.
33. Are there other investment-related program requirements?					
Yes. For instance, the fund is encouraged to provide training to workers on economic and financial matters and to give economic development.	No.	Yes. For instance, a fund is encouraged to provide education to workers on economic and financial matters and give priority to community and regional economic development.	Yes. For instance, the fund is encouraged to emphasize worker ownership, economic education and empowerment of workers, and corporate social responsibility.	No.	Yes. The Workers Investment Fund is encouraged, for instance, to promote economic awareness and empowerment of workers.
V. RESTRICTIONS ON INVESTMENT					
34. Is a fund restricted from investing in certain firms or sectors?					
No.	No.	Yes. A fund is restricted from investing is natural resource industries (e.g. fishing, forest products, mining), the financial sector, land development and retails.	Yes. A fund is restricted from investing is natural resource industries (e.g. agriculture, mining, oil and gas), the financial sector land development and retail.	Yes. No more than 15% or a fund's total investment can go towards publicly-traded enterprises.	No.

Table 1. *(Continued)*

Québec	Federal Government	British Columbia	Manitoba	Ontario	New Brunswick
35. How much can a fund invest in a single business?					
No more than 5% of the fund's total capital (or up to 10% under special circumstances) at the time of an investment.	The lesser of $15 million or 10% of fund capital at the time of an investment.	No more than $5 million per company for a period of two years.	No more than 10% of total fund capital at the time of an investment.	No more than $10 million or 10% of fund capital at the time of an investment, whichever is less.	No more than $10 million or 10% of fund capital at the time of investment, whichever is less.
36. Is a fund restricted as to its controlling share in a business?					
No.	No.	Yes. Majority control is not permitted except under special circumstances (e.g. worker buyouts or financial distress).	No. Majority control is encouraged if it facilitates worker buyouts/owners hip.	Yes. A fund may not have "control", but the definition is broad and permits majority ownership.	No

Note: Saskatchewan (1992), Nova Scotia (1994), and Prince Edward Island (1992) are similar to Part X.3 of the Federal Income Tax Act.

and Lerner. Like the covenants found in limited partnership agreements, these covenants and policies may vary over time and with changing economic conditions.

The third source of restrictions is the legislation under which the LSVCC is formed. Each of the provincial (and federal) enactments that allow for the creation of LSVCCs impose restrictions on the fund that are in many respects more onerous than those found in limited partnership agreements. These restrictions, which are set out in Table 1, affect both the supply side (the flow of funds to entrepreneurial firms) and the demand side (the demand by entrepreneurial firms for LSVCC capital) of the market. Unlike contractually negotiated covenants in limited partnership arrangements, these restrictions are not the product of informed bargaining between arm's length commercial parties, but reflect the objectives of the relevant legislature. Moreover, LSVCC statutes change very little (and in most pertinent respects, not at all) over time with changing economic conditions. The rigidity of the LSVCC statutory governance mechanism limits the ability of both supply and demand sides of the market to react to changing economic conditions by altering pertinent contractual arrangements. This is in sharp contrast to the governance of private limited partnership organizations, in which changes are observed over time in response to changing conditions of demand and supply (Gompers and Lerner, 1996).

As noted earlier, pursuant to LSVCC legislation, LSVCCs are typically formed with multiple objectives, although the principal motive has been to expand the pool of venture capital (Osborne and Sandler, 1998). These statutorily specified objectives are indicated in items #4 and 33 in Table 1. The extent to which goals other than profit maximization are pursued in practice varies from one province to another. For example, in Quebec, the legislative goals are pursued quite vigorously. However, a number of funds incorporated in other provinces have publicly stated that (despite their broad statutory mandates) they will pursue profit maximization to the exclusion of other objectives (Halpern, 1997; MacIntosh, 1994). Osborne and Sandler state that in Ontario (where more than half of all venture capital investments by dollar value are made), there is essentially no consideration of objectives other than profit maximization (Osborne and Sandler, 1998).

Another difference from private funds arises in the investor lock-in period. As indicated in item 12 of Table 1, the lock-in period is seven years in Manitoba, and eight years in all other jurisdictions except Quebec (in which the shares must be held until retirement). Individuals withdrawing prior to the elapse of this period will lose their LSVCC tax credits (although not the deductability of the contribution, if it was made through an RRSP). By contrast, private fund investors are typically locked in for 10 years. The shorter horizon for LSVCC funds and the ability of investors to make demand redemptions force the fund to maintain liquidity against the event of redemptions. As noted earlier, this is partly responsible for the overhang of uninvested funds referred to earlier. Because the overhang is invested in low risk instruments such as treasury bills and bank deposits, we would expect that LSVCC funds will have both lower risk and return when compared to other types of funds. The longer duration of private funds and the inability of investors to made demand redemptions not only allows for investment of all the contributed capital, but also provides more breathing room to bring investee firms to fruition and more flexibility in exiting.

Other features of the legislative structure depart from contractual arrangements observed in private funds, and are likely to adversely affect performance. In four provinces (Table 1, item 15) there is a limit on the amount of funds raised in any given year, at a threshold (in the range of CAN$20–40 million) that may prevent the exploitation of economies of scale associated with venture capital investing. Further, in response to the common practice of placing up to half (and in some cases more) of a fund's capital in treasury bills and similar low risk instruments (again, the problem of "overhang"), all jurisdictions now require that an LSVCC invest a certain portion of its capital contributions in eligible businesses within one to three years of receipt (Table 1, items 28, 30–31). This constraint can have the effect of forcing the fund to invest in inferior businesses if an investment deadline looms (the violation of which would result in severe penalties).

LSVCCs are also geographically constrained; typically a majority of the salaries and wages paid by the fund (or assets or employment) must be within the sponsoring province (Table 1, item #26). This limits the businesses that can be vetted for investment purposes, and may also impose a constraint on any relocation of the business as it grows and/or participation in follow-on investments. In Ontario (the province in which the majority of LSVCC investments are made), the fund cannot acquire "control". However, this constraint may be more apparent then real, since control is defined as the ability to "determine the strategic operating, investing and financing policies of the corporation or partnership without the co-operation of another person".[10] The provincial administrators take the view that this does not prohibit a shareholding in excess of 50%. A similar prohibition against control in B. C. is defined in the traditional manner, thus excluding majority ownership and limiting a B. C. fund's governance options.

In addition, the timing of LSVCC capital contributions differs from that of private funds. In a private fund, the fund will only be able to secure commitments from investors when the underlying investment fundamentals are favorable. These committed funds are drawn down if and when needed. By contrast, because most LSVCC capital consists of RRSP contributions, LSVCC funding is concentrated in the first three months of the calendar year (i.e. immediately prior to the cut-off date for claiming the tax benefits associated with the contribution for the previous

[10] *Community Small Business Investment Funds Act*, S.O. 1992, c. 18, s.1(3).

calendar year). These funds are received immediately from investors, rather than being drawn down as needed. This gives rise to highly lumpy receipts by the LSVCCs, and tends to divorce capital raising from the underlying fundamentals of the investment market.

In sum, the legislative, contractual and governance structures of LSVCC funds lead us to hypothesize that the LSVCC is an inferior form of venture capital organization that should have high agency costs and low returns relative to private venture capital funds. We briefly consider the performance of LSVCCs in Section 5.

5. THE PERFORMANCE OF LSVCCs

The marketing materials of the LSVCCs typically stress the tax advantage of the investment, rather than the investment return. Thus, for example, under the heading "Why Invest with Us", the first item on the B. E. S. T. Fund website is "Tax Savings".[11] An elaborate chart indicates the nature of the combined federal and provincial tax savings for individuals in various tax brackets. The second item is "investment performance", which consists of describing returns as "above average", without any actual performance figures or any indication of the definition of "average".

Data mining in marketing materials is common among LSVCCs.[12] For example, on its website, the Crocus Fund[13] does not present figures related to individual performance, but rather presents LSVCC average performance for 1 month and 1 year and compares this performance to various market benchmarks. While these figures show the LSVCC index outperforming other indices (i.e. incurring smaller losses), a full presentation of performance compared to these same market benchmarks with five and ten year returns shows gross underperformance (as we document below). The Crocus Fund website also stresses LSVCCs comparatively low volatility, which we confirm below. This low volatility may well benefit the investor, but is artificially manufactured to give the LSVCCs a comparative advantage over other forms of investment in attracting investment capital.

As this section will make clear, it is not surprising that marketing efforts have focused as little as possible on investment returns, since these have been extremely

[11] See http://www.bestcapital.ca/why_invest.htm#2

[12] The typical LSVCC report on the Internet does not meet AIMR's Performance Presentation Standards; see http://www.aimr.com

[13] "Not Just a Pretty Tax Credit" at http://www.crocusfund.com/advisor/printconcept14.html

poor. Figure 9 and Table 2 present a fuller account of LSVCC performance over the past 10 years than can be found in any LSVCC marketing materials.[14]

The poor returns of LSVCC funds over the past 10 years is striking. While the return to the LSVCC index over the 1992–1999 period was 28%, it was 160% for the TSE 300 Index, 180% for the Globle Canadian Small Cap Peer Index, and 650% for the U.S. VC Index (as computed by Peng, 2001).[15] Figure 9 shows that LSVCC funds have even underperformed 30-day treasury bills.

While we do not have data comparing the performance of the LSVCCs with their Canadian private fund counterparts, Brander et al. (2002) find that LSVCC performance has lagged private fund performance in both a statistically and economically significant manner.

Since venture capital is a risky asset class, a priori one would expect the LSVCCs to exhibit an average beta significantly in excess of one. However, as Table 2 shows, the average LSVCC beta (measured in respect of funds for which 3 years of data was available) is only 0.3782. The distribution of betas across the LSVCCs, and the returns in each beta category for the 1-month, 3-month, 6-month, 1-year, 3-year, 5-year and 10-year period ending 6/1/2002 are presented in Table 2.

The low average beta is very surprising. We have noted that LSVCCs hold some of their contributed capital in cash (e.g. treasury bills and bank deposits), which can be expected to lower the beta. However, Table 2 indicates that in any given year, on average only about 20% of total capital remains uninvested. This would seem to be far too little to account for the observed beta. If the average portfolio has 20% of its capital in zero beta instruments, this suggests that the beta of the other assets is still only 0.47. Since 90% of the investments of the LSVCCs are in start-up and expansion financing, a very risky asset class, this does not seem possible.

The low beta appears to be an artefact both of infrequent valuations and valuation practice. LSVCC shares do not trade publicly. Hence, there is no opportunity for the public market to price the shares, and hence no real measure of the volatility of a given fund's assets. Rather, betas are determined from the net asset value (NAV) reported periodically (usually quarterly) by each fund. The infrequent valuations create problems in calculating betas that are similar to those encountered in thinly traded stocks; the paucity of data points leads to a tendency to understate betas.

[14] The data presented may in theory, exhibit survivorship bias because of our inability to obtain data for defunct LSVCCs. If any such bias exists, it will serve to *overstate*, rather than understate LSVCC performance. More to the point, however, to date no LSVCCs have been wound up.

[15] The US VC Index value from Peng (2001) is not available for 2000 and 2001. Peng referred the authors to http://www.ventureeconomics.com for a somewhat comparable U.S. VC performance statistic for 2002, as indicated in Figure 9. The Venture Economics Statistic, however, is not computed with the same degree of accuracy as done by Peng (2001).

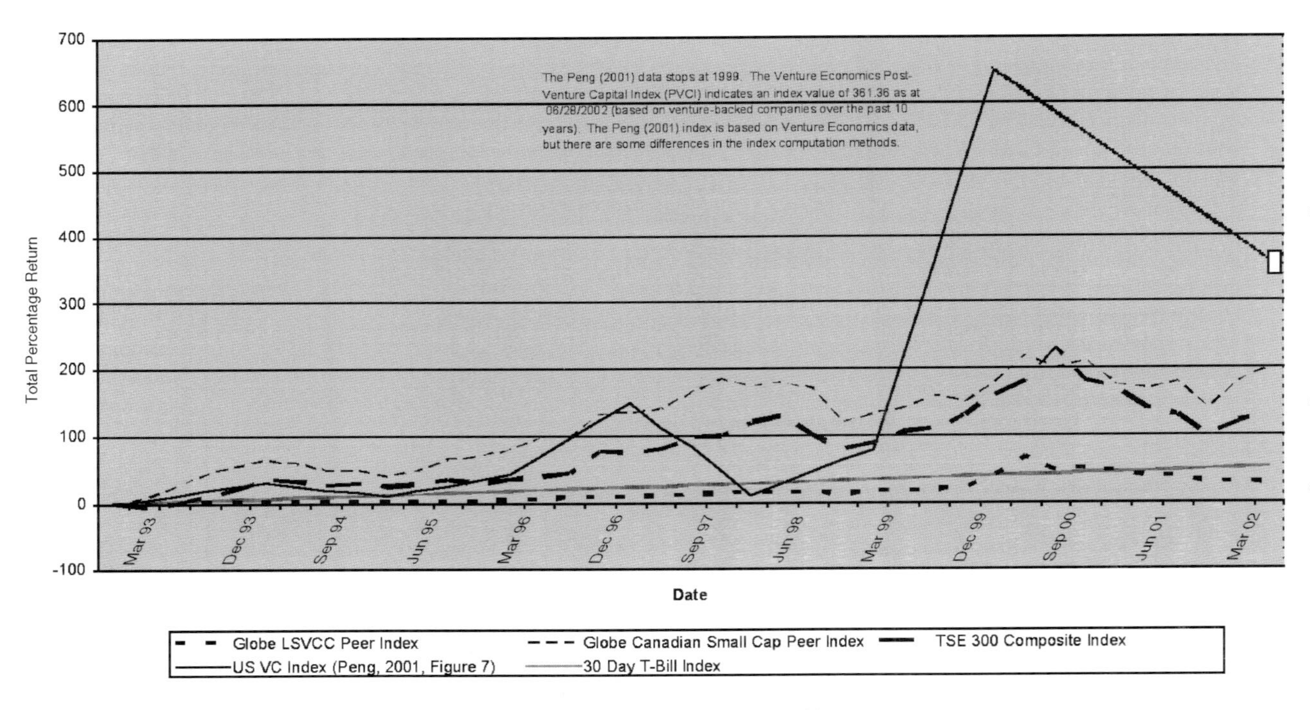

Figure 9. Selected indices 1992–2002.

Table 2. Summary statistics: Returns by characteristics of Labour-Sponsored Investment Funds (LSIFs)

	Number of funds	Average left hand column values	Average 1-month return	Average returns (annualized) to period ending 1 June 2002					
				Average 3-month return	Average 6-month return	Average 1-year return	Average 3-year return	Average 5-year return	Average 10-year return
LSVCF average	50		−1.530	−3.000	−4.480	−10.410	1.830	1.950	2.890
Nesbitt Burns Cdn Small-Cap Index			−0.050	8.910	27.500	9.560	10.400	5.730	10.510
−0.2 < 3-Year Beta < 0	3		−1.070	−1.637	−3.173	−4.810	−1.740	0.445	–
0 < 3-Year Beta < 0.2	2		−0.735	−1.480	−0.835	−2.585	−0.855	1.650	–
0.2 < 3-Year Beta < 0.4	4	Average Beta of All LSVCFs = 0.378	−0.753	−2.435	−2.085	−9.348	4.600	8.137	7.110
0.4 < 3-Year Beta < 0.6	5		−3.386	−4.236	−5.284	−18.006	1.814	1.966	–
0.6 < 3-Year Beta < 0.8	1		−3.490	−5.320	−3.010	−23.780	−10.840	−5.130	−1.340
0.8 < 3-Year Beta < 1	1		−7.800	−12.630	−11.720	−11.760	13.060	6.800	–
1 < 3-Year Beta < 1.2	1		−2.120	−7.800	−17.790	−36.560	−15.000	−7.600	–
3-Year Beta Not Available	33		−1.363	−2.745	−3.781	−7.071	–	–	–
VC Fund Early Stage Focus	18		−0.847	−1.951	−2.077	−7.316	5.820	4.033	–
VC Fund General Technology Focus	8		−1.797	−3.617	−4.913	−11.720	7.257	5.020	–
VC Fund Specific Technology Focus	15		−1.089	−2.824	−3.888	−6.739	4.430	2.000	–
VC Funds part of VC Firm with >1 Fund	38		−1.956	−3.376	−5.038	−11.809	0.316	1.171	2.885
Age < 1 Year	15		−0.857	−1.039	–	–	–	–	–
1 Year < Age < 2 Years	6		−1.322	−4.603	−3.842	−7.872	–	–	–
2 Years < Age < 3 Years	10		−2.194	−3.166	−3.745	−6.591	–	–	–
3 Years < Age < 4 Years	1	Average Age of All LSVCFs = 44.180 (Months)	0.500	0.400	−0.570	−3.240	7.270	–	–
4 Years < Age < 5 Years	1		−0.300	−0.830	0.950	−1.450	−0.720	–	–
5 Years < Age < 6 Years	4		−2.277	−1.640	−3.503	−17.813	6.040	3.920	–
6 Years < Age < 7 Years	1		−2.120	−7.800	−17.790	−36.560	−15.000	−7.600	–
8 Years < Age < 9 Years	6		−3.560	−5.508	−7.023	−11.495	1.328	1.843	–
Age > 9 Years	6		−1.584	−3.668	−1.682	−11.092	−1.976	3.456	2.885

Federal	14		−1.334	−1.998	−3.164	−7.523	−1.213	−0.624	−0.096
Ontario	29		−1.582	−2.851	−2.814	−6.403	0.207	0.994	–
British Columbia	2		−5.255	−9.560	−5.545	−16.625	7.865	5.105	3.555
Saskatchewan	2		0.030	−0.290	−0.170	−1.245	3.635	–	–
Manitoba	2		−0.600	−1.410	−0.810	−2.750	−2.125	1.100	–
Quebec	1		0.000	−7.000	−7.000	−11.400	8.600	8.100	–
New Brunswick	1		0.000	0.000	0.000	–	–	–	–
Assets < $20 million	18	Average Assets of All LSVCFs = 2002 $Can 79.798 (Excluding Quebec) = $Can 168.202 (including Quebec)	−1.399	−2.215	−3.486	−9.912	2.590	2.735	–
$20 million < Assets < $40 million	9		−0.968	−1.561	−1.881	−3.869	7.270	–	–
$40 million < Assets < $60 million	2		−0.675	−2.390	−3.235	−10.730	−8.660	4.000	–
$60 million < Assets < $80 million	9		−1.687	−3.648	−4.341	−9.866	4.600	3.560	–
$80 million < Assets < $100 million	1		−0.550	−0.080	−4.040	−7.640	8.460	5.300	–
Assets > $100 million	11		−2.905	−5.623	−6.034	−14.454	−0.704	1.394	2.457
Cash[a] <33%	13	Average Cash of All LSVCFs = 0.351	−1.402	−3.952	−5.331	−9.475	−2.356	1.594	8.100
33% < Cash < 66%	11		−2.293	−3.104	−3.892	−13.312	1.711	3.021	–
Cash > 66%	5		−0.290	−1.276	−1.026	−5.803	8.460	5.300	–
Equity[a] < 33%	10	Average Equity of All LSVCFs = 0.375	−0.450	−1.516	−1.452	−4.113	4.110	3.545	–
33% < Equity < 66%	18		−2.194	−4.152	−5.526	−13.630	0.516	2.513	8.100
Equity > 66%	1		−0.900	−1.990	−2.570	−4.050	−3.530	2.200	–
Bonds[a] < 33%	17	Average Bonds of All LSVCFs = 0.274	−1.751	−2.952	−3.197	−11.739	2.096	3.358	–
33% < Bonds < 66%	9		−1.509	−3.052	−6.859	−9.684	−4.277	0.120	–
Bonds > 66%	3		−0.517	−2.043	−0.157	0.830	–	–	–

"–" means that returns are not systematically publicly reported, or the fund was recently introduced so the data are not available for the period.

Data sources: www.globefund.com, www.morningstar.ca

[a] Security allocations not known for 21 of the 50 LSVCFs. Other/unknown category not reported.

We believe that there is another reason, however, for the artificially low betas. The NAV from which betas are calculated is the price at which new buyers buy in, and current holders cash out. It is set periodically by each LSVCC fund. The LSVCCs have an incentive to artificially reduce reported volatility in order to attract purchasers, particularly given that many purchasers of LSVCCs are undiversified (many holding only the shares of a single LSVCC in their retirement portfolios).[16] Low volatility is frequently a selling point for the LSVCCs, as it is on the Crocus Fund web cite, discussed above. Since valuations of private companies are inherently subjective and subject to wide confidence intervals, smoothing is not difficult to achieve. The ability to artificially smooth NAVs gives the LSVCCs an advantage over many other asset classes, including mutual funds, whose NAVs are subject to market determination.

What of the fact that valuations are typically required to be performed by an "independent" valuer? While the use of an independent valuer would initially appear to limit the extent to which management can massage NAVs, the nominally independent valuer has an incentive to bow to management pressure in order to secure future valuation work with the fund or other funds. Management is free to call the shots, so to speak, because the controller (the sponsoring union) and management have a commonality of interest. Each wants to secure the greatest number and dollar value of contributions (particularly where the union's remuneration takes the form of a percentage of NAV). The shareholders, many of whom are undiversified, share the interest of management and the valuer in smoothing NAVs (although it is likely that most, being unsophisticated investors, are unaware of this practice, and, lacking control, would have great difficulty in opposing it even if they found it disagreeable).

Table 2 also indicates performance by the age of the VC fund. There is no support for the proposition that older LSVCCs generate greater returns.[17]

As discussed in Section 4, LSVCCs have geographic restrictions on the location of their investee firms (see also item 26 in Table 1). While the worst short-term performance has been in B. C., the best long-term performance has

[16] Osborne and Sandler report that "A survey of FSTQ [Solidarity] shareholders undertaken in 1989 indicated that 45 percent of the shareholders invested for the first time in their lives in an RRSP when they acquired shares of FSTQ, and that 39 percent of the shareholders had only one RRSP (consisting of shares of FSTQ): Sorecom Inc., for the Fonds de solidarite des travailleurs du Quebec, June 1989, unpublished". Osborne and Sandler, 1998, at 559 (note 216).

[17] Data on LSVCC fund manager experience have not been compiled. It is unclear whether VC fund manager experience should be defined to include experience only in private equity investing, or to also include related experience. Education, network contacts, syndication arrangements, etc., could also be considered.

also been in B. C. This is attributable to differences across the funds within B. C.[18] Similarly, differences across other jurisdictions are attributable to the variance in performance across the funds. As indicated in Table 1 (see Section 4 above), LSVCC legislation is quite similar in the various jurisdictions; differences in performance across jurisdictions cannot be explained only by reference to differences in the legislation across jurisdictions. It is nevertheless interesting to note that the best long-term performance is observed in Saskatchewan and British Columbia.

LSVCC performance by asset sizes in Table 2 generally indicates that the larger funds had worse short-term performance in the most recent year. Performance over periods of more than a year is best among funds in the mid-range of assets (CAN$40–80 million).

Table 2 also presents performance results by current security allocations (between bonds, equity and cash or cash equivalents). It has been noted by some industry analysts that LSVCCs hold bonds in part to realize a book value return (interest reported in financial statements from fixed income investments) in order to attract new capital to their funds.[19] As discussed in Section 2, LSVCCs are required to keep a percentage of their assets in liquid securities. The most interesting finding from Table 2 is that the LSVCCs with the best long-term performance are those that currently have more than 66% of their assets in cash, funds with less than 33% of their assets in equity, and with less than 33% of their assets in bonds. One interpretation of this finding is that the better LSVCC fund managers have put a higher proportion of their assets in liquid securities in the current market environment. Further research is warranted.

In sum, LSVCCs are an asset class with low returns, artificially low betas and significant restrictions on ownership. A typical LSVCC investor invests for the tax savings and not the economic returns. As discussed in Section 3, investors cannot withdrawal their invested capital for a period of 8 years. This has serious implications for the ability of the market to discipline LSVCC managers. The corporate governance mechanisms detailed in Table 1 suggested we should expect LSVCCs to have inferior returns. The evidence in Figure 9 and Table 2 is supportive.

[18] The Working Opportunity Balanced Fund created in January 1992 has outperformed the Working Opportunity Growth Fund created in January 2000. The same venture capital firm runs these two funds. The Working Opportunity Balanced Fund has had the highest returns since inception (7.11%).

[19] This was noted by Mary Macdonald of Macdonald & Associates, Ltd. (the company that collects venture capital data for the Canadian Venture Capital Association Annual Reports) at a University of Toronto lecture in 1998.

6. CONCLUSION

Labour-Sponsored Venture Capital Corporations (LSVCCs) are a unique investment vehicle that enables individuals to make investments of up to CAN$5,000 (CAN$3,500 in some jurisdictions) in venture capital and receive significant tax savings. Unlike venture capital limited partnerships, LSVCC governance mechanisms are partly dictated by statute. Thus, for example, LSVCCs must reinvest capital contributions within a limited amount of time in entrepreneurial firms, or face fines (or even revocation of their licence to operate as a LSVCC). LSVCC statutes also do not change over time or vary according to the economic environment; in the U.S., by contrast, privately negotiated limited partnership agreements change over time and across economic environments. This has been hailed as a key component of the success of the U.S. venture capital industry (Gompers and Lerner, 1996, 1999).

The very low LSVCC returns over the past 10 years relative to other investments in Canada are striking. In Figure 9 we showed that LSVCCs have had lower returns than the Globe Canadian Small Cap Peer Index, the TSE 300 Composite Index and the U.S. VC index. The return to the LSVCC index over the 1992–1999 period was 28%, but 160% for the TSE 300 Index, 180% for the Globe Canadian Small Cap Peer Index, and 650% for the U.S. VC Index (as computed by Peng, 2001). LSVCCs even underperformed 30-day treasury bills.

Our empirical findings on LSVCC performance are generally consistent with the theoretical research of Kanniainen and Keuschnigg (2000, 2001), Keuschnigg (2002), Keuschnigg and Nielsen (2001, 2002a, b) on public policy towards venture capital. Simple tax breaks towards venture capital will not necessarily facilitate successful entrepreneurial finance. The very low LSVCC returns relative to comparable investments in Canada cements this view. Our related research indicates that LSVCC portfolios are significantly larger than their non-LSVCC counterparts in Canada (Cumming, 2001), and LSVCCs have crowded-out other types of private equity in Canada (Cumming and MacIntosh, 2001b). The Canadian experience with LSVCCs is highly suggestive that similar structures should not be adopted in other countries.

REFERENCES

Amit, A. R., Brander, J., & Zott, C. (1997). Venture capital financing of entrepreneurship in Canada. In: P. Halpern (Ed.), *Financing Innovative Enterprise in Canada* (pp. 237–277). University of Calgary Press.

Amit, R., Brander, J., & Zott, C. (1998). Why do venture capital firms exist? Theory and Canadian evidence, *Journal of Business Venturing*, *13*, 441–466.

Black, B. S., & Gilson, R. J. (1998). Venture capital and the structure of capital markets: Banks vs. stock markets, *Journal of Financial Economics*, *47*, 243–277.

Brander, J. A., Amit, R., & Antweiler, W. (2002). Venture capital syndication: Improved venture selection vs. the value-added hypothesis, *Journal of Economics and Management Strategy* (forthcoming).

Canadian Venture Capital Association (1978–2002). *Venture Capital in Canada: Annual Statistical Review and Directory*, Toronto.

Cumming, D. J. (2000). *The Convertible Preferred Equity Puzzle in Canadian Venture Capital Finance*, Working Paper. University of Alberta. Available on www.ssrn.com

Cumming, D. J. (2001). *The Determinants of Venture Capital Portfolio Size: Empirical Evidence*, Working Paper. University of Alberta. Available on www.ssrn.com

Cumming, D. J., & MacIntosh, J. G. (2001a). Venture capital investment duration in Canada and the United States, *Journal of Multinational Financial Management*, *11*, 445–463.

Cumming, D. J., & MacIntosh, J. G. (2001b). *Crowding Out Private Equity: Canadian Evidence*, Working Paper. University of Alberta and University of Toronto. Available on www.ssrn.com

Cumming, D. J., & MacIntosh, J. G. (2003a). Venture capital exits in Canada and the United States, *University of Toronto Law Journal*, *53*, 101–200. Available on www.ssrn.com

Cumming, D. J., & MacIntosh, J. G. (2003b). The extent of venture capital exits: Evidence from Canada and the United States. In: L. D. R. Renneboog, & J. McCahery (Eds), *Venture Capital Contracting and the Valuation of High-Tech Firms* (forthcoming), Oxford University Press. Available on www.ssrn.com

Cumming, D. J., & MacIntosh, J. G. (2003c). A cross-country comparison of full and partial venture exits, *Journal of Banking and Finance*, *27*, 511–548. Available on www.ssrn.com

Department of Finance (Canada) (1996). 1996 Budget, Budget Plan, Annex 5, Tax Measures: Supplementary Information and Notice of Ways and Means Motions (March 6).

Francis, B., & Hasan, I. (2001). Venture capital-backed IPOs: New evidence, *Journal of Financial Services Research* (forthcoming).

Gompers, P. A. (1998). Venture capital growing pains: Should the market diet? *Journal of Banking and Finance*, *22*, 1089–1102.

Gompers, P. A., & Lerner, J. (1996). The use of covenants: An empirical analysis of venture capital partnership agreements, *Journal of Law and Economics*, *39*, 463–498.

Gompers, P. A., & Lerner, J. (1998). What drives venture fundraising? *Brookings Proceedings on Microeconomic Activity*. Opt cit. National Bureau of Research Working Paper 6906 (January 1999).

Gompers, P. A., & Lerner, J. (1999). *The venture capital cycle*, Cambridge: MIT Press.

Gompers, P. A., & Lerner, J. (2001). *The Money of Invention: How Venture Capital Creates New Wealth*, Cambridge: Harvard Business School Press.

Halpern, P. (Ed.) (1997). *Financing Growth in Canada*, University of Calgary Press.

Jeng, L. A., & Wells, P. C. (2000). The determinants of venture capital funding: Evidence across countries, *Journal of Corporate Finance*, *6*, 241–289. Available on www.ssrn.com

Kanniainen, V., & Keuschnigg, C. (2000). The optimal portfolio of start-up firms in venture capital finance. CESifo Working Paper No. 381, *Journal of Corporate Finance* (forthcoming).

Kanniainen, V., & Keuschnigg, C. (2001). Start-up investment with scarce venture capital support. CESifo Working Paper No. 439. Posted on www.ssrn.com

Keuschnigg, C. (2002). Taxation of a venture capitalist with a portfolio of firms. University of St. Gallen Working Paper.

Keuschnigg, C., & Nielsen, S. B. (2001). Public policy for venture capital, *International Tax and Public Finance*, *8*, 557–572.

Keuschnigg, C., & Nielsen, S. B. (2002a). Tax policy, venture capital and entrepreneurship, *Journal of Public Economics*, *87*, 175–203.

Keuschnigg, C., & Nielsen, S. B. (2002b). Start-ups, venture capitalists, and the capital gains tax. University of St. Gallen and Copenhagen Business School Working Paper.

Macdonald, M. (1992). *Venture Capital in Canada: A Guide and Sources*, Toronto: Canadian Venture Capital Association.

MacIntosh, J. G. (1997). Venture capital exits in Canada and the United States. In: P. Halpern (Ed.), *Financing Innovative Enterprise in Canada* (pp. 279–356). University of Calgary Press.

MacIntosh, J. G. (1994). Legal and Institutional Barriers to Financing Innovative Enterprise in Canada. Monograph prepared for the Government and Competitiveness Project, School of Policy Studies, Queen's University, Kingston, Discussion Paper 94–10.

Peng, L. (2001). Building A Venture Capital Index. Yale Center for International Finance Working Paper. Available on www.ssrn.com

Sahlman, W. A. (1990). The structure and governance of venture capital organizations, *Journal of Financial Economics*, *27*, 473–521.

Sorenson, O., & Stuart, T. (2001). Syndication networks and the spatial distribution of venture capital investments, *American Journal of Sociology*, *106*, 1546–1588.

New Venture Investment: Choices and Consequences
A. Ginsberg and I. Hasan (editors)

Chapter 9

'New' Stock Markets in Europe: A 'New' Exit for Venture Capital Investments

FABIO BERTONI[a,**] and GIANCARLO GIUDICI[b,*]

[a] *Dipartimento di Ingegneria Gestionale, Politecnico di Milano, P.zza L. Da Vinci 32, 20133 Milano, Italy*
[b] *Facoltá di Economia, Universitá Tor Vergata, Roma, Italy*

ABSTRACT

The number of new listings in continental Europe has considerably grown in the late 1990s thank to 'new stock markets' (NMs) designed for fast-growing and high-tech firms willing to raise new finance. European NMs should represent an important way-out for private equity and venture capital. In this work, we study a large sample of 575 IPOs listed between 1996 and 2001 on the major European NMs. We compare the characteristics of venture-backed IPOs with other new listings, focussing on the ownership structure before and after the IPO. We highlight that VCs maintain a relevant stake in NM companies even after the listing.

1. INTRODUCTION

It is common belief that efficient IPO markets and stock exchanges increase the level of venture capital investing, and therefore innovation and employment. Conversely,

* Corresponding author. Tel.: +39-02-23992793; fax: +39-02-23992710.
** Tel.: +39-02-23992738; fax: +39-02-23992710.
E-mail addresses: fabio.bertoni@polimi.it (F. Bertoni), giancarlo.giudici@polimi.it (G. Giudici).

the lack of capital is often cited as one of the major impediments to innovation in enterprises.

In this vein, the competitive advantage of U.S. innovative companies during the 1990s over their European counterparts has been explained by a number of determinants, among which the different development of financial markets, and the scarce vivacity of venture capital (VC) and stock exchanges. As a consequence, European countries have been eager to promote private equity investments, and the growth of stock exchanges. The efforts seem to have been successful, considering that in 2000 more companies listed on European stock exchanges that in the U.S. Much of the merit is due to the birth of the 'New Stock Markets' established for small but fast-growing firms (the EASDAQ in Belgium, the *Neuer Markt* in Germany, the *Nouveau Marché* in France, the *Nuovo Mercato* in Italy, just to mention the largest). At the same time, record levels of private equity investments in growth companies have been achieved both in the U.S. and in Europe.

European 'New Markets' (NMs) provide interesting insights on the relationship between stock markets and venture capital. In fact, it is worth analyzing if the establishment of these new exchanges had any effect on venture capital investments, and exploring the characteristics of venture-backed NM IPOs. Giudici and Paleari (2002) already highlighted that NM companies raise a significant amount of new capital, if compared to the existing equity resources. They show that retail investors gave a relevant valuation to intangible capital, by purchasing shares at a price significantly higher than the firms' book value. Therefore, they argue that stock markets (and not private equity investors) have been the real financiers of NM companies in most cases, this confuting the traditional belief that professional private equity investors face most of the risk in technology-based companies life-cycle, before the listing on stock exchanges.

In this work, we aim at exploring the role of venture capitalists in NM companies before, during and after the listing. We try to point out if relevant differences characterize venture-backed IPOs vs. other companies, and European VC-backed IPOs vs. their U.S. counterparts.

We consider a large sample of 575 firms listed on European 'New Markets' from 1996 to 2001, built by Giudici and Roosenboom (2002). Indeed, we find that VC-backed IPOs are significantly smaller than other IPOs. They exhibit lower sales and earnings, but they also raise less equity capital. On the contrary, they are not significantly younger and more underpriced.

We compare our statistics with the numbers collected by Ljungqvist and Wilhelm (2003) for U.S. IPOs from 1996 to 2000. The comparison highlights many analogies (NM companies are not significantly younger than U.S. companies and the initial mean underpricing is surprisingly similar) but also some peculiarities (NM companies are smaller and less profitable; insider ownership is concentrated in the hands of CEOs and the presence of VCs is less frequent and significant, secondary sales occur more frequently at the IPO).

We find that venture capitalists completely exit the investment in a very few cases. On the average, they sell only 16.29% of their shares at the IPO (corresponding to about 5% of the company equity capital). The fraction of the investments liquidated is larger, the older the firm, the larger the current sales, but the smaller the accounting value of the assets. Technology companies are more easily exited by VCs than others. The evidence seems to suggest that: (i) VCs retained shares in companies in which their marginal contribution to the creation of value might be significant also after the listing (i.e. young companies, with a scarce capability to generate cash, characterized by further growth opportunities); (ii) they tended to sell shares of the smallest companies in their portfolio; and (iii) they took advantage of the market euphoria towards technology stocks.

Our findings challenge the traditional view that going public coincides with private equity investors' exit from the firm and strengthen the hypothesis that the listing on European NMs is not considered a stage subsequent to private equity financing, but a further relevant 'public' source of funds alongside venture capital. The role of private VCs in NMs is perceived as strategic even after the IPO: their permanence provide a 'certification effect' strengthened by lock-up provisions.

The remainder is organized as follows. The topic of venture capital financing is briefly recalled in Section 2. The venture capital markets in Europe and in the U.S. are compared in Section 3. Stock exchanges dedicated to fast-growing technology companies (European NMs and their main U.S. counterpart, the NASDAQ as well as the British AIM/Techmark) are analyzed in Section 4. Section 5 contains the empirical analysis on the IPO sample and finally, Section 6 presents some concluding remarks.

2. VENTURE CAPITAL FINANCING

Technology requires R&D investments, and R&D activity needs to be financed. Most of the times, the entrepreneurs' personal savings are not sufficient to cover the investments above; therefore, outside finance must be collected. However, innovation is risky and financiers do not like to engage in risky investments if no adequate contractual provisions are at work.

Financial constraints are often considered one of the main impediments to start-ups and high-technology firms seeking to expand and grow (Giudici and Paleari, 2000; Gompers and Lerner, 1999; Himmelberg and Petersen, 1994; Manigart and Struyf, 1997; Moore, 1994). Most of the times the entrepreneurs' personal wealth is not sufficient to fund new start-ups neither managerial experience is available. Even if a founder had enough capital, he probably would not be ready to invest all his property into a project, since, in the case of failure, he would lose everything. Without being offered some kind of insurance, he would prefer not to invest.

Hence, the access to finance capital is fundamental for fast-growing innovative firms, in order to sustain marketing and R&D expenditures. Nevertheless, several factors increase the cost of capital for young innovative firms compared to mature companies.

Uncertainty, information asymmetry and the risk of failure in developing new technologies are higher than in traditional firms (Binks et al., 1992; Westhead and Storey, 1997). This raises the costs related to external financing, as stated by the agency theory (Jensen and Meckling, 1976) and by the pecking order theory (Donaldson, 1961; Myers, 1984). Potential problems concerning risk and information asymmetry include the moral hazard and the adverse selection externalities. The entrepreneur usually has a superior knowledge of the future prospects of the project to be financed. He can use this knowledge to reduce his effort, and maximize his own utility, not the project's value. On the other hand, financiers are not able to discriminate among good and bad projects. Moreover, young firms have a short track record and a few public documents available to outsiders, this raising the cost of collecting information (Binks and Ennew, 1996).

Debt financing may not be available, because of the high risk and the lack of collaterizable assets. However, since interest payments would slow down the expansion of a young firm anyway, a loan is not a good financing instrument for innovative growing firms.

The resort to equity financing provided by closed-end funds and venture capitalists (corporate or business angels) is often the only solution to raise capital out of inside entrepreneurs (Gompers and Lerner, 1999; Sahlman, 1990). In fact, young firms are usually too small to be financed through bond or equity issue on capital markets.

Private equity investors negotiate complex contracts and covenants with entrepreneurs to mitigate conflict of interests, engage in active monitoring and place valuable managerial competencies at growing small firms' disposal. Advisory activities by the venture capitalists are very important and raise the chances for the survival of the firm, since entrepreneurs often have no business experience. Baker and Gompers (2003) find that venture capital backing improves firm outcomes also in the long run, reducing significantly the failure rate. Venture capitalists' stakes in the equity capital have a relevant image benefit, and provide a sort of 'certification effect' on the firm's quality (Megginson and Weiss, 1991).

Venture capital financing allows to separately allocate cash flow rights, voting rights, board rights, liquidation rights and other control rights, as the solution to conflicts of interest or agency problems between investors and entrepreneurs (Kaplan and Strömberg, 2003). For example, often venture capitalists purchase a combination of common equity, preferred equity and convertible bonds, and obtain veto rights, as to take over the firm's control in case of entrepreneurs' failure or opportunistic behaviour (Cornelli and Yosha, 1998; Hellmann, 1998). Capital is infused at stages corresponding to significant developments in the life of the company, following the business plan objectives. Stage financing limits the venture capitalist's losses in case of default and represents a threat of abandonment in the short run. Redemption covenants provide the venture capitalists with the means to extract the original investment from an unsuccessful company, as well as a credible threat of withdrawal over the entrepreneur.

Venture capitalists raise finance from outside investors and identify investment opportunities and projects. The returns flow as capital gains upon completion of the project. Seed capital is the first type of financing a newly founded company might want to secure, in order to fund R&D and commercial expenditures; start-up investments are targeted at companies gearing up to produce and market their products. Finally, in the expansion stage investing, the company has to fund growth opportunities, enlarging its manufacturing and distribution capacity, as well as engaging in external acquisitions.

Most venture capitalists exit from their investments in one of four ways: (i) sale after the company completes an initial public offering; (ii) sale of shares pursuant to an acquisition of the portfolio company; (iii) redemption of the venture capitalist's shares pursuant to contractual options; and (iv) write off, in case of unsuccessful ventures.

Venture funding is believed to have a positive impact both on creating jobs and boosting capital markets (Black and Gilson, 1998) and on innovation. Hellmann and Puri (2000) highlight a positive relationship between the market success of innovator firms and the type of financing obtained (in particular whether they obtain venture capital or not). Kortum and Lerner (2000) estimate that venture capital accounts for 15% of recent industrial innovation in the U.S.

Yet, the presence of efficient Exchanges dedicated to start-ups and growth firms is necessary to provide investors with an exit. In fact, Jeng and Wells (2000) show that a strong IPO market is the main force behind venture capital, especially later stage investments.

3. VENTURE CAPITAL INVESTMENTS: USA VS. EUROPE

Venture capital is by no means the main source of capital neither in the U.S. nor, to a higher extent, in Europe. In the U.S., the venture capital industry invested about $240 billion from 1990 to 2000 (NVCA, 2002), while companies listed on the New York Stock Exchange and the NASDAQ during the same period raised equity in seasoned issues for more than $230 billion and $380 billion respectively

(NASDAQ, 2002). The difference is even more significant if we take into account initial offerings and the issues of debt.

While in the U.S. the term venture capital is linked exclusively to equity-related investments in start-ups, or in high growth companies, European statistics refer to a more general concept of venture capital, that includes any commitment to unlisted companies at any stage, from seed investments to replacements, buy-outs and turn-around operations, the bulk of the activity being later-stage investments. Notwithstanding, in Europe, the venture capital industry is even less developed than in the U.S. From 1990 to 2000, €85 billion have been raised by professional private equity investors in Europe. Besides, considering only 1998 and 1999, companies listed on the major EU exchanges issued equity for more than €200 billion (EVCA, 2002; FESE, 2002).

Despite its small volumes, venture capital has helped to create world-wide many successful innovative multinational companies, particularly in high-tech industries. The European Venture Capital Association (2002) reports that over the period 1991–1995 venture-backed European companies experienced exceptional growth rates outperforming those of top 500 European companies. On average, revenues rose by 35% annually, twice as fast as the top established companies. Staff numbers increased by an average of 15% per year over the same period, and only by 2% for benchmark top companies. Investments in plant, property and capital equipment grew by an average of 25% annually. In 1995, R&D expenditure represented on average 8.6% of total sales compared to 1.3% for the top companies.

Historically venture capital activity has been significant only in the U.S. and in the U.K., but quite weak in other major European economies (Germany, France, and Italy). This evidence has been attributed to several determinants, and it has been invoked as one of the key distinctions between bank-based vs. stock-market-based financial systems (Allen and Gale, 1999). Yet, the gap has been reduced in the most recent years.

Figure 1 shows the ratio between venture capital investments and GDP in 2000, in the U.S. and in some European countries. The U.S. economy was still leading with 1.24%, but in Europe the incidence of investments on domestic wealth has significantly increased, as in 1998 the mean percentage ranged from 0.15% to 0.30%.

Recent years were the biggest ever for venture capital investments both in the U.S. and in Europe. Figure 2 shows the evolution of venture capital activity in the two areas. It is evident the impressive growth of total investments in the late 1990s.

In Europe, venture capital activity traditionally concerns non-innovative sectors, later development stages or investments without control and monitoring. Yet, 2000 saw the largest amount of money ever invested by European private equity and venture capital firms: funds invested in private equity totaled a record €35 billion in 10,440 companies, an increase of 39% on the 1999 stock. Amount invested in

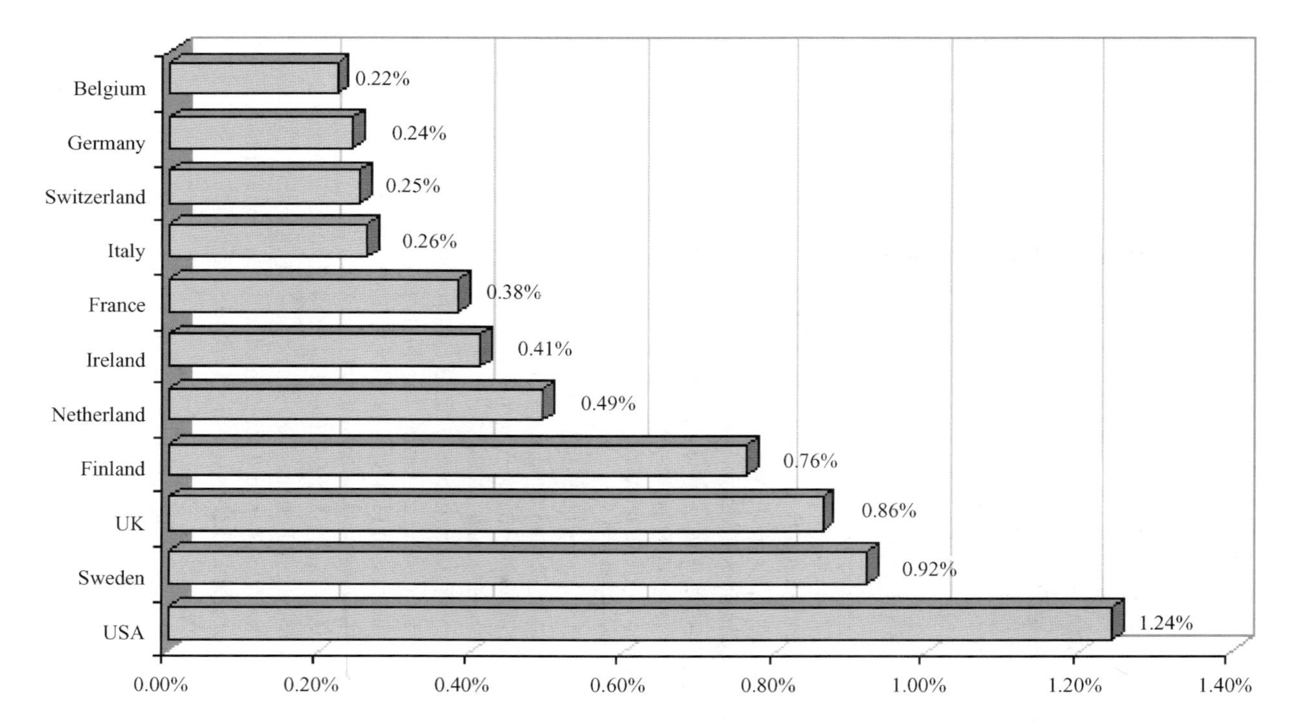

Figure 1. Venture capital investments in Europe and in the U.S., as percent of GDP in 2000. *Source:* European Venture Capital Association (2002), National Venture Capital Association (2002).

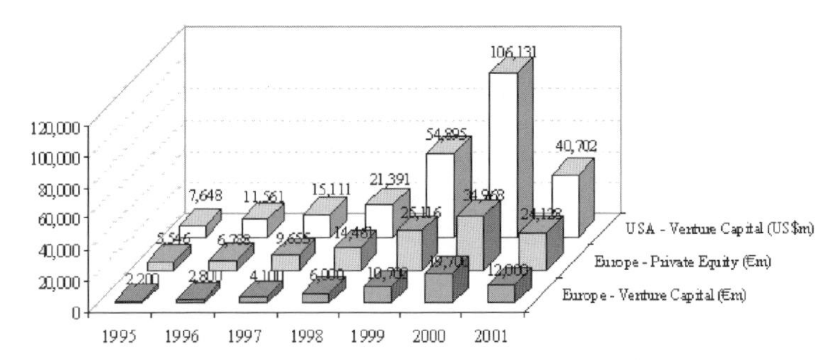

Figure 2. Annual venture capital investments in Europe (data in € million) and in the U.S. (data in $ million) from 1995 to 2001. The amounts of all private equity investments in Europe (data in € million) are also reported. *Source:* European Venture Capital Association (2002), National Venture Capital Association (2002), PricewaterhouseCoopers (2002).

early stage (seed and start-up) more than doubled to 19% of the total. Expansion investments totaled 37% of the amount, and buyout 41%. Capital invested in European high-technology companies totaled €11.5 billion in 2000, up 68% from 1999 (PricewaterhouseCoopers, 2002). Corporate investors significantly increased their appetite for risky start-ups, doubling their investments from €2.4 billion in 1999 to €4.8 billion in 2000. In detail, €19.7 billion went into pure venture capital stages, up 84% from 1999.

In the U.S., in 2000, a record $145 billion has been reached in private equity investing, with venture capital funds originating almost 72% of the total amount ($104 billion). Technology-related investments totaled $85 billion, up 91% from 1999 (PricewaterhouseCoopers, 2002).

Interestingly, the difference between funds raised and invested has been increasing in Europe, this suggesting a growing capability to channel investments outside the continent. For the first time since 1995, in 2000 in Europe, the largest amount of funds raised (24%) came from pension funds, with banks contributing 22% and insurance companies 13%. U.K. continued to lead Europe in fund raising with €17.7 billion, and €13.2 billion invested. France followed with €7.5 billion raised and €5.3 billion invested. Germany came third with €6.1 billion raised and €4.8 billion of amount invested. Italy was fourth contributing with €2.9 billion raised and €3.0 billion invested.

The numbers have changed radically in 2001 both in Europe and in the U.S. The capital invested in technology companies went down astonishingly (−35% in Europe) and Internet-related investments marked a downward spiral, declining to the lowest level in two years. Expansion stage companies received a greater attention in 2001 compared to 2000. In Europe, new private equity investments accounted 31%

less than during 2000 (€24 billion). Early stage ventures have fallen 68% by value, although the amount invested in pure VC (€12 billion) has been larger than in 1999.

In the U.S., venture capital investments in 2001 totaled $40.7 billion, down 61% from 2000. About $35.4 billion have been invested in technology companies, almost six times higher than in Europe. In 2000, technology VC investments in the U.S. were almost nine times greater than in Europe, so there is some evidence that Europe is catching up.

4. STOCK MARKETS FOR TECHNOLOGY FIRMS: NASDAQ VS. EUROPEAN 'NEW MARKETS'

Stock markets provide the source for new capital, but they also allow private equity investors to divest their stakes when the firm is mature. The main risk faced by investors and venture capitalists is the risk of not getting their funds back. Thus, a viable exit mechanism is extremely important to the development of venture capital (Jeng and Wells, 2000). First, it provides a financial incentive for equity-compensated managers to expend effort. Second, it gives the managers a call option on control of the firm, since venture capitalists relinquish control rights at the listing (Black and Gilson, 1998). Third, it allows venture capitalists to recycle both financial and non-financial resources from successful companies to early stage business and start-ups.

Initial public offerings (IPOs) allow private companies to create the floating capital required by the exchange admission rules (Ritter, 2002). IPOs may have numerous advantages for a company and its entrepreneurs. New equity capital may be raised, and more easily the managers can engage in stock-financed acquisitions and expansions. Yet, the transition from privately held to publicly traded company imposes substantial and costly changes in the firm's organization. Moreover, prior to go public, intermediates acting as consultants and underwriters have to be hired. The contractual right to initiate an IPO ('demand registration right') is commonly discussed between the entrepreneur and the venture capitalists (Halloran et al., 2000). A threat to invoke such an option could be used by the venture capitalist to coerce the entrepreneur into pursuing an efficient business strategy.

The most relevant problem IPO firms face is a substantial information asymmetry between insiders (controlling entrepreneurs and managers) and investors about the company value. Analysts frequently use comparable multiples to come up with a preliminary price range, to value a firm going public, and rely on forecasts of the next years' accounting numbers.

Technology-based innovative companies are expected to face several troubles when going public. The fact that only scant information on R&D and other technology activities is publicly disclosed by innovator firms compounds the information problem of investors when evaluating IPOs. The information asymmetry is more relevant, the higher the intensity of firms' intangible assets, thus affecting financing and the going public process (Lev, 2000). Maksimovic and Pichler (2001) show that in an emerging industry the timing of external financing and the choice between public and private equity financing depends on the technological risks in the industry. Firms that go public early may gain advantage from beginning full-scale operations before their rivals, but risk being displaced from more efficient rivals during a period of technological change. Moreover, they can provide valuable information to potential competitors. Finally, technology-based companies are almost never profitable in the short run.

The remarks above suggest that traditional stock markets are often inadequate to host small but fast-growing firms, the most relevant problems being tight listing requirements, high business risk and potential illiquidity. Therefore, many countries established specific stock exchanges for fast-growing and high-tech firms. On one hand, the listing requirements in this 'new markets' are less severe, on the other hand several warranties protect investors from the insiders' opportunistic behavior.

In the 1990s, Europe has been eager to reply the success of the hi-tech and 'new economy' enterprises in the U.S. European lateness has been blamed on the scarcity of venture capital investments and private equity active investors, but in particular on the absence of high-growth and high-tech segments such as the NASDAQ (European Commission, 1994). As a reaction, European governments sustained the birth of 'new stock markets' for hi-growth and hi-tech companies. In 1996, referring to the model of the NASDAQ, the EASDAQ market was founded in Belgium, to host young companies willing to obtain international audience from investors. The market has been directly taken over by the NASDAQ and changed its name in Nasdaq Europe in 2001.

In 1996, the French *Nouveau Marché* was born, and in 1997 the German *Neuer Markt* as well. In 1999, the London Stock Exchange opened a new specific segment for hi-tech firms, TechMark, although a market for small caps (Alternative Investment Market, AIM) was already at work. Italy in 1999 established the *Nuovo Mercato*. In 2000, the Spanish *Nuevo Mercado* opened to high-growth firms. Other 'new' stock markets have been established in the Netherlands (*Nieuwe Markt*), in Belgium (Euro.NM Belgium), in Sweden (*Nya Marknaden*), in Finland (*NM-List*), in Switzerland (SWX New Market), in Ireland (ITEQ), in Denmark (KVX Growth), in Portugal (*Novo Mercado*) and in Greece (New Market NEHA).

The major European exchanges agreed to launch the Euro.NM network, grouping the German, French, Italian, Belgian and Dutch 'new stock markets'. However, the Euro.NM initiative failed to establish operating links between the member markets, and the network was disbanded on 31 December 2000, leaving full autonomy to the single NMs. The increasing number of listing on the German, French and Italian markets, compared to the EASDAQ decline, seemed to favour the paradigm of several national 'new markets', instead of a single pan-European exchange. Yet, the negative markets momentum after April 2000 and the reports of several failures among NM companies raised several questions, and even caused the crisis of some NMs (for example, the German *Neuer Markt* closed in 2003).

The basic differences between NMs and main exchanges are the listing requirements.[2] Generally, they are less binding, allowing young but not yet profitable firms to be listed and issue new capital. Investors' protection is favored through the commitment to regularly disclose information, to assure research coverage and to lock-up inside equity (generally up to one year). Therefore, private equity investors have to sell their shares at the IPO, or retain them for a longer period. This provision assures that insiders will not sell opportunistically their shares immediately after the listing.

The market trading rules are designed to provide liquidity and often firms are required to appoint a 'specialist' (an intermediate displaying bids on the exchange book).

The minimum accounting value of the equity ranges from €1 million (Euro.NM Belgium) to €1.65 million (SWX New Market). The Euro.NM Belgium, the SWX New Market and the *Neuer Markt* require a minimum age equal to three years, while one year old companies may list on the other markets. The Spanish *Nuevo Mercado* is the only one accepting exclusively profitable companies paying dividends. The other markets list companies even with losses, but with an ambitious business plan and relevant growth opportunities. The EASDAQ (NASDAQ Europe) requires either that the IPO company equity has a book value equal to at least €10 million and gross income equal to at least €1 million, or a book value equal to at least €20 million, or a market capitalization equal to €20 million, with sales larger than €50 million. The U.K. Techmark is a segment of the main board; therefore the listing requirements are not specific, and firms are admitted according to their business activity (for example, IT and software companies are automatically listed on the Techmark).

The capital owned by outsiders must represent 20/25% of the total equity, albeit in some cases exceptions are tolerated. Remarkably, most markets require that at least 50% of the IPO shares must be newly issued. This should boost IPO firms to make new investments and grow.

Interestingly, the going public process is not the same in all the exchanges considered. However, NMs generally adopted the U.S. model, i.e. book building

---

[2] A survey of the listing requirements imposed by NMs, compared with main exchanges, is contained in Giudici and Roosenboom (2002).

Table 1. Statistics about 'New' stock markets around continental Europe, as at June 2002, and comparison with the NASDAQ and Techmark/AIM

Stock market	Country	Market capitalization[a]	Market capitalization/ GDP (%)	Listed companies
Neuer Markt	Germany	38,361	1.39	288
Nuevo Mercado	Spain	11,827	1.61	12
Nouveau Marché/ Euronext	France	10,266	0.52	164
Nuovo Mercato	Italy	9,014	0.64	45
NASDAQ Europe (EASDAQ)	Belgium	4,240	n.s.[b]	44
SWX New Market	Switzerland	1,869	0.58	15
ITEQ	Ireland	1,293	0.57	8
KVX Growth	Denmark	1,198	0.50	13
Nya Marknaden[c]	Sweden	513	0.15	16
NM-List	Finland	403	0.16	17
Nieuwe Markt/ Euronext	The Netherlands	379	0.07	11
New Market NEHA	Greece	98	0.06	3
Euro-NM Belgium/ Euronext	Belgium	57	0.02	11
NASDAQ	USA	2,548,027	20.54	3,883
TechMark/AIM	UK	449,417	28.26	907

Source: Federation of European Stock Exchanges (2002), NASDAQ (2002).
[a] Data in € million.
[b] The ratio is not significant, because the majority of listed companies are not incorporated in Belgium.
[c] Nya Marknaden is not a regulated market, although it shares the trading platform provided by the Stockholm Exchange.

with open price. The public offering is reserved to retailers; meanwhile institutions are allocated shares, on the basis of the book building process. A price range is contained in the prospectus, while the final offer price is decided after the collection of institutions' bids.

Table 1 reports some basic statistics about 'new stock markets' in Continental Europe, as at June 2002. A comparison with the NASDAQ and with the British AIM/Techmark is also reported. It is evident that European NMs are still far from the U.S. and U.K. figures. Yet, the German *Neuer Markt* reached a remarkable market capitalization, which is comparable to the other NMs combined, and before the downfall of 2000 and 2001 represented about 6% of German GDP. The Italian and Spanish markets have been able to achieve a significant dimension in a few months. The French *Nouveau Marchè* stands out by the number of listed companies, while the other exchanges have minor dimensions. The Nasdaq Europe experienced a decrease in the number of listed firms, especially in 2001.

5. A SURVEY ON VENTURE-BACKED IPOs ON EUROPEAN NMs

In this section, we explore the role of venture capital investors in NM IPOs. We consider a large sample of 575 IPO companies, listed on the major European NMs (the German *Neuer Markt*, the French *Nouveau Marché*, the Italian *Nuovo Mercato*, the EASDAQ/NASDAQ Europe, the Euro.NM Belgium and the Dutch *Nieuwe Markt*) from 1996 to 2001, built by Giudici and Roosenboom (2002). This sample comprises all new listings on the markets above, excluding only spin-offs, financial companies and firms already listed in other exchanges.

Table 2 splits the sample by market and year of listing.

The largest number of IPOs (314, equal to 54.6% of the sample) went public on the German *Neuer Markt*. The French *Nouveau Marché* hosted 149 IPOs. Thirty-nine companies listed in Italy (*Nuovo Mercato*). Fourty-six firms opted for the EASDAQ/NASDAQ Europe. The Dutch and Belgian Markets hosted 14 and 13 IPOs, respectively. Many sample companies went public in 1999 and 2000 (178 and 213, respectively), confirming that this was a good period for new issues in Europe. Conversely, only 20 companies listed in 2001 as a consequence of the world-wide bearish market momentum.

Table 2 also highlights the number of IPOs listed during the same period on the NASDAQ (2,392) and AIM (883) markets,[3] confirming the vivacity of U.S. and U.K. IPO markets compared to continental Europe.

Table 3 reports some basic statistics about the sample IPO companies. The mean age is equal to 12.35 years. At the IPO the mean accounting value of the assets is equal to €25.582 million. Sales on the average account for €30.156 million, while earnings (EBITDA) amount to €1.615 million. The mean fraction of equity held by venture capitalists before the IPO is equal to 10.65%, while it decreases to 8.65% after the IPO (due both to the dilution effect of newly issued shares and to sales of secondary shares). Total IPO proceeds are large, if compared to the book value of existing assets (mean value €54.198 million). At the IPO both primary and secondary shares may be offered. In our sample, newly issued capital on the average (€45.994 million) is almost twice the accounting value of the assets (the listing requirements force IPO companies to issue new shares). Secondary shares (proceeds amount to €8.204 million on the average) occur in 374 cases (65% of the sample companies).

[3] Techmark is not considered in Table 2, since it is a segment of the main exchange.

Table 2. The sample: Initial Public Offerings on European 'New Stock Markets' and comparison with the NASDAQ and AIM (Techmark is not considered since it is a segment of the main exchange), from January 1996 up to December 2001

Stock Market	Country	1996	1997	1998	1999	2000	2001	Grand total
Neuer Markt	Germany	–	10	38	124	131	11	314 (54.6%)
Nouveau Marché (Euronext)	France	14	17	40	30	43	5	149 (25.9%)
Nuovo Mercato	Italy	–	–	–	6	29	4	39 (6.8%)
EASDAQ (Nasdaq Europe)	Belgium	4	13	12	12	5	–	46 (8.0%)
Nieuwe Markt (Euronext)	The Netherlands	–	3	8	1	2	–	14 (2.4%)
EuroNM.Belgium (Euronext)	Belgium	–	1	4	5	3	–	13 (2.3%)
Total European sample	–	18 (3.1%)	44 (7.7%)	102 (17.8%)	178 (30.9%)	213 (37.0%)	20 (3.5%)	575 (100%)
Total Nasdaq	USA	680 (28.4%)	494 (20.7%)	273 (11.4%)	485 (20.3%)	397 (16.6%)	63 (2.6%)	2,392 (100%)
Total AIM	UK	145 (16.4%)	107 (12.1%)	75 (8.5%)	102 (11.6%)	277 (31.4%)	177 (20.0%)	883 (100%)

Table 3. Descriptive statistics about the sample IPO companies (575 IPOs listed on European 'New Stock Markets' between January 1996 and December 2001)

	Mean value	Median value
Assets accounting value (€ million)[a]	25.582	10.574
Sales (€ million)[a]	30.156	11.496
Current EBITDA (€ million)[a]	1.615	1.111
Company age (years)	12.35	9.00
Employees	196	105
Fraction of equity capital held by venture capitalists before IPO (%)	10.65	0.00
Fraction of equity capital held by venture capitalists after IPO (%)	8.65	0.00
Offer size (€ million)	54.198	28.223
Newly issued capital (€ million)	45.994	23.782
Secondary shares offered (€ million)	8.204	2.859
Floating equity capital after the IPO (%)	29.38	28.00
First-day return (%)	+34.50	+10.00

Source of the data: Giudici and Roosenboom (2002).
[a] Accounting value, as from the balance sheet at the year of the listing.

The floating capital after the IPO on the average is equal to 29.38%. Remarkably, we report a significant first-day return compared to the IPO offer price, equal to 34.50%.

Not surprisingly, the characteristics of NM companies are significantly different from those of companies listed on primary European exchanges during the same period. Giudici and Roosenboom (2002) highlight that NM companies are significantly younger (they report that the mean age for IPOs on established European exchanges is 37 years) and less profitable, although they relatively collect more equity capital, with respect to their existing assets. NM companies are also more underpriced (13.07% is the first day return during the same period in European established markets).

Compared to U.S. IPOs, NM companies are not significantly younger (Ljungqvist and Wilhelm, 2003, hereafter L&W, report a mean company age equal to 13.6 years for their U.S. sample covering 1996–2000), but still smaller and less profitable (in the US, mean revenues account for $208 million and the book value of equity accounts for $475.5 million). In the U.S., secondary sales occur in only 27.47% of the IPO companies. The initial underpricing in the U.S. is surprisingly similar: 33.66%.

Table 4 splits the sample between venture-backed IPOs (258) and other companies (317). VC backed IPOs are defined as companies participated by venture capitalists (closed-end funds, corporate VCs) prior to the IPO. They represent 44.87% of the total IPO sample, while in the L&W's analysis they represent 50.85% of the panel.

Table 4. Comparison between venture-backed IPOs and other companies (respectively 258 and 317 IPOs listed on European 'New Stock Markets' between January 1996 and December 2001). VC-backed IPOs are defined as companies participated by venture capitalists (closed-end funds, corporate VCs) before the IPO

	Venture-backed IPOs	Other IPOs
Sample size	258	317
Assets accounting value (€ million)[a]	21.073*	34.693*
Sales (€ million)[a]	19.508***	38.529***
Current EBITDA (€ million)[a]	0.478***	2.540***
Company age (years)	11.9	12.7
Employees	163***	224***
Fraction of equity capital held by CEOs and other insiders before IPO (%)[b]	52.15***	73.11***
Fraction of equity capital held by CEOs and other insiders after IPO (%)[b]	38.03***	52.73***
Fraction of equity capital held by industrial companies before IPO (%)	4.73***	12.93***
Fraction of equity capital held by industrial companies after IPO (%)	3.60**	9.35**
Floating equity capital after the IPO (%)	30.71**	28.31**
Offer size (€ million)	44.694*	61.933*
Newly issued capital (€ million)	35.745*	54.336*
First-day return (%)	+36.27	+33.05

Note: The asterisks (*), (**), (***) denote that the difference between the two samples is significant at the 90%, 95% and 99% level, respectively.

Source of the data: Giudici and Roosenboom (2002).

[a] Accounting value, as from the balance sheet at the year of the listing.

[b] Other insiders comprise executive directors, non-executive directors and employees.

VC-backed IPOs are significantly smaller (in terms of assets and employees), they report lower sales (€19.508 million vs. €38.529 million) and earnings (€0.478 million vs. €2.540 million), but—interestingly—they are not significantly younger (11.9 years vs. 12.7 years). Issue proceeds are lower for venture-backed IPOs, as well as newly issued capital (raised through the issue of primary shares). Yet, the proportion of funds raised with respect to the existing assets is about the same for the two samples. Not surprisingly, the fraction of equity capital owned by insiders (CEOs, executive and non-executive directors, employees) and by industrial partnering companies is lower in venture-backed IPOs. Also the floating capital on the average is smaller. Remarkably, VC-backed IPOs are not significantly more underpriced than other offerings (36.27% vs. 33.05%); this result contrasts with the U.S. evidence, where during the same period VC-backed IPOs have been more severely underpriced (Loughran and Ritter, 2003).

Table 5 deeply investigates upon the composition and evolution of venture-backed companies' ownership structure around the IPO. The stakes held by venture capitalists, banks, insurance and other financial companies, industrial partnering companies, insiders (CEOs, executive and non-executive directors, employees), are measured.

Stakes ultimately held by CEOs are on the average larger than reported in L&W's sample (29.90% vs. 20.55%). Yet in the aggregate insiders on the average own 52.15% of the pre-IPO equity (see also Table 4), lower than reported in the U.S. (59.94%). This suggests that officers and employees' ownership is less common in NM companies than in the U.S., and voting power is more concentrated at the CEO level.[4] On the contrary, in Europe VCs seem to own fewer shares prior to the IPO in participated companies than in the U.S. (30.34% vs. 40.36% reported by L&W).

In Europe on the average 4.91% of the equity capital is put on the market by venture capitalists at the IPO, representing 16.29% of their stakes. In 100 cases (38.76% of the sample VC-backed companies) VCs do not sell any share at the IPO. These results seem to contrast with the common belief that the IPO represents the primary exit opportunity for private equity investors.

Even less shares are sold by insiders (CEOs put on the market only 4.42% of their shareholding, executive directors only 4.00%, non-executive directors only 2.73%). In more than 50% of the sample companies insiders do not sell shares at the IPO. Secondary shares also occur among banks (3.44%), financial and insurance companies (2.45%), and corporate partnering companies (1.98%).

The evidence is consistent with the trend detected in the U.S. by L&W who document that secondary sales by both insiders and VCs at the IPO have become less frequent in 1999 and 2000 (they report that VCs sell shares in only 14.51% of the sample VC-backed companies).

After the IPO, insiders maintain a relevant fraction of the equity capital even in venture-backed companies (in aggregate 38.03%, see also Table 4). They maintain the majority of the equity capital (>50%) in 86 companies (33.3% of the sample VC-backed companies).

We finally looked at the determinants of VC selling at the IPO. We considered several potential explanatory variables. We introduced the log of the company age (LN_AGE), the log of current reported sales (LN_SALES), the log of the accounting

[4] L&W's statistics refer to their total sample, including non-venture-backed companies. Easily the mean shareholding held by CEOs in U.S. venture-backed companies is even lower, this strengthening the significance of the difference between U.S. and Europe.

Table 5. Composition of inside shareholding and venture capital investments in NM venture-backed IPOs

	Shareholding before the IPO (%)		Shares sold at the IPO (%)[a]		Shareholding after the IPO (%)	
	Mean value	Median value	Mean value	Median value	Mean value	Median value
Venture capitalists	30.34	27.30	4.91 (16.29)	2.20 (9.42)	19.20	16.47
Banks	1.18	0.00	0.28 (3.44)	0.00 (0.00)	0.66	0.00
Insurance/other financial companies	3.41	0.00	0.46 (2.45)	0.00 (0.00)	2.23	0.00
Industrial companies	4.73	0.00	0.20 (1.98)	0.00 (0.00)	3.60	0.00
Chief executive officers	29.90	24.95	1.33 (4.42)	0.00 (0.00)	22.01	18.17
Executive directors	15.16	6.62	0.78 (4.00)	0.00 (0.00)	10.93	4.72
Non-executive directors	4.38	0.00	0.37 (2.73)	0.00 (0.00)	3.06	0.00
Employees	2.71	0.00	0.03 (1.11)	0.00 (0.00)	2.03	0.00
Others/untracked	8.19	1.64	0.24 (2.85)	0.00 (0.00)	5.57	1.26
Floating capital	–	–	–	–	30.71	28.60

Note: Shareholding before the IPO is measured as a percentage of the capital outstanding before the IPO. Secondary shares sold at the IPO are measured both as a percentage of the total capital outstanding before the IPO and (in parentheses) as a percentage of the equity capital held before the IPO by each category. Shareholding after the IPO is measured as a percentage of the capital outstanding after the IPO. Sample: 258 VC-backed IPOs listed on European 'New Stock Markets' between January 1996 and December 2001.

Source of the data: Giudici and Roosenboom (2002).

[a] Percentage of the total capital outstanding before the IPO and (in parentheses) percentage of the equity capital held before the IPO by each category.

value of total assets (LN_ASSETS) and the log of the capital raised through the issue of primary shares at the IPO (LN_CAP_R). We hypothesize that VCs should be more willing to exit the investments in mature firms, with a significant capability to generate cash through their operating activity. Firms raising more capital at the IPO should need less financial assistance from VCs and be more easily exited.

Then we introduce the ratio between the market capitalization at the IPO and the book value of assets (M_B), the VCs' ownership level at the IPO (VC_STAKE) and the IPO share price volatility measured in 50 days after the listing (VOL), as a proxy of the company risk. VCs holding more shares in a company should have more incentives to sell at the IPO, as well as they could consider the risk of the business in deciding how many shares should be sold, or the growth opportunity expressed in the firm's evaluation.

We take into account other control variables: the stock market index performance in 50 days prior to the issue of the IPO prospectus (MRKT_INDEX) and the index volatility during the same period (MRKT_VOL). Finally, we introduce two dummy variables (TECH_DUMMY and INTERNET_DUMMY) characterizing technology stock (corresponding to 144 companies among 258) and Internet stock (54 companies), as classified in Giudici and Roosenboom (2002). VCs may be more willing to sell shares if the market momentum is favorable, in order to take advantage of investors' optimism, also given the high-tech and Internet euphoria experienced in the late 1990s and in 2000.

Since the dependent variable (the fraction of VC shares sold at the IPO) takes values only between 0% and 100%, we adopted a Tobit model.[5] Table 6 reports the estimation results and the variables' marginal effects.

Model (1) highlights that the company age, the reported sales and the total assets are significant determinants of VC selling at the IPO. Not surprisingly, VC sales are larger, the older the company and the larger the current sales. Venture capitalists do not exit investments in young IPO companies, with scarce capability to generate cash from the operating activity. Interestingly, VC sales are larger, the lower the accounting value of the assets. This may indicate that the book value of assets is not considered by VCs as a proxy of the firm's maturity. Indeed, VCs seem to be more willing to exit investments in smaller companies, and to maintain shares in larger firms. It is reasonable that the VCs' bargaining power is weakened after the IPO: by retaining shares, the VCs—like any other institutional investor—are exposed to the risk that the company is not able to obtain adequate attention from the market

[5] In this case, OLS estimates would be inconsistent, because of the thresholds characterizing the dependent variable (we cannot observe if VCs short sell or even purchase IPO shares). Maximum likelihood estimates have been derived with LIMDEP software.

Table 6. The determinants of VC selling at the IPO: maximum likelihood estimation of the Tobit model

Variable	(1)	(2)	(3)	(4)
Constant	−0.254 (−0.157)	−0.142 (−0.088)	0.213 (0.131)	0.243 (0.149)
LN_AGE	0.100*** (0.061)	0.094*** (0.058)	0.088*** (0.054)	0.089*** (0.054)
LN_SALES	0.061*** (0.037)	0.062*** (0.038)	0.058*** (0.035)	0.059*** (0.036)
LN_ASSETS	−0.049** (−0.030)	−0.057** (−0.035)	−0.060** (−0.037)	−0.063** (−0.039)
LN_CAP_R			−0.017 (−0.010)	−0.016 (−0.010)
M_B			−0.009* (−0.006)	−0.009* (−0.005)
VC_STAKE				0.094 (0.057)
VOL				−0.732 (−0.448)
MRKT_INDEX				−0.071 (−0.044)
MRKT_VOL				−2.630 (−1.611)
TECH_DUMMY	0.060* (0.037)		0.065* (0.040)	0.074** (0.045)
INTERNET_DUMMY		−0.049 (−0.030)		
Log likelihood function	−91.08	−91.90	−86.52	−85.53
Sample size	258	258	258	258

Notes: The dependent variable is the fraction of VC shares sold at the IPO. The independent variables are: the log of the company age (LN_AGE), the log of current sales at the IPO (LN_SALES), the log of the accounting value of total assets (LN_ASSETS), the log of the new equity capital raised at the IPO (LN_CAP_R), the market/book ratio (M_B), the level of VCs shareholding before the IPO (VC_STAKE), the IPO share volatility, measured in 50 trading days after the IPO (VOL), the market index performance and volatility measured in 50 trading days prior to the issue of the IPO prospectus (MRKT_INDEX and MRKT_VOL). The dummy variables TECH_DUMMY and INTERNET_DUMMY are equal to 1 if the company is classified by Giudici and Roosenboom (2002) as tech-stock or specifically as internet-stock, respectively. Estimated marginal effects of independent variables are shown in parentheses. Sample: 258 VC-backed IPOs listed on European 'New Stock Markets' between January 1996 and December 2001.

The asterisks (*), (**), (***) denote that the coefficient are significantly different from zero at the 90%, 95% and 99% level, respectively.

or research coverage. This risk is evidently larger in small firms and pushes VCs to divest their shares. In other words, VCs prefer to maintain shares in large companies, characterized by larger liquidity and lower costs of information updating.

VC sales are more frequent in technology-based firms, this providing support to the hypothesis that during the period considered VCs took advantage of the market euphoria towards high-tech stocks. On the contrary, model (2) shows that similar evidence is not detected by separately considering the sample of Internet companies.

In models (3) and (4), the significance of the age, sales, assets value and technology dummy variables is confirmed and the estimation coefficients are quite stable. No significant relationship is detected among other variables, except for the ratio between the equity capitalization and total assets (M_B). VC sales are larger, the lower the growth opportunities expressed by the market evaluation; the evidence is consistent with the hypotheses that VCs prefer to retain shares in companies with larger market capitalization (like any other institutional investor trading on the stock exchanges), and companies characterized by growth opportunities to be further exploited (i.e. companies in which the marginal contribution of VCs to the creation of new value is more significant).

Interestingly, VCs seem not to be influenced by the short-term market momentum and by the relative size of the pre-IPO investments.

6. CONCLUDING REMARKS

The birth of 'New Stock Markets' (NMs) around Europe significantly contributed to the increase of the number of new listings, and to attract technology firms to the stock markets.

In this work, we explored the characteristics of NM companies, focussing on the role of venture capitalists. Not surprisingly, NM companies are younger, smaller and less profitable than companies listed on the 'main' exchanges. Compared to U.S. IPO companies, they are not significantly younger, but their size and profitability are smaller. The presence of venture capitalists in the equity capital prior to the IPO is less frequent than in the U.S. Inside ownership is more concentrated in the hands of CEOs, while in the U.S. it is more dispersed among executive directors and employees. At the IPO, secondary sales occur more frequently in Europe than in the U.S.

Focussing on the sample of VC-backed companies, we highlighted that they are not significantly younger than other companies, although their size and reported sales are smaller. Consistently with the U.S. evidence, we showed that VC sales at the IPO are lower than it could be expected. On the average, VCs sell 16.29% of their shares. If we consider that lock-up provisions are compulsory in NM companies, this result demonstrates that VCs often commit themselves not to exit their investment neither at the IPO nor in the following months. Hence, do really NMs represent a 'new' exit for venture capital?

Giudici and Paleari (2002) state that NM companies are far from being later-stage enterprises, willing to finance their expansion. On the contrary, NMs are characterized as 'markets for projects' in which ambitious business plans are sold to the public. In this sense, NMs do not represent primarily an exit occasion for private equity investors, but a 'public venture capital market' financed by retailers. This work confirms such hypothesis. Indeed we find that: (i) VCs retain shares in companies in which their marginal contribution to the creation of value may be significant also after the listing (i.e. young companies, with a scarce capability to generate cash, characterized by further growth opportunities); (ii) they tend to sell shares of the smallest companies in their portfolio; and (iii) they took advantage of the market euphoria towards technology stocks during the 1990s.

Therefore, we conclude that NMs play two distinct roles. First, they provide a partial exit for venture capitalists and private equity investors. Second, they provide an important source of finance alongside (and not subsequently to) venture capital in the phase of start-up.

Our results contribute to shed new light on VC activity in Europe and the characteristics of growth companies listed on NMs. In the future, it will be interesting to verify if the advent of NMs in Europe will contribute to the general economic progress and in particular to promote venture capital, technological innovation and employment.

Europe is willing to fill the gap with the U.S. economy by replying the successful contamination among finance and innovation. The 'new economy' euphoria, coinciding with the Internet diffusion, contributed to a glorious growth of NMs in Europe. Yet, investors have been generally disappointed by the IPOs market return. In a few months, several companies went default, because of financial distress, or company frauds. The reputation of NMs deeply suffered from this bad momentum. Yet, no one doubts about the opportunity to establish specific segments for growth companies.

Even more, in the long term it is possible that a large single (or more) pan-European market for high-growth and high-tech stock will emerge. Anyway, we are convinced that a strong 'home-bias effect' will persist in the next years and integration among European stock exchanges is still far to come, despite the efforts of the European Commission.[6] European companies seem not to be willing to go public on a pan-European market. Rather, they prefer first to list on their domestic

[6] During the 2002 summit in Barcelona, EU governments adopted the recommendation of the 'Lamfalussy Committee', towards a further harmonization of listing procedures across EU stock exchanges, and the issue of a 'EU passport' for IPO companies. Yet, no specific directives have been targeted to the establishment of a unique European stock exchange.

market, and then—the largest companies—to cross-list in foreign countries, where they can reap major visibility.

We believe that in the future new growing opportunities will be generated by emerging technologies such as wireless and third-generation mobile platforms, where Europe has a clear lead over the U.S., and other areas such as life science, security software, optical technology, interactive television. Yet, we think that stock markets should clearly distinguish between financing the growth of established companies and financing over-optimistic business plans. To this extent, the role of professional institutional investors, like VCs, is fundamental.

ACKNOWLEDGMENTS

The authors acknowledge research assistance from Andrea Randone as well as financial support from Cofinanziamento MIUR and CNR Ageozia 2000 Progetto Giovani Ricercatori. Many thanks to Peter Roosenboom for sharing the IPO database.

REFERENCES

Allen, F., & Gale, D. (1999). Diversity of opinions and financing of new technologies, *Journal of Financial Intermediation*, *8*(1), 68–89.

Baker, M., & Gompers, P. A. (2003). The Determinants of Board Structure at the Initial Public Offering, *Journal of Law and Economics* (forthcoming).

Binks, M., & Ennew, C. (1996). Growing firms and the credit constraint, *Small Business Economics Journal*, *8*, 17–25.

Binks, M., Ennew, C., & Reed, C. (1992). Information asymmetries and the provision of finance to small firms, *International Small Business Journal*, *11*(1), 35–46.

Black, B., & Gilson, R. (1998). Venture capital and the structure of capital markets: Banks vs. stock markets, *Journal of Financial Economics*, *47*(3), 243–277.

Cornelli, F., & Yosha, O. (1998). *Stage Financing and the Role of Convertible Debt*, IFA Working Paper No. 253.

Donaldson, G. (1961). *Corporate Debt Capacity: A Study of Corporate Debt Policy and the Determination of Corporate Debt Capacity*, Harvard Business School, Division of Research.

European Commission (1994). *Research into the Financing of New Technology based Firms (NTBSFs)*, Paris: Final Report.

EVCA European Venture Capital Association (2002). Yearbook, http://www.evca.com

FESE Federation of European Stock Exchange (2002). Statistics Database, http://www.fese.be

Giudici, G., & Paleari, S. (2000). The provision of finance to innovation: A survey conducted among Italian technology-based small firms, *Small Business Economics Journal*, *14*, 37–53.

Giudici, G., & Paleari, S. (2002). R&D financing and stock markets. In: M. Calderini, P. Garrone, & M. Sobrero (Eds), *Corporate Governance, Market Structure and Innovation*, Edward Elgar.

Giudici, G., & Roosenboom, P. G. J. (2002). Pricing Initial Public Offerings on 'New' European Stock Markets, European Financial Management Association Conference (26–28 June). London, UK.

Gompers, P. A., & Lerner, J. (1999). *The Venture Capital Cycle*, MIT Press.

Halloran, M. J., Benton, L. F., Gunderson, R. V., Del Calvo, J., & Kintner, T. W. (2000). *Venture Capital and Public Offering Negotiation*, Aspen Law and Business.

Hellmann, T. (1998). The allocation of control rights in venture capital contracts, *Rand Journal of Economics*, *29*(1), 57–76.

Hellmann, T., & Puri, M. (2000). The interaction between product market and financing strategy: The role of venture capital, *The Review of Financial Studies*, *13*(4), 959–984.

Himmelberg, C. P., & Petersen, B. C. (1994). R&D and internal finance: A panel study of small firms in high-tech industries, *Review of Economics and Statistics*, *76*, 38–51.

Jeng, L. A., & Wells, P. C. (2000). The determinants of venture capital funding: Evidence across countries, *Journal of Corporate Finance*, *6*(3), 241–289.

Jensen, M. C., & Meckling, W. H. (1976). Theory of the firm: Managerial behavior, agency costs and ownership structure, *Journal of Financial Economics*, *3*, 305–360.

Kaplan, S. N., & Strömberg, P. (2003). Financial contracting theory meets the real world: An empirical analysis of venture capital contracts, *Review of Economic Studies*, *70*, 281–316.

Kortum, S., & Lerner, J. (2000). Assessing the contribution of venture capital to innovation, *Rand Journal of Economics*, *31*, 674–692.

Lev, B. (2000). *New Accounting for a New Economy*, Working Paper, Stern Business School, NY.

Ljungqvist, A. P., & Wilhelm, W. J. (2003). IPO pricing in the Dot-com bubble: Complacency or incentives? *Journal of Finance*, *58*, 723–752.

Loughran, T., & Ritter, J. R. (2003). *Why Has IPO Underpricing Changed Over Time?* Working Paper.

Maksimovic, V., & Pichler, P. (2001). Technological innovation and initial public offerings, *Review of Financial Studies*, *14*(2), 459–494.

Manigart, S., & Struyf, C. (1997). Financing high-technology start-up in Belgium: An explorative study, *Small Business Economics Journal*, *9*, 125–135.

Megginson, W. L., & Weiss, K. A. (1991). Venture capitalist certification in initial public offerings, *Journal of Finance*, *46*, 879–903.

Moore, B. (1994). Financial constraint to the growth and development of small high technology firms. In: A. Hughes, & D. Storey (Eds), *Finance and the Small Firms*, London: Routledge.

Myers, S. (1984). The capital structure puzzle, *Journal of Finance*, *39*, 572–592.

NASDAQ (2002). *The NASDAQ-AMEX Fact Book and Company Directory*, http://www.nasdaq.com

NVCA National Venture Capital Association (2002). Venture Capital Yearbook, http://www.nvca.com

PricewaterhouseCoopers (2002). *Money for Growth – The European Technology Investment Report 2001*, http://www.pwcmoneytree.com

Ritter, J. R. (2002). Investment banking and securities issuance. In: G. Constantinides, M. Harris, & R. Stulz (Eds), *Handbook of the Economics of Finance*, North-Holland.

Sahlman, W. (1990). The structure and governance of venture capital organizations, *Journal of Financial Economics*, *27*, 473–521.

Westhead, P., & Storey, D. J. (1997). Financial constraints on the growth of high-tech small firms in the U.K. *Applied Financial Economics*, *7*, 197–201.

New Venture Investment: Choices and Consequences
A. Ginsberg and I. Hasan (editors)

Chapter 10

Post-Issue Performance of Hot IPOs

ANNIKA SANDSTRÖM[a] and JOAKIM WESTERHOLM[b,*]

[a] *Swedish School of Economics and Business Administration, Helsinki, Finland*
[b] *School of Business H69, University of Sydney, NSW 2006, Australia*

ABSTRACT

This study is the first attempt to analyze the operating performance of Finnish initial public offerings. The objective is to detect if there is a relationship between the timing of the IPO and the post-issue operating performance of the IPO firm and to assess if this is an explanation for long-run price underperformance observed for IPO companies. We study a sample of 73 IPOs, which includes practically all new listings during the years 1984–2000. We detect a long-run underperformance in returns that is related to the activity of the IPO markets, industry clustering of IPOs and the operating performance of the IPO firms. We conclude that the operating performance of the IPO companies is strongest the year before the IPO and consistently weakens over the five years following the IPO.

1. INTRODUCTION

Initial public offerings [IPOs] with large initial returns have a distinguishing feature of clustering in time and industry. For example, in the late 1990s the high tech-, and web infrastructure companies dominated the booming market for IPOs and many of the newly listed firms experienced phenomenal price gains, implying signs of underpricing or "irrational exuberance" in the markets. This kind of speculative

*E-mail address: j.westerholm@econ.usyd.edu.au (J. Westerholm).

investing invoked an intense discussion among academics and practitioners. Many observers argued that these so-called "dot-coms" are unlike anything to come before, and should be valued differently applying a new set of metrics.

Different type of IPO-market patterns have been extensively analyzed in the last two decades. Empirical regularities from IPO research suggest that the number of successful offerings of IPO shares appear to follow cyclical patterns with the resulting waves often disproportionately populated with firms in particular industries. A number of studies have documented the over-reacting initial price behavior as well as the subsequent long-run stock price underperformance of the IPO firms.[1] While the Internet IPO boom of 1998–2000 was extraordinary in nature, a significant body of academic research suggests that nearly all IPO markets over the past 40 years have experienced a number of similar recurrent features: an impressive first-day "pop" and market under-performance over a longer-term time horizon. Several recent papers argue that investor optimism about issuers can explain the long-run underperformance.

Ritter (1991) and Loughran and Ritter (1995) suggest that investors are periodically overoptimistic about the earnings potential for young growth companies. They document that underperformance is concentrated among relatively young growth companies, especially those going public in high volume years. They interpret their long-run underperformance results as evidence that firms take advantage of transitory windows of opportunity by issuing equity when, on average, they are substantially overvalued.[2] The finding that stocks of firms issuing new equity significantly underperform the market in the long-run suggest that firms issue equity when their stocks are overpriced and that stock markets underreact to the lemon problem.

Teoh, Welch and Wong (1998) identify earnings management by issuers as a potential source of investor optimism. In particular, they suggest that investors may overpay for IPOs if they focus on exaggerated earnings that are not representative of the long-run economic earnings power of the issuer. Although IPO issuers subsequently experience relatively poor post-issue operating performance (Jain & Kini, 1994), equity analysts do not appear to fully anticipate this (Rajan & Servaes, 1997).

To date, relatively few studies have examined the operating performance of IPO firms, especially in the context of the cyclical behavior of IPO markets. The operating performance of newly issued companies should be a primary concern for investors in relation to short term price swings around and after the issue date.

---

[1] For more details see Ritter (1984, 1991, 2002). For an extensive perspective on Finnish IPO experience, see Westerholm (2003).

[2] In the equity pricing literature, a window of opportunity is defined as a period when issuers can raise capital at favourable terms. There is some disagreement, however, as to whether the window is due to time-varying asymmetric information (see e.g. Bayless & Chaplinsky, 1996 and Choe, Masulis & Nanda, 1991 or investor optimism (see e.g. Loughran & Ritter, 1995).

If the stock price changes are considered to be forward looking and correct, then the post-issue operating performance would improve. Also, if raising additional equity by means of IPO should improve the firm's performance, then the previously documented long-run underperformance is in clear conflict with this presumption.

Jain and Kini (1994) examine the post-issue performance of 682 IPOs issued in the U.S. during 1976–1988 and find significant decline in the operating performance measured using various accounting ratios. They also find that sales growth and capital expenditure of IPOs exceed that of an industry-matched sample of firms. Jain and Kini conclude that a declining sales growth and/or reduction in capital expenditures cannot explain the observed decline in operating performance and indicate that IPO firms time their offerings to coincide with peak operating performance. Helwege and Liang (1996) who introduce the question of the relation between cyclical patterns in IPO markets and firm operating performances, compare 575 U.S. firms that went public in the hot issue year 1983 with firms that went public during the cold issue year 1988. They find that over half of the volume during the hot issue period with large initial returns is concentrated to four industries. The two set of firms they use have similar operating performance, but stock returns are worse for firms that went public in the hot market, which they regard consistent with investor over optimism in hot markets, but not with the asymmetric information models.

We now raise the following question: is the post-issue operating performance of IPO firms consistent with the stock price effect observed during the time of the IPO as well as the during the subsequent five years? Especially, is it possible to find patterns in the succeeding operating years of the IPO firm, which link to specific time periods of the IPO cycles? The Helsinki Stock Exchange (HEX) studied in this paper offers two episodes of frequent and large price gains leading to an ever-increasing number of listings in what seems to be "hot issue markets" (Ritter, 1984).[3]

This analysis extends the international evidence on initial public offerings (IPOs) to new issues in Finland between 1984 and 2000. The Helsinki stock exchange [HEX] has experienced quite a few "hot" IPOs. Particularly the telecommunications and technology sectors of the market have experienced tremendous growth and Finland ranks first in a world wide comparison of growth in market capitalization.[4] We find that the initial return for the sample of IPOs is 27% in excess of the first

[3] Ibbotson and Jaffe (1975) initially studied the clustering-phenomenon in more detail and found that clustering is often preceded by periods with high initial returns, suggesting that 'hot issue' markets trigger IPO waves. Since their 'hot-issue-markets' are defined as periods in time in where IPOs have conspicuously high initial returns instead of markets with abnormal IPO-volume, we distinguish between this phenomenon by calling markets with high IPO-concentrations "*IPO- waves*" and reserve "*hot-issue markets*" for periods with high initial returns.

[4] Meridian Securities (2002).

day market return and that the studied market exhibit strong cycles both in terms of pricing and volume. We also find that the return on the IPO stocks significantly underperform the sector and the market index during the five years after the IPO. We isolate three significant factors related to this underperformance, the activity of the IPO markets, industry clustering of IPOs and the operating performance of the IPO firms.

The remainder of the chapter is organized as follows. Section 2 presents the empirical analysis including a description of the data and investigations of stock market response, industry clustering, market cycles, liquidity and operating performance. Section 3 summarizes the findings and presents the conclusions.

2. EMPIRICAL ANALYSIS

2.1. Data sources and sample description

Our sample of IPOs issued on the Helsinki Exchange are collected for the period 1984–2000 using the records of HEX Group and previous studies by Keloharju (1992). Data on offering dates, number of shares issued and daily returns are from the information services provided by Bloomberg L. P. The sources of accounting information used in this study are Bloomberg and the yearly issues of the factbook "Pörssitieto" by G. Kock (1984–1996). When data in these primary sources were unavailable, supplementing data were collected directly from the prospectuses and annual reports of the issuing firms. The main source for price data of the IPOs prior the 1990s is the database compiled at the Swedish School of Economics, as the Bloomberg database do not include material from this time period.

Sample firms used in this study meet the following selection criteria; data on operating performance is available from the year prior the IPO to five years after; data on pricing and share turnover from the listing and onwards. In addition, we accept only issues that can be classified as "pure" IPOs, e.g. any equity offerings conducted simultaneously with transfers between separate lists on the HEX are not included. A total of 73 firms from a full set of 150 equity offerings satisfy these criteria. As in all type of data limitation, it is possible that the data selection criteria may have caused and inadvertent survivorship bias. We do not expect the sample selection to affect our results. The larger of the companies not included are left out because they are not strictly speaking IPOs, while a few small issues with a minimal stock market impact are left out because there is simply no data available on these issues.

Table 1 summarizes the number of IPOs, the amount of capital raised and the average initial return per year. We classify 1988, 1989, 1999 and 2000 as high activity years (highlighted in table), based on the yearly relative IPO-activity measure. This measure is the ratio of IPOs during the year to the total number of

Table 1. IPO issuance by year

Year	Nr of total issues[a]	Selected Nr of IPOs (sample)	Relative activity	Size (EURO m, sample)	Average initial return (%, sample)
1984	5	2	3	23.5	28.3%
1985	3	–	2	–	–
1986	4	–	3	–	–
1987	7	1	5	2.1	117.9%
1988	**43**	**8**	**29**	**139.5**	**30.3%**
1989	**21**	**7**	**14**	**162.8**	**3.7%**
1990	–	–		–	–
1991	–	–		–	–
1992	–	–		–	–
1993	–	–		–	–
1994	5	3	3	347.6	−0.2%
1995	4	4	3	300.9	1.2%
1996	3	2	2	85.9	18.7%
1997	11	7	7	399.3	14.5%
1998	10	5	7	1359.6	18.5%
1999	**19**	**19**	**13**	**625.7**	**37.8%**
2000	**15**	**15**	**10**	**488.2**	**27.0%**
Total	150	73	100	3,935.2	
Average				357.7	27.1%
Median				300.9	18.7%

Note: Number of IPOs, Total Proceeds Raised and Average Initial Return.
[a]Total set includes a number of equity issues of which some are not classified as "pure IPOs." In the remainder of this study, these issues are used only in determining the industry clustering and the IPO activity measures. The EURO was close to par with the U.S. dollar at the end of the sample period.

IPOs during 1984–2000 and shows a value higher than 10% for the selected years. A total of 67% of our sample IPO firms went public during these years.[5]

2.2. Stock market response

The initial stock price reaction of each individual firm as well as the following long-run price development are presented in Appendix. We measure the initial

[5] During 1988 a strong demand in the banking and investment sector opened the window for 43 IPOs that raised a combined EURO 507.7 million in capital. The very large gross proceeds in 1998 is explained by the privatisations of the state owned companies Fortum Oyj and Sonera Oyj.

return as the relative change in price less the market and alternatively the business sector index return as outlined in Eq. (1).

$$\text{Initial Return} = \frac{\text{CLOSING PRICE}_t - \text{IPO PRICE}_{t-1}}{\text{IPO PRICE}_{t-1}} - \frac{\text{INDEX}_t - \text{INDEX}_{t-1}}{\text{INDEX}_{t-1}} \quad (1)$$

Our results show that the initial stock price return is significantly positive as reported in Table 1. The mean level of initial return is 27.1%, with a median level of 18.7%. On a yearly basis we find that there is a distinctive variation in the mean level of initial returns with the highest level at 117.9% observed in year 1987 and the lowest level of initial return of −0.2% observed in year 1994.

The selected sample shows thus the traditional signs reflecting "hot issue markets." In Table 1 there are some indications that high initial returns precede high volume years. Meanwhile, the descriptive statistics in Table 2 show that initial returns are still significantly higher for high volume years (28% compared to 17% on average). This finding can be partly explained by the extraordinary returns during 1999 and 2000. The mean initial return by industry is 13% for financial companies, 13% for industrial companies, −1% for basic materials, 16% for consumer products, 21% for telecommunications and 65% for computer technology and software sectors. The majority of IPOs from the last two groups occurred mainly just in these "high tech" years. Further, the proceeds were smaller in the high volume years, reflecting the smaller average size of the IPO firms. The finding of IPO firms being initially highly priced might reflect investors having expectations of high earnings growth in the future and they appear to value firms going public based on the expectation that earnings growth will continue.

Next, we analyze the long-run price performance of our sample by comparing the firm performance to the business sector and to the market as a whole. We calculate the abnormal return compared to the business sector index and the market

Table 2. Proceeds and size of firms that issued in high and low volume years

Year		High volume	Low volume	Test statistic
Size EURO million	Mean	28.9	105.0	2.48[a]
	Median	14.0	44.0	
Initial return (%)	Mean	28.4%	17.1%	−0.79
	Median	5.4%	4.7%	
Number of firms		49	24	

Note: Significance tests for differences in mean values are based on two-sample *t*-test.
[a] Denotes significance on 1% level.

index as follows:

$$\text{Long} - \text{Run Price Performance} = \frac{\text{PRICE}_t - \text{PRICE}_{\text{IST\_TRADINGDAY}}}{\text{PRICE}_{\text{IST\_TRADINGDAY}}} - \frac{\text{INDEX}_t - \text{INDEX}_{\text{IST\_TRADINGDAY}}}{\text{INDEXT}_{\text{IST\_TRADINGDAY}}} \quad (2)$$

The return from the closing market price on the first day of public trading is measured to the market price on the relevant anniversary date. Long-run performance is reported as abnormal return for periods from 1 to 5 years after the offering. We test if the mean of the abnormal returns is significantly different to the return of the sector and market indexes as a result of long-run underperformance documented in earlier studies such as Loughran and Ritter (1995).

Table 3 offers statistics on the returns on the sample IPOs compared to the business sector- and all share index returns. For all 4 years after the IPO the returns are lower for the IPO firms than for the benchmark indices. The reported differences are significant on 1% level. In year 5 an improvement is detectable: the difference in return compared to the business sector index is no longer significant. The IPO firms still underperform the all share index return in year 5.

The mainstream literature on long-term performance focuses on long-term underperformance. Because underperformance is an anomalous phenomenon, many authors search for explanations based on financial market imperfections. The explanations put forward can mainly be placed into three groups. The first identifies the existence of underperformance and provides behavioral and expectations-based explanations for the phenomenon. Ritter (1991), Rajan and Servaes (1997) among others argued that firms go public when investors are over-optimistic about the growth prospects of IPO companies. Investors overpay initially but mark prices down as more information becomes available and hence the expected long-run returns decrease with the decrease in initial investor sentiment. The second group provides explanation using the agency cost hypothesis. For example Jain and Kini (1994) found a significant positive relation between post-IPO performance and equity retention by the original shareholders. The third group explains under-performance as a mis-measurement. The risk mis-measurement hypothesis proposes that long-run underperformance may be due to a failure to adjust returns for time-varying systematic risk. No empirical evidence has been found for this hypothesis by Ritter (1991), Keloharju (1993) and Ljungqvist (1997). They tried to adjust for risk but still found that the newly listed firms underperform. Also the possibility, that statistical inference conducted using traditional testing methods or imperfect benchmarking cause poor long-run returns, remains in discussion.

Table 3. Long-run price performance

Year	Year 1 excess return compared to		Year 2 excess return compared to		Year 3 excess return compared to		Year 4 excess return compared to		Year 5 excess return compared to	
	Sector	Market	Sector	Market	Sector	Market	Sector	Market	Sector	Market
Aver.	0.24	0.05	−0.53	−0.55	−1.35	−0.92	−1.03	−1.14	−0.14	−0.91
Std. dev.	1.71	1.55	1.36	0.91	3.26	1.54	1.77	1.74	0.73	1.48
t-Value	-	-	−2.34[a]	−2.79[a]	−2.20[a]	−3.31[a]	−2.51[a]	−3.40[a]	−1.16	−2.72[a]
N	50	73	38	61	23	45	16	34	11	24

[a] Denotes significance on 1% level.

More recently, however, attention has shifted from underperformance to long-term performance in general, including search for other than financial market imperfections as explanations for under- or outperformance. In our study we are not directly interested in explaining long-run underperformance. Rather, we are interested in identifying measurable firm- and market characteristics at the time of the IPO that are related to long-run performance in a systematic way. The characteristics that we examine are based on theoretical considerations, based on both previous research on IPOs and more general theory of the firm.

2.3. Industry clustering

The basic argument for the overvaluation hypothesis, documented by Ritter (1991) and Loughran and Ritter (1995) is that a temporary overvaluation of a certain industry's stock drives an increased number of IPOs to the market. We start by testing whether the long-term performance of IPOs is concentrated mainly in the clustered IPOs and define a "clustering" variable as follows:

$$\text{CLUST} = \frac{\text{IPOs in same sector in 15-month period around the offering}}{\text{IPOs in the same sector (1984–2000)}} \quad (3)$$

The sector is defined with six different industry categories: (1) Financial (all financial companies, including banks, investment companies etc.); (2) Industrial (including machinery, construction, etc.); (3) Basic materials (energy oil etc.); (4) Consumer products (non-cyclical and cyclical); (5) Telecom (Telecommunications equipment, media, internet and communication); and (6) High Tech (computer technology and software).[6] The 15 month period around the offering is symmetric: it lasts from seven month before the offering-month until the seven month after the offering month. We divide by the total number of IPOs in our observation period to have a measure of clustering, not merely of intensity. The cluster measure is expected to be positively related to initial return and negatively related to post-issue return if the temporary overvaluation hypothesis by Ritter (1991) is supported.

In Table 4 we group the sample IPOs into quartiles according to their cluster measure. As expected we observe higher returns in the quartiles with a higher industry concentration. A graphical analysis of the long-run cumulative excess return for these quartiles over the five years after the IPO reveals that the long-run

[6] For the purposes of this study, we use a narrowed grouping of the industries reported by HEX which we define "sector." HEX identifies 16 branches including many similarly featured groups which against our narrow data set would end up to a too broad classification.

Table 4. Industry clustering in IPOs

Full set quartile	Cluster range	Nr of firms in cluster (full set)	Average cluster value	
Q1	0–13.2	46	7.2	
Q2	13.2–26.5	44	19.3	
Q3	26.5–39.7	24	32.3	
Q4	39.7–52.9	36	46.7	
Sample				
Quartile	Cluster range	Nr of firms in cluster (sample)	Average cluster value	Initial return (%)
Q1	0–12.8	22	6.5	10
Q2	12.8–25.7	21	18.7	14
Q3	25.7–38.5	13	29.5	7
Q4	38.5–51.4	17	46.0	70

Note: Comparison of industry clustering in sample population and the full set.
Q1 = low industry clustering, Q4 = high industry clustering.

performance is better for the IPOs from industries with low clustering when comparing the quartiles to the market index. In section 2.9 the industry cluster variable is included in regressions with both the initial return and the post-issue return as dependent variables (Figure 1).

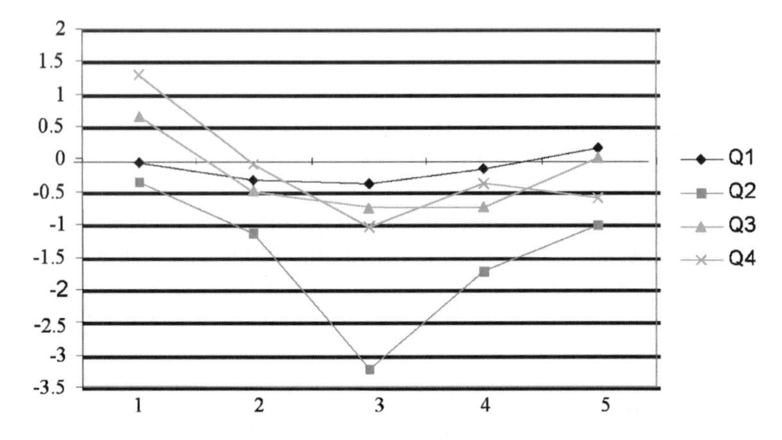

Figure 1. Long-run performance of cluster portfolios. The long-run cumulative excess return for quartiles of industry clustered IPOs compared to the sector index.

2.4. Liquidity

Recent studies e.g. Eckbo and Norli (2000) and Westerholm (2001) find a negative relationship between the price performance of IPO firms and their liquidity measured by transaction costs. This could be explained as follows. Higher liquidity justifies a higher price for liquid IPO stock in comparison to less liquid IPOs in the initial offering while the subsequent high liquidity in the aftermarket leads to a lower expected return on liquid securities compared to less liquid securities. The studies above also found that IPO stocks are more liquid, have lower transaction costs, than matching companies and thus the long-run price underperformance can be explained as by a lower liquidity premium on IPO stock in comparison to the market on average. We include transaction costs measures and turnover rate measures such as the bid-ask spread [BAS], turnover rate [τ] and amortized spread [τ?BAS], to gauge the relationship between liquidity and aftermarket price and operating performance for IPO firms. The turnover rate is aggregated trade by trade and the bid-ask spread is measured at the closing of the market each day after the IPO. Daily measures are aggregated to yearly averages. We estimate the following relationships:

$$\text{Initial Return} = \text{BAS}_t + \tau_t + \text{IPO size}_t \tag{4}$$

$$\text{Cumulative Excess Return}_t = \text{BAS}_t + \tau_t + \text{IPO size}_t \tag{5}$$

Table 5 describes the development in the liquidity measures over the investigated years. The closing spread level for IPO stock is remarkably high 5.5% on average

Table 5. Liquidity

Year	Average size of IPOs ? million	Average relative bid-ask spread	Average yearly turnover rate	Average yearly amortized spread
1992		0.0161	9.5037	0.4836
1993		0.0552	3.6706	0.2024
1994	115.9	0.0623	2.7495	0.1714
1995	75.2	0.0725	2.2612	0.1638
1996	43.0	0.0890	3.4135	0.3038
1997	57.0	0.0342	3.9638	0.1356
1998	271.9	0.0585	7.0263	0.4108
1999	31.7	0.0346	15.2084	0.5260
2000	32.5	0.0721	20.5595	1.4823
Average	62.0	0.0549	7.5952	0.4311
Median	32.5	0.0585	3.9638	0.3038

Note: The liquidity measures are equally weighted yearly averages for all IPOs in the sample.

with a lower level of 3.5% during one of the more active year; 1999. The turnover rate (number of shares traded to number of shares outstanding per year) is remarkably high for IPO stock considering that the yearly turnover rate for the Finnish market in total has been between 30 and 70% during the investigated years. This might imply that the stocks are very liquid but in fact with the high transaction costs it appears that a very large fraction of the market value, 43% on average when converted to a yearly measure, has been paid in transaction costs. A large part of this unusually high activity around the exchange introduction can be contributed to unwinding of short term positions taken by the underwriter and short term speculators. In any case it seems inevitable that the prices have had to fall to compensate for these high transaction costs.

In Table 6 the relationship between initial return and bid ask spread and turnover rate as well as the relationship between cumulative abnormal return and the bid ask spread and turnover rate have been estimated. A higher bid ask spread is related to lower initial return and lower cumulative abnormal return. Thus decreasing liquidity of the sample IPOs over time appears to result in lower returns since prices adjust to lower price levels to compensate for the illiquidity. The turnover rate does not have a significant relationship to return while the IPO size is negatively related to initial return and positively related to cumulative abnormal return. This is expected since smaller IPOs tend to have larger initial returns due to scarce supply of shares while in the longer term larger IPOs more rarely experience drastic downward adjustments in price post-IPO due to lower risk and higher liquidity. A liquidity premium effect in IPOs is thus supported; for the investigated sample prices adjust to an on average decreasing liquidity. See Amihud (2002) for a discussion on the effects of unexpected changes in liquidity on contemporaneous stock returns.

Table 6. liquidity vs. initial return and cumulative excess return

	Dependent variable 1: initial return	Dependent variable 2: cum. excess return$_t$
Intercept (*t*-value)	0.379[a] (17.53)	0.248 (7.24)
Bid-Ask Spread [BAS] (*t*-value)	−0.720[a] (−4.23)	−0.844 (−3.36)
Turnover rate [τ] (*t*-value)	0.002 (1.28)	0.003 (1.11)
IPO size [LGSIZE] (*t*-value)	−0.045[a] (−10.79)	0.048 (3.71)
n	5054	5054
Adjusted R^2	0.0365	0.0056

[a] Denotes significance on 1% level.

2.5. Operating performance changes

We next turn to examining the long-run operating performance of the IPOs. Among the empirical studies focusing on operating performance changes, two distinguishing hypotheses have been proposed to explain the observed stock and operating performance decline of firms. The hypothesis in Jensen and Meckling (1976), which posits that the interests of managers and other stockholders become less closely aligned as manager's stakes decrease and ownership becomes more disperse, is probably the most studied one. Also the window of opportunity hypothesis, which argues that corporate managers decide on equity offerings depending on favorable economic conditions is also a widely studied argument. Myers and Majluf (1984) argue that if managers are better informed than outside investors, firms are more likely to issue equity when the equity is overvalued. Thus, the announcement of an equity offering conveys negative information about firm value. The long-run underperformance is therefore interpreted as evidence that firms take advantage of transitory widows of opportunity by issuing equity when, on average, they are substantially overvalued.

Despite general acceptance of the efficiency paradigm in its various strengths, investment bankers and issuers alike persist in spending great efforts and money in finding the right "window of opportunity" in the market, where temporary "mispricings" would optimize the proceeds of their planned financial operations. In a more general fashion, the asymmetric information hypothesis argues that corporate managers posses superior information about their firms than outside investors. According to models in this category, hot issue markets occur when better quality firms are pulled into the equity market due to lower information asymmetry costs. The explanation is however in conflict with often observed poor long-run operating performance for IPOs issued during active markets, Helwege and Liang (1996).

To reconcile these different positions, we investigate the post-issue performance of the sample IPOs and relate the findings to the earlier presented variables; initial pricing, long-term stock price performance and industry clustering. The objective is to detect if there is a relationship between the timing of the IPO and the post-issue operating performance of the IPO firm and to determine whether the stock market valuation impact is consistent with subsequent operating performance of IPO firms.

Measures of operating performance

Of primary interest in this study are changes in *Return on Asset*. We use three different definitions of this measure and additionally we analyze the *Profitability* and *Operating Cash Flows to Total Assets*.

For each of these five measures, the value for the year preceding the IPO (−1) and the year of the IPO (year 0) are compared to the measures from *the years 1, 3 and 5* following the IPO.

(1) *Net Return on Total Assets* is calculated by dividing net income (losses) minus total cash preferred dividends by average assets. The net return on total assets is an accepted measure of the efficiency of the utilization of corporate assets to generate net income for common stockholders. (2) *Operating Return on Assets*, is calculated as earnings before interests and taxes (EBIT) divided by average assets. This is a measure of the utilization of corporate assets to generate operating revenue. The (3) *Total Assets Turnover Ratio*, a common gross performance measure, is calculated by dividing net sales by total assets. Total assets turnover ratio is an expression of the efficiency of utilization of corporate resources for generation of sales revenue. (4) *Operating Return on Sales* (the EBIT-margin) is earnings before interest and taxes divided by sales. Finally, the (5) *Operating cash flow to Total Assets* equals earnings before interest and taxes, depreciation and amortization expenses (EBITDA) minus capital expenditures, divided by total assets. Operating cash flows deflated by assets is a useful measure, as it is a primary component in net-present-value (NPV) calculations used to value a firm. As argued in Barber and Lyon (1997) it is also a "cleaner" measure of operating performance for two reasons; (1) earnings include interest expense, special items, and income taxes, which can obscure operating performance; and (2) operating cash flows represent the economic benefits generated by a firm, and as a pre-tax measure, they are unaffected by the changes in tax status or capital structure that accompany equity offerings. In addition, the operating cash flow to total assets should be the most reliable measure since it is the least dependent on accounting practices.

We use operating performance measures to directly test for changes in performance of the IPOs and also to ascertain the sources of poor long-run stock market performance documented. The deflation of financial statement items by Total Assets has been established by Lewellen and Huntsman (1970) as a suitable technique for controlling some of the statistical aberrations commonly present in corporate performance measures.

Evidence on operating performance changes

This section reports the results for operating performance for the IPO firms in the sample during the seven year period of our analysis. Table 7 exhibits descriptive statistics on the selected measures, while Tables 8 and 9 report the median changes in each respective measure and the results of the tests of significance. The differences between the mean and median values indicate the skewness of accounting

Table 7. Descriptive statistics on operating performance measures

Return on Assets	−1	0	1	3	5
Average	9.3	6.5	0.2	3.6	2.5
Median	7.5	6.5	4.2	4.5	1.2
MIN	−88.2	−61.8	−146.9	−20.8	−10.3
MAX	47.2	45.7	37.8	17.3	17.0
St.dev	19.1	16.8	24.9	6.3	5.4
Nr of observations	65	71	71	36	21
Operating return on assets	**−1**	**0**	**1**	**3**	**5**
Average	17.5	12.2	5.7	7.0	6.9
Median	12.8	11.1	6.4	9.3	5.8
MIN	−96.7	−60.5	−103.6	−20.4	−1.8
MAX	103.7	96.0	100.7	24.9	26.5
St.dev	27.0	21.1	25.0	8.3	6.2
Nr of observations	64	69	69	33	18
Asset turnover	**−1**	**0**	**1**	**3**	**5**
Average	1.49	1.23	1.13	0.99	0.81
Median	1.20	1.07	1.07	0.96	0.91
MIN	0.03	0.00	0.07	0.08	0.06
MAX	4.74	3.98	3.12	3.57	1.81
St.dev	1.03	0.75	0.65	0.74	0.52
Nr of observations	65	71	70	36	21
EBIT-margin	**−1**	**0**	**1**	**3**	**5**
Average	11.4	8.4	3.6	10.1	7.9
Median	10.4	9.3	6.4	7.1	5.9
MIN	−37.8	−105.3	−110.0	−21.0	−22.7
MAX	46.0	48.2	58.4	60.7	31.2
St.dev	13.5	19.4	21.6	14.7	11.7
Nr of observations	63	68	68	33	18
Cash flow to assets	**−1**	**0**	**1**	**3**	**5**
Average	8.5	8.2	−0.8	3.8	4.3
Median	10.6	7.3	0.8	3.6	3.3
MIN	−68.2	−57.3	−89.3	−10.0	−28.1
MAX	45.6	222.4	50.6	24.6	23.7
St.dev	18.0	31.4	21.4	7.6	10.5
Nr of observations	66	69	59	32	20

Note: The table reports yearly measures in Return on Assets (ROA), Operating Return on Assets, Asset Turnover, EBIT-margin and Cash flow to Assets.

ROA = (net income-total cash, preferred, dividends)/Avg.Assets × 100. Operating-ROA= EBIT/Avg.Assets × 100. Asset turnover = Net Sales/Avg.Assets. EBIT-margin = Earnings before interest and taxes/Net Sales × 100. Cash flow to Assets = (EBITDA (Earnings before interest, taxes, depreciation and amortization)−Capital Expenditures)/Avg.Assets × 100.

Table 8. firm characteristics in the years following the year prior to the IPO

Return on Assets	−1–0	−1–1	−1–3	−1–5
Median change, all firms (%)	−0.09[b]	−0.31[b]	−0.58[b]	−0.32[a]
Nr of observations	65	65	31	16
Median change, hot firms (%)	−0.19[b]	−0.42[b]	−0.83[b]	−0.83[a]
Nr of observations	44	44	11	11
Median change, cold firms (%)	0.10	−0.15[a]	−0.18[a]	0.03
Nr of observations	21	21	20	5
Operating return on assets	**−1–0**	**−1–1**	**−1–3**	**−1–5**
Median change, all firms (%)	−0.10[b]	−0.32[b]	−0.16[b]	−0.34[b]
Nr of observations	62	63	28	13
Median change, hot sample (%)	−0.22[b]	−0.55[b]	−0.73[a]	−0.44[a]
Nr of observations	43	43	8	8
Median change, cold sample (%)	0.09	−0.07[a]	−0.09[b]	−0.25i
Nr of observations	20	21	20	5
Asset turnover	**−1–0**	**−1–1**	**−1–3**	**−1–5**
Median change, all firms (%)	−0.07[b]	−0.11[b]	0.00[b]	0.19
Nr of observations	63	63	30	15
Median change, hot sample (%)	−0.18[b]	−0.17[b]	0.01	0.24[a]
Nr of observations	44	44	11	11
Median change, cold sample (%)	0.03	−0.08[a]	0.00[a]	−0.04
Nr of observations	21	21	20	5
EBIT-margin	**−1–0**	**−1–1**	**−1–3**	**−1–5**
Median change, all firms (%)	−0.04[b]	−0.37[b]	−0.15[a]	−0.45[b]
Nr of observations	62	63	28	13
Median change, hot sample (%)	−0.13[b]	−0.63[b]	−0.67i	−0.53[a]
Nr of observations	42	42	8	8
Median change, hot sample (%)	0.03	−0.09[a]	−0.01i	−0.05
Nr of observations	21	21	20	5
Cash Flow to Assets	**−1–0**	**−1–1**	**−1–3**	**−1–5**
Median change, all firms (%)	−0.15[b]	−0.67[b]	−0.72[b]	−0.60[b]
Nr of observations	55	57	28	13
Median change, hot sample (%)	−0.25[b]	−1.00[b]	−0.72[b]	−0.65[b]
Nr of observations	34	36	8	8
Median change, cold sample (%)	0.11	−0.27[b]	−0.72[b]	−0.55i
Nr of observations	21	21	20	5

Note: The table reports yearly changes in Return on Assets, Operating Return on Assets, Asset Turnover, EBIT-Margin and Cash flow to Assets. Changes are expressed as the median values of the differences in respective operating measure compared to the same measure *the year before the IPO*. The significance is determined according to Wilcoxon signed rank test of the reported differences.

[a] Denotes significance on 5% level and i on 10% level.

[b] Denotes significance on 1% level.

Table 9. Firm characteristics in the years following the year of the IPO

Return on assets	0–1	0–3	0–5
Median change, all firms (%)	−0.23[b]	−0.51[b]	−0.44[a]
Nr of observations	70	36	21[a]
Median change, hot firms (%)	−0.28[b]	−0.89[b]	−0.65
Nr of observations	46	13	13[a]
Median change, cold firms (%)	−0.10[b]	−0.09[a]	−0.30[a]
Nr of observations	24	23	8[a]
Operating return on assets	**0–1**	**0–3**	**0–5**
Median change, all firms (%)	−0.18[b]	−0.22[b]	−0.35[a]
Nr of observations	68	33	18[a]
Median change, hot firms (%)	−0.22[b]	−0.71i	−0.41
Nr of observations	44	10	10i
Median change, cold firms (%)	−0.16[b]	−0.21[b]	−0.29
Nr of observations	24	23	8[a]
Asset turnover	**0–1**	**0–3**	**0–5**
Median change, all firms (%)	−0.09[b]	−0.11[b]	−0.09[a]
Nr of observations	71	36	21[a]
Median change, hot firms (%)	−0.11[b]	−0.13[b]	−0.10
Nr of observations	46	13	13[a]
Median change, cold firms (%)	−0.06[b]	−0.08[b]	−0.09
Nr of observations	24	23	8[a]
EBIT-margin	**0–1**	**0–3**	**0–5**
Median change, all firms (%)	−0.15[b]	−0.23[b]	−0.14[a]
Nr of observations	67	33	18
Median change, hot firms (%)	−0.24[b]	−0.09	−0.07
Nr of observations	43	10	10
Median change, cold firms (%)	−0.12[b]	−0.23[b]	−0.14[a]
Nr of observations	24	23	8
Cash flow to assets	**0–1**	**0–3**	**0–5**
Median change, all firms (%)	−0.46[b]	−0.64[b]	−0.55[a]
Nr of observations	65	33	18[a]
Median change, hot firms (%)	−0.83[b]	−0.60[b]	−0.64
Nr of observations	41	10	10[a]
Median change, cold firms (%)	−0.26[b]	−0.69[b]	−0.29[a]
Nr of observations	24	23	8i

Note: The table reports yearly changes in Return on Assets, Operative Return on Assets, Asset Turnover, EBIT-Margin and Cash flow to Assets. Changes are expressed as the median value of the differences in respective operating measure compared to the same measure *the year of the IPO*. The significance is determined according to Wilcoxon signed rank test of the reported differences.

[a]Denotes significance on 5% level and i on 10% level.

[b]Denotes significance on 1% level.

data why we use non-parametric tests and analyze medians rather than means. We use the Wilcoxon signed-rank test to test for significant changes in each of the variables for all time windows. This procedure tests whether the median difference in variable values between the pre and post-IPO samples is zero. Conclusions are then made based on the standardized test statistic Z, which for samples of at least 10 observations approximately follows a standard normal distribution.

Table 7 reports the raw ROA, OpROA, Asset Turnover, EBIT-Margin and Cash Flow-ratios. The median value of ROA in year -1 is 7.5 which decreases in a rather steady pace to reach 1.2 in year 5. Similar patterns are found also in the other operating performance variables. In year 3 the ROA, OpROA and EBIT-margin show some improvement from this negative pattern. The median operating cash flow to Total Assets decreases most during the first year after the IPO and remains at a lower level than before the IPO during year 3 and 5.

Table 8 reports changes with performance in year -1 as the benchmark. Compared to the year of the IPO as well as the first, third and fifth year after, all issuing firms exhibit significant declines in all selected performance variables. Only the Asset Turnover shows some improvement in year 5. For example, the median change in the ratios of ROA of -9%, -31%, -58% for the years 0,1 and 3 represent significant declines at 1% level. The change of -32% in year 5 is significant on 5% level. The decreasing median operating cash flow compared to the year before the IPO is significant on 1% level for all years. The largest annual change in the other ratios seem to occur from year -1 to year $+1$ which may indicate towards managers attempting to window-dress by overstating pre-IPO performance. Comparing the hot firms to cold firms[7] reveals that the cold firms seem not to exhibit as strong decline and do not give as significant results as the hot firms. For instance, the ROA for the cold sample turns positive in year 5 and the operating cash flow for the cold sample decreases significantly only years 1 and 3, while the measures decrease significantly all years for the hot sample.

Table 9 report the results of the test for the changes from year 0 to year 1, 3 and 5. Similarly as in Table 8 the results of the independent analyses of the selected financial performance measures are all found to have a significant reduction in the years following the IPO. Results of the analysis of hot and cold segments are somewhat inconclusive as the levels of significance vary from segment to segment. However again, the cold firms do not seem to exhibit as strong decline as the hot firms do. There is consistency in median changes in operating cash flow to total assets when the changes are measured against the year before the IPO. The sample

[7] For brevity we call the high activity IPO periods defined in section 2.1 for hot firms, and the low activity IPO firms for "cold firms."

shows significant decreases on 1% level during all years for all IPOs, only the cold sample does not decrease significantly in year 5.

It appears that the years of recession in Finland 1990–1994 do have some impact in the highly negative results, we obtain. To control for the variation in operating performance of firms during these years, we adjusted the sample by excluding firms issued during the 1980s. According to our expectations, the new sample including 55 firms listed during the 1990s exhibit somewhat softer declines during all time frames both compared to Year –1 and Year 0. However, the changes are still negative and highly significant. Up to year 5 there is some modest signs of turnaround in some measures but this examination is not motivated due to the fewer number of observations available in the later years.

The findings of the operating performance changes are similar to earlier findings of i.e. Jain and Kini (1994) and could be explained by the window of opportunity hypothesis suggested by Ritter (1991) and Loughran and Ritter (1995). One interpretation is given by Jensen's (1986) free cash flow theory, that predict a general decline in the profitability of firms which conduct equity offerings. He argues that there are important divergences of interest between managers and shareholders that might induce managers to issue equity, retain excess cash flow in the firm and use the cash flow for value reducing activities, such as investment in negative net-present-value (NPV) projects. This problem is especially acute for firms with few positive-NPV investment opportunities, which might involve the "hot firms" with more significant declines. As the market heats up, some of these firms may have gone public for opportunistic reasons, to extract surplus from sentiment investors. However, it should be noted that these raw measures of operating performance can be exposed to industry-wide effects beyond the manager's control.

2.6. Cross-sectional determinants of stock returns

So far, our analysis has focused separately on the stock market performance and changes in operating performance of IPO firms relative to their pre-IPO levels. We now turn to the question of relationship between the initially presented explanations for long-run underperformance and initial returns (market cyclicality, clustering and liquidity), as we simultaneously analyze the operating performance measures against the stock price behavior.

A descriptive table and correlation analysis of the variables is presented in Table 10. The results of multivariate regression with initial return as dependent variable are reported in Table 11. A set of variables may act as potential proxies for ex-ante uncertainty about market performance; operating performance variables (ROA, Operating ROA, Asset Turnover, Operating Return on Sales); volume (IPO

Table 10. Descriptive statistics and correlation analysis of variables

Variable	N	Mean	Std. dev	Variance	MIN	MAX
Return	5054	0.038	0.817	0.668	−1.000	32.248
CAR	5054	0.292	1.060	1.124	−1.024	14.927
ROA	5054	2.813	13.260	175.840	−146.910	45.668
Op. ROA	5054	0.066	0.148	0.022	−1.036	1.007
Asset turnover	5054	0.919	0.740	0.548	0.000	4.648
EBIT margin	5054	6.152	14.681	215.540	−110.000	60.656
Am spread	5054	0.037	0.294	0.086	0.000	7.508
Industry clustering	5054	25.263	14.773	218.240	2.703	51.351
Ln (size)	5054	2.570	1.490	2.220	−0.078	6.960

Variable	Return	CumRe	ROA	Op ROA	Ass. TO	EBIT m	Am Spr	Cluster
Return	1							
Cum. Return	0.024	1						
ROA	0.002	0.061	1					
Op. ROA	0.009	0.030	0.858	1				
Asset Turnover	−0.010	0.095	0.155	0.246	1			
EBIT margin	−0.037	−0.027	0.520	0.521	0.033			
Am Spread	−0.012	−0.027	0.006	0.015	0.086	−0.029	1	
Ind. Clustering	−0.022	0.037	−0.154	−0.146	−0.266	−0.003	0.002	1
Ln (Size)	0.027	0.066	0.047	0.028	0.085	0.059	0.002	0.207

Note: The table provides descriptive statistics and correlation analysis of the monthly observations for the investigated variables. The first panel reports, Number of observations, Mean, Standard Deviation, Variance, Minimum and Maximum observations and the second panel reports correlation coefficients for the variables.

Table 11. Estimation of the relationship between high initial returns and ex-ante uncertainty about market performance

Dependent variable: initial return	Seasoning and measures HOT dummy	Operating performance Relative IPO activity
Intercept (t-value)	0.269** (10.73)	0.273** (12.09)
ROA_t (t-value)	0.001 (0.82)	0.002** (2.19)
Op. ROA_t (t-value)	0.005 (0.07)	−0.088 (−1.12)
Asset $turnover_t$ (t-value)	−0.008 (−1.06)	−0.008 (−1.32)
Oper. Ret. on sales t (t-value)	−0.003** (−5.37)	−0.002** (−4.08)
Size of IPO (t-value)	−0.047** (−10.13)	−0.053** (−10.93)
Hot/Cold Dummy (t-value)	−0.211** (−11.8)	–
Relative IPO activity (t-value)	–	−1.415** (−8.58)
Industry Clustering (t-value)	0.008** (10.55)	0.011** (8.56)
n	5054	5054
Adjusted R^2	0.0995	0.1079

Note: The following two equations are estimated for the time series over the whole cross-section of IPO stocks the year before the IPO, the year of the IPO and the year after the IPO. Initial Return = ROA_t + Op.ROA_t + Asset Turnovert + $Size_t$ + IPOActivity $Dummy_t$ + Industry $Clustering_t$. (12) Initial Return = ROA_t + Op.ROA_t + Asset $Turnover_t$ + $Size_t$ + Relative IPO $Activity_t$ + Industry $Clustering_t$. (13) The equation is estimated using pooled cross sectional and time series data with a autocorrelation and heteroskedasiticity corrected variance-covariance matrix.

size); seasoning (hot/cold) and as an alternative the activity of the IPO market; and Industry Clustering. A positive relationship between initial return and seasoning as well as between initial return and firm operating performance is expected. The basic assertion about issue size is that larger IPOs should experience less underpricing. However, the effect of possible market movement as a result of rising demand between the date of fixing of the offer price and the first trading day, might influence of percentage offered to public, and thus the initial return. This would increase the initial return in smaller IPOs with high demand relative to larger IPOs. The estimations are made with Operating performance measures for the year prior to, the year of and the year after the IPO. The estimations include only one of the operating performance variables, ROA, Operating ROA, Asset Turnover, Operating return on sales in each estimation due to expected high correlation between the measures.

$$\begin{aligned}\text{Initial Return} = {} & \text{ROA}_t + \text{Op.ROA}_t + \text{Asset Turnover}_t \\ & + \text{Operating Return on Sales}_t + \text{Size}_t + \text{IPO Activity Dummy}_t \\ & + \text{Industry Clustering}_t. \qquad (6)\end{aligned}$$

$$\text{Initial Return} = \text{ROA}_t + \text{Op.ROA}_t + \text{Asset Turnover}_t + \text{Operating Return on Sales}_t + \text{Size}_t + \text{Relative IPO Activity}_t + \text{Industry Clustering}_t. \quad (7)$$

In the pooled time series and cross-sectional estimations reported in Table 11, the size variable is significant and takes the expected negative sign; smaller IPOs have higher initial returns. The seasoning variables are significant and take surprising negative signs. When size difference between hot and cold markets is picked up the by the size variable there appears to in fact be a negative relationship between seasoning and initial return. Industry clustering is positively related to initial returns as expected. The general key measure of profitability and performance; operating margin emerges as having a strong negative relationship with initial return indicating that highly priced IPO-firms seem to end up in a negative transformation of the operating environment. Return on Assets has a weaker positive relationship with initial return indicating that overly optimistic run-ups at the IPO are on average followed only by modest increases in net revenue generated by the better capitalized company after the IPO.

In the following we take a new approach in that we regard the IPO return as a combination of the initial return and the post issue return. This way Eq. (8) describes the total return that investors would be expected to be interested in and measures of operating performance, liquidity, IPO market activity and industry clustering that investors would observe in the market. The dependent variable, cumulative excess return, is the cumulative monthly return on the IPO stock in excess of the market index, calculated as the cumulative monthly percentage change from the IPO price less market return. We use independent variables lagged by one month to more closely replicate a realistic market situation where ex-post available data is used in the estimations. Since Eq. (8) is estimated in levels we have to be aware of a possible non-stationarity in the time series (Table 12).

$$\text{Cumulative excess return}_{i,t} = \text{ROA}_{i,t-1} + \text{Amortized. Spread}_{i,t-1} + \text{Relative IPO Activity}_{i,t-1} + \text{Industry Clustering}_{i,t-1} + \ln(\text{Size of IPO}_{i,t-1}) \quad (8)$$

In relation to cumulative excess returns and in order of explanatory power the relative IPO activity variable is significantly negative, the industry clustering variable is significantly positive and the return on assets variable is significantly positive while the size and the liquidity variables are insignificant. Thus the total return on an IPO is higher during less active IPO markets in general, when the

Table 12. Estimation of the impact of industry clustering, market cycles, liquidity operating performance and issue size on the intermediate to long-term price performance of IPOs

Dependent variable: Cumulative excess return$_t$	
Intercept (t-value)	0.209** (4.53)
ROA$_{t-1}$ (t-value)	0.007** (5.39)
Amortized Spread$_{t-1}$ (t-value)	0.018 (0.73)
Relative IPO Activity$_{t-1}$ (t-value)	−3.612** (−8.48)
Industry Clustering$_{t-1}$ (t-value)	0.021** (6.37)
Ln (Size of IPO$_{t-1}$) (t-value)	0.010 (0.73)
n	5054
Adjusted R^2	0.0549

Note: The table reports the estimated coefficients for the equation: Cumulative Excess Return$_t$ = ROA$_{t-1}$ + Am. Sprd$_{t-1}$ + Relative IPO Activity$_{t-1}$ + Industry Clustering$_{t-1}$ + ln (IPO Size$_{t-1}$) The Cumulative excess return is the monthly cumulative return on the IPO stock since the initial public offering in excess of the market index. The equation is estimated using cross sectional and time series data with a autocorrelation and heteroskedasiticity corrected variance-covariance matrix.

IPO belongs to a set of highly industry clustered IPOs and when return on assets is higher.

3. CONCLUSIONS

This study documents the overall stock market and operating performance of 73 Finnish initial public offerings during the period 1984–2000. Consistent with the majority of prior research, we find initial returns for the sample of IPOs that have averaged 27% in excess of the first day market return as well as long-term returns that significantly underperform the sector and the market index during the five years after the IPO.

Analyzing the changes in operating performance, we find a sharp statistically significant decrease in several different accounting variables for up to five years following the IPO. Diverse *Return on Assets* -variables, the *Profitability* and the *Operating Cash Flows to Total Assets* show declines, both in comparison of the year before and the year of the IPO. For most variables, the levels seem to be strongest the year before the IPO.

This suggests that any projected growth opportunities implicit in the initial valuation fail to materialize subsequently. These results are not consistent with asymmetric information models of IPO pricing and rather support behavioral theories based on investor overconfidence. However, when the firms were disaggregated into high-activity and low-activity panels, according to the IPO activity in the market in the year of listing, the data suggested that the performance is different between the two categories.

We consider a number of possible explanations for the poor subsequent performance. We conclude that industry clustering, many IPOs issued in the same industry segment during a period, is related to positive IPO returns. Liquidity of IPOs is significantly related to the performance of the IPO stock. It appears that traders in IPO stock are prepared to accept very high transaction costs since the turnover rate is much higher than for the market on average despite the higher than average transaction costs; This should affect the price levels for IPO stock adversely. Another factor that should affect the price levels of IPO stock adversely is the poor operating performance of the IPO companies.

Finally, the study examined in a simultaneous test the impact of the suggested factors and found that the size and liquidity effects are less important than three significant factors related to IPO underperformance, the activity of the IPO markets, industry clustering of IPOs and the operating performance of the IPO firms at the time of listing. The factors are consistently related to firm post-issue performance both measured as return on stock and as operating performance.

In light of the technology-heavy character of the recent IPO period, it appears that even if investors are in general better off investing in IPOs during less active IPO markets, the firm belonging to a group of many other new issues in the same industry as well as having a better than average history of operating performance predicts a good future for the company.

REFERENCES

Amihud, Y. (2002). Illiquidity and stock returns: Cross-section and time-series effects, *Journal of Financial Markets*, *5*, 31–56.

Barber, B. M., & Lyon, J. D. (1997). Detecting abnormal operating performance: The empirical power and specification of test statistics, *Journal of Financial Economics*, *43*, 341–372.

Bayless, S., & Chaplinsky, M. (1996). Investment banking, reputation, and the underpricing of initial public offerings, *Journal of Finance*, *37*, 955–957.

Choe, H., Masulis, R., & Nanda, V. (1991). Common stock offerings across the business cycle: Theory and evidence, *Journal of Financial Intermediation*, *7*, 69–90.

Eckbo, E., & Norli, O. (2000). Leverage, liquidity and long-run IPO returns, Unpublished Working Paper, Amos Tuck School of Business, Dartmouth, USA.

Helwege, J., & Liang, N. (1996). Initial public offerings in hot and cold markets, Working Paper at the Federal Reserve Bank of New York, USA.

Ibbotson, R. G., & Jaffe, J. F. (1975). Hot issue markets, *Journal of Finance*, *30*, 1027–1042.

Jain, B., & Kini, O. (1994). The post-issue operating performance of IPO firms, *Journal of Finance*, *49*, 1699–1726.

Jensen, M. C., & Meckling, W. M. (1976). The theory of the firm: Managerial behavior, agency costs and ownership structure, *Journal of Financial Economics*, *3*, 305–360.

Jensen, M. C. (1986). Agency costs of free cash flow, corporate finance and takeovers, *American Economic Review*, *76*, 323–329.

Keloharju, M. (1992). Essays on initial public offerings, The Helsinki School of Economics and Business Administration.

Keloharju, M. (1993). Winner's curse, legal liability and the long-run price performance of initial public offerings in Finland, *Journal of Financial Economics*, *34*, 251–277.

Lewellen, W. G., & Huntsman, B. (1970). Managerial pay and corporate performance, *American Economic Review*, *60*, 70–720.

Ljungqvist, A. (1997). Pricing initial public offerings: Further evidence from Germany, *European Economic Review*, *41*, 1309–1320.

Loughran, T., & Ritter, J. R. (1995). The new issues puzzle, *Journal of Finance*, *50*, 23–51.

Myers, S. C., & Majluf, N. S. (1984). Corporate financing and investment decisions when firms have information that investors do not have, *Journal of Financial Economics*, *13*(2), 187–221.

Rajan, R., & Servaes, H. (1997). Analysts following on initial public offerings, *Journal of Finance*, *52*, 507–529.

Ritter, J. R. (1984). The 'hot issue' market of 1980, *Journal of Business*, *57*, 215–240.

Ritter, J. R. (1991). The long-run performance of initial public offerings, *Journal of Finance*, *46*, 3–28.

Teoh, S.-H., Welch, I., & Wong, T. J. (1998). Earnings management and the long-run market performance of initial public offerings, *Journal of Finance*, *53–56*, 1935–1974.

Westerholm, J. (2001). The importance of liquidity in initial public offerings: Findings from the Finnish stock market, *The Journal of the Economic Society of Finland*, *54*, 135–145.

Westerholm, J. (2003). Do exchange listings bring the desired improvement in liquidity? Observations from the Nordic markets for venture capital and IPOs, *Research in Banking and Finance*, *3*, 331–347.

APPENDIX

Investigated IPOs

Company	IPO date dd/mm/yyyy	Size M euro	Initial return Exc market	Return 1 year Exc market	Return 3 year Exc market	Return 5 year Exc market
Tietoenator Oyj	01/06/1984	5.0	0.4323	0.0112	−0.5747	−1.0135
Skop-Rahoitus Oyj	17/12/1984	18.5	0.0258	−0.0326	−0.7126	−0.9056
Olvi Oyj	29/06/1987	2.1	0.7754	0.1906	−0.1458	0.4570
Chips Abp	10/02/1988	9.8	−0.0279	−0.1669	−0.2562	0.3578
Leo Longlife Plc Oyj	17/06/1988	2.7	0.2432	−0.1569	−0.2393	0.0734
OP Sijoitus Oyj	05/07/1988	5.0	0.1293	−0.1552	−0.1126	−0.4849
Interavanti Oyj	11/07/1988	21.0	0.2314	−0.1866	−0.1337	−0.4650
Hackinan Oyj	19/07/1988	14.1	0.4414	−0.2042	−0.1951	−0.2889
Viatek Oyj	16/08/1988	1.8	0.7431	−0.3673	−0.2240	−0.6503
Vaisala Oyj	20/09/1988	1.7	−0.0694	−0.3147	−0.1466	2.6943
SKOP Bank Oyj	22/11/1988	83.3	0.1743	−0.1520	−0.3053	−0.7444
Interbank Oyj	12/04/1989	0.9	0.1767	−0.1961	0.0142	0.0377
Hartwall Oyj	07/06/1989	10.8	0.0334	−0.1046	0.4169	0.0954
OKO Bank Oyj	26/06/1989	45.4	0.2732	0.0485	0.0211	−0.4410
Nobiscum Oyj	14/08/1989	2.3	0.0497	−0.3299	−0.3476	−0.9665
Stromsdahl Oyj	14/08/1989	3.4	−0.1045	−0.3775	−0.2397	−0.5553
Lemminkäinen Oyj	15/08/1989	24.2	−0.2447	−0.0594	−0.1355	−0.6553
Rautaruukki Oyj	08/09/1989	75.9	0.0066	−0.0300	0.1421	0.1734
Raute Oyj	27/09/1994	7.7	0.0101	−0.6378	−1.2500	−3.4954

Kemira Oyj	10/11/1994	191.7	−0.0036	−0.0693	−0.3815	−3.9580
Espoon Sähkö Oyj	24/11/1994	148.2	−0.0136	0.1696	0.5666	−3.2624
Nokian Renkaat Oyj	01/06/1995	26.6	−0.0132	0.6071	5.0175	−3.0358
Suunto Oyj	14/06/1995	8.4	0.0092	0.2236	−0.8395	
Rauma Oyj	27/06/1995	139.9	0.0380	0.1795	−1.0768	
Neste Oyj	27/11/1995	126.0	0.0156	0.0554	−0.3570	
KCI Konecranes Intern. Oyj	27/03/1996	82.4	0.2143	0.8220	−1.3744	−2.7828
Kauppakaari Oy	22/07/1996	3.5	0.1598	−0.5329	−1.7527	−1.9377
PK Cables Oyj	03/04/1997	18.3	0.5242	0.1454	−3.8748	
Nordic Aluminium Oyj	24/04/1997	25.6	0.0649	−0.6315	−4.2885	
Kyro Oyj	09/06/1997	37.2	0.4108	−0.3296	−4.6505	
Rocla Oyj	17/06/1997	13.6	0.0179	−0.0414	−4.4630	
Elcoteq Network Oyj	26/11/1997	101.2	0.0189	−0.8581	−0.9469	
Janko Pöyry Group Oyj	02/12/1997	87.5	−0.0349	−0.5785	−2.2959	
Metsä Tissue Oyj	09/12/1997	115.9	0.0102	−0.7544	−3.3722	
A-rakennusmies Oyj	30/04/1998	23.0	0.1940	−0.4658	−1.1698	
Sponda Oyj	01/06/1998	138.4	0.1279	−0.6073	−1.3804	
JOT Automation Group Oyj	15/09/1998	50.8	0.1452	−1.4996	−1.3192	
Sonera Yhtymä Oyj	10/11/1998	1053.2	0.4012	0.5029	−1.2881	
Rapala Normark Oyj	04/12/1998	94.2	0.0560	−1.6510	−1.1879	
Janton Oyj	11/03/1999	27.6	0.0464	−1.1859	−1.3693	
Marimekko Oyj	12/03/1999	3.0	−0.0366	−1.8619	−1.3214	
TJ Group Oyj	15/03/1999	4.0	0.5544	6.5165	0.0000	
Eimo Oyj	23/03/1999	22.0	−0.0229	0.7615	−1.3022	
Teleste Oyj	30/03/1999	74.0	0.0192	0.3829	−1.2503	
Stonesoft Oyj	12/04/1999	14.0	0.0270	8.0552	−1.1092	

APPENDIX *Continued*

Investigated IPOs

Company	IPO date dd/mm/yyyy	Size M euro	Initial return Exc market	Return 1 year Exc market	Return 3 year Exc market	Return 5 year Exc market
Nedecon-Network Devel. Oyj	01/06/1999	2.0	0.7715	−1.3490		
Technopolis Oyj	08/06/1999	10.0	−0.0434	−1.4537		
Biohit Oyj	23/06/1999	9.0	0.1918	−0.8484		
Perlos Oyj	23/06/1999	213.0	0.2463	1.1821		
Sanitec Oyj	06/07/1999	88.0	0.1540	−1.0479		
TH Tiedonhallinta Oyj	06/09/1999	39.0	−0.0334	−1.3169		
Sysopen Oyj	27/09/1999	4.0	0.5493	−0.8625		
Tieto-X Oyj	28/09/1999	13.0	−0.0458	−0.7431		
Liinos Oyj	08/10/1999	31.0	0.0732	−1.0896		
Proha Oyj	15/10/1999	9.0	−0.1901	1.8346		
Aldata Solution Oyj	22/10/1999	7.0	0.1659	5.9765		
F-Secure Oyj	04/11/1999	50.0	2.5520	−0.3592		
Comptel Oyj	09/12/1999	7.0	2.2127	0.3989		
Basware Oyj	29/02/2000	8.0	3.1850	−0.2613		
Satama Interactive Oyj	15/03/2000	59.0	0.9109	−0.5143		
Saunalahti Oyj	10/04/2000	23.0	−0.2422	0.2233		
EQ Online Oyj	14/04/2000	69.0	−0.0049	−0.5267		
Etteplan Oyj	27/04/2000	10.0	−0.0461	0.0559		
Tekla Oyj	22/05/2000	21.0	0.0698	−0.0039		

Wecan Electronics Oyj	22/05/2000	27.0	0.0509	0.0387
Iocore Oyj	29/05/2000	9.0	0.1067	−0.2987
Digital Open Network Env. Oyj	19/06/2000	14.0	−0.1472	−0.3129
Biotie Therapies Oyj	29/06/2000	21.0	0.0858	0.2753
Tecnomen Oyj	30/06/2000	104.0	−0.0174	−0.2814
Okmetic Oyj	03/07/2000	45.0	−0.0936	0.2760
Beltton -Yhtiöt Oyj	09/10/2000	9.0	0.0541	0.0687
Vacon Oyj	14/12/2000	33.0	0.1559	0.7557
SSH Comm. Security Oyj	19/12/2000	36.0	−0.0174	−0.4043

New Venture Investment: Choices and Consequences
A. Ginsberg and I. Hasan (editors)

Chapter 11

Is Accounting Information Relevant to Valuing European Internet IPOs?

PIETER KNAUFF[a], PETER ROOSENBOOM[b]
and TJALLING VAN DER GOOT[c,*]

[a] *School of Management, University of Amsterdam, Amsterdam, The Netherlands*
[b] *Department of Finance, Eramus University, Rotterdam, The Netherlands*
[c] *Department of Finance, University of Amsterdam, Amsterdam, The Netherlands*

ABSTRACT

This paper investigates the relevance of accounting information to valuing European Internet IPOs during the years 1998–2000. We show that market value is negatively related to earnings in the Internet bubble period before April 2000. This is consistent with an Internet firm's start-up expenditures being considered as assets, not as costs. Furthermore, we find that free float is value relevant in the bubble period. Underwriters and issuers restricted the supply of shares at the IPO. This drove up market prices as investors were eager to buy Internet IPO shares. In the post-bubble period, sales are positively associated with market value. This indicates that investors started to value market share after the Internet bubble burst. Our results remain qualitatively similar under various robustness checks.

Profits were for wimps
The Economist (May 12th, 2001)

* Corresponding author. Tel.: +31-20-525-4171.
E-mail addresses: vdgoot@fee.uva.nl (T.v.d. Goot); proosenboom@fbk.eur.nl (P. Roosenboom).

1. INTRODUCTION

This paper examines European Internet firms that have gone public and experienced sky-high market valuations in comparison with traditional non-Internet companies during the bubble years 1998–2000. It has been argued that the accounting information of firms in nascent, technology-based industries such as the Internet industry is of limited value to investors. Hand (2000a) quotes: "For most companies there at least some widely agreed upon yardsticks: book value, current earnings, projected earnings growth. Internet companies have no tangible assets ... little or nothing in the way of earnings and their future growth is impossible to predict reliably. So investors can't use their customary yardsticks." (Net stock rules: Masters of a parallel universe, *Fortune*, 6/7/99). In the same way, European equity fund managers were quoted saying: "Press coverage, not profits, can determine the attractiveness of a company" (Resources aplenty as equity groups pile in, *Financial Times*, 12/2/99). Doubts about the value relevance of accounting information led analysts to use non-accounting information to value Internet IPOs. For example, European Internet Service Providers (ISPs) were valued on the basis of subscriber numbers rather than revenues: "In the absence of much in the way of revenues, let alone profits, ISPs queuing for a flotation have to be judged to a large extent on subscriber numbers." (ISPs bolster Europe's Internet IPOs, *Financial Times*, 3/6/00). In this paper, we investigate whether the claim that accounting information is of limited use to valuing European Internet IPOs can be substantiated.

We note that many European Internet companies report losses because they heavily invest in customer base creation and brand development. These investments are either immediately expensed in income statements or capitalized and amortized. Hand (2000a) reports an inverse relationship between market value and earnings for U.S. Internet firms during the period of 1997 to mid 1999. Losses therefore appear to enhance, not reduce market value because investors recognize that losses arise because of strategic expenditures by management, not poor performance. In this paper, we examine whether a similar relationship can be found between market value and losses for European Internet companies. Additionally, we investigate whether the negative pricing of losses persists beyond the bursting of the Internet bubble in April 2000.

Our contribution to the existing literature is threefold. First, prior empirical research on the value relevance of accounting information for Internet firms has been dominated by U.S. studies (Core, Guay and Van Buskirk, 2001; Hand, 2000a, b). To our knowledge, this study is the first to examine the value relevance of accounting information for European Internet companies. In particular, we analyze a sample of 138 European Internet companies that went public during the years 1998–2000. The countries included are Belgium, Finland, France, Germany, Italy, the Netherlands, Norway, Spain, Sweden, Switzerland, and the United Kingdom.

Second, we analyze the relevance of accounting information to valuing European Internet firms at the time of the Initial Public Offering (IPO), both before and after the Internet bubble burst in April 2000. The IPO is an important event for Internet companies. Schultz and Zaman (2001) find that Internet firms go public to raise money and to grab market share through takeovers and strategic alliances. Going public thus yields an important advantage to Internet firms vis-à-vis competitors and potential entrants in the Internet sector. In addition, there are large differences in the amount and type of information available for IPO firms vs. publicly traded Internet companies. For example, reliable web traffic measures are generally unavailable for start-up Internet firms at the time of their IPO, particularly in Europe. Investors therefore need to rely on the information disclosed in the prospectus when valuing European Internet IPOs. Following Bartov, Mohanram and Seethamraju (2001), we examine the value drivers underlying Internet IPOs from two perspectives: the offer price and the stock price at the end of the first trading day. The underwriter and the issuing firm set the offer price. This allows us to examine the determinants of the offer price that are considered to be key by the most informed parties in the IPO market. The stock price at the end of the first trading day indicates how small investors perceive the value drivers of European Internet IPO firms.

Third, we extend our analysis to non-accounting information that is available at the time of the IPO. We examine free float (i.e. the percentage of shares being sold in the IPO) and the percentage of post-IPO shares held by the largest shareholder. We expect to find that free float is negatively related to market value. If fewer shares are sold at the IPO, the supply of shares is restricted. This may drive up market prices as investors scramble for Internet IPO shares (Bartov, Mohanram and Seethamraju, 2001; Hand, 2000b). This effect is expected to be especially important during the Internet bubble period. We expect to find a positive relationship between market value and the percentage of post-IPO ownership of the largest shareholder. The largest owner is the party closest to the Internet firm. If the largest owner decides to retain a large fraction of the post-IPO shares, this may signal positive news to investors about the value of the Internet firm (Schultz and Zaman, 2001).

The remainder of the chapter is organized as follows. In Section 2, we discuss prior research. Section 3 presents descriptive statistics. Section 4 contains the methodology. Section 5 presents empirical results. Section 6 concludes the chapter.

2. A REVIEW OF INTERNET VALUATION

The main empirical finding of papers on the valuation of U.S. Internet firms is that typical accounting information still plays a dominant role in explaining the cross-sectional variation in market valuations (Core, Guay and Van Buskirk, 2001; Hand, 2000a, b). Nonetheless, web traffic (e.g. number of visitors to the web site, page views) has additional explanatory power beyond that of standard accounting measures for specific types of Internet companies, such as e-tailers and content and portals (Demers and Lev, 2000; Rajgopal, Kotha and Venkatachlan, 2000; Trueman, Wong and Zhang, 2000). In this section, we discuss several U.S. studies on the valuation of Internet firms. We refer to Ofek and Richardson (2001) for a survey of market efficiency in the Internet sector.

Hand (2000a) finds that basic accounting data is value relevant in a non-linear way. Analyzing 167 pure-play Internet firms from 1997 to mid 1999, he finds that Internet firm's market values are linear and increasing in the book value of equity, but concave and increasing (decreasing) in negative (positive) net income. Internet firm's market values are reliably positively related to selling and marketing expenses. The losses of Internet firms resulting from high marketing expenditures are found to be negatively related to the stock price, which is consistent with the argument that large marketing costs are intangible investments and not period expenses. The larger the losses, the higher the market values. Hand (2000a) argues that investors seem to recognize that losses reflect strategic expenditures for customer base creation and market share, not poor performance. Hand (2000a) also investigates a sample of 116 Internet firms at the time of their IPO. Again, he finds that market value at the IPO, whether evaluated using the first-day closing price or final offer price, is a linear and positive function of the book value of equity, but concave and increasing (decreasing) in negative (positive) net income.

In a companion study, Hand (2000b) investigates the importance of supply and demand forces above economic fundamentals. Hand (2000b) finds that Internet market values are correlated with proxies for demand and supply forces in the form of public float and institutional ownership. After controlling for accounting data and web traffic, his results show that Internet firm's market values are negatively related to public float and positively related to institutional ownership. The negative relation between free float and market value indicates that stock prices are higher if the supply of freely traded shares in the hands of the general public is restricted. Second, the positive relationship between institutional ownership and Internet firm market value is consistent with Internet firms benefiting from the market stabilizing force resulting in less volatile stock prices and a lower cost of capital provided by greater institutional ownership. Interestingly, Hand (2000b) reports that market values are reliably related to only one out of four measures of web traffic—the number of unique visitors. Web traffic only explains a small fraction of the cross-sectional variation in market values. Instead, accounting variables such as book value of equity and earnings overwhelmingly determine the market value of Internet firms.

Core, Guay and Van Buskirk (2001) examine the explanatory power and stability of a regression model of market values on traditional accounting variables for a large sample of firms over the past 25 years. They investigate how equity valuation changed in the recent New Economy sub-period of 1996–1999. Overall, their results suggest that traditional explanatory variables such as earnings and book value of equity, remain applicable to firms in the New Economy period, but that there is greater variation remaining to be explained by uncorrelated omitted factors. This shows that accounting information remains important in the New Economy period.

Trueman, Wong and Zhang (2000) are unable to find a significant positive association between bottom-line net income and the stock price in their sample of e-commerce companies and portals/communities. However, they do find a positive and significant association between gross profits and stock prices. This finding can be explained by the fact that bottom-line net income is often reduced by large non-recurring costs or costs that analysts and investors consider being investments rather than expenses. They argue that gross profit, in contrast, reflects the firm's current operating performance and is viewed as a more stable benchmark for future profitability. Trueman, Wong and Zhang (2000) also investigate web traffic measures. They find that web traffic measures (i.e. the number of visitors to the site and page views) have incremental explanatory power above accounting information such as earnings. The importance of web traffic measures declines when the components of earnings are examined instead of bottom-line net income, suggesting that some of the value implications of web traffic measures are already captured by the components of earnings. Moreover, the significance of web traffic does not mean that accounting information is irrelevant to valuing Internet firms. According to Ofek and Richardson (2001), web traffic is likely to give a better idea of the growth opportunities of the firm.

Rajgopal, Kotha and Venkatachlan (2000) also find a positive relation between market-to-book ratios and web traffic after controlling for accounting information. They analyze a sample of 86 Internet firms from the sub-sectors ISPs, content/portals and e-commerce. They assume that web traffic is driven by managerial choices. While other studies see web traffic as an exogenous variable, they use it as an endogenous variable. They show that web traffic increases if the firm engages in strategic alliances with America Online, media visibility and marketing expenditure. The value-relevance of web traffic therefore does not come from fundamental links between traffic and revenues but possibly from future growth potential through network effects and customer relationships.

While the aforementioned studies examine possible value drivers over the years 1998–1999, Demers and Lev (2000) compare the role of various types of information in explaining stock prices before and after the Internet shakeout in the spring of 2000. They study a sample of 84 Internet companies over the period February 1999 to May 2000. In 1999, marketing expenditures and product and development costs are positively related to price-to-sales ratios, implying that the market sees marketing and R&D costs as intangible assets instead of current expenditures. In 2000, however, this relation disappears suggesting that the stock market is no longer willing to capitalize extraordinary expenditure as intangible assets. Demers and Lev (2000) also investigate the importance of web metrics (the attraction of new visitors to the site, retention of visitors at the site and the ability to generate repeat visits). Contrary to recent doubts about the relevance of web traffic caused by the Internet shakeout, they report that web metrics are still relevant to valuing Internet firms in 2000.

Bartov, Mohanram and Seethamraju (2001) examine the valuation of 100 U.S. Internet IPOs from 1996 to June 1999. They distinguish between the viewpoint of the underwriter and the issuing firm vs. the viewpoint of small investors. The viewpoint of the underwriter and issuing firm is captured by the final offer price. They find that offer prices are increasing in sales, sales growth and positive cash flows, but decreasing in free float and negative cash flows. They argue that negative cash flows are viewed as strategic investments. Earnings seem to play no role in valuing Internet IPOs. Bartov, Mohanram and Seethamraju (2001) capture the perspective of small investors by the market price on the first trading day. They report that first-day market prices are increasing in book value of equity, sales growth and the positive information gathered by the underwriter in the bookbuilding process, but decreasing in free float. Again, earnings do not play a role in valuing Internet IPOs. In the next section, we describe our sample of European Internet IPOs.

3. DATA AND DESCRIPTIVE STATISTICS

3.1. Sample selection

Following Hand (2000a) we define an Internet company as a company which obtains the majority of its revenues (>51%) through or because of the Internet. We use three sources to collect an initial sample of 185 Internet companies that went public on a European stock exchange between January 1, 1998 and December 31, 2000. Before January 1998 and after December 2000, there are few Internet firms going public in Europe. Internet companies are selected from: (i) all traded companies on European stock markets for young, high growth companies in Belgium (*Euro.NM Belgium*), France (*Nouveau Marché*), Germany (*Neuer Markt*), Italy (*Nuovo Mercato*) and the Netherlands (*NMAX*), and from the London Stock Exchange (*Alternative Investment Market* and *Official List*) and *EASDAQ* (now *NASDAQ Europe*); (ii) all members of the *Bloomberg European Internet Index*; and (iii) through several interviews with merchant bankers from Kempen & Co.

Table 1. Geographical distribution of European Internet IPOs

Country	*N*	Percentage
Belgium	2	1.4
Finland	3	2.2
France	27	19.6
Germany	64	46.4
Italy	9	6.5
The Netherlands	1	0.7
Norway	1	0.7
Spain	1	0.7
Sweden	5	3.6
Switzerland	3	2.2
United Kingdom	19	13.8
EASDAQ	3	2.2
Total	138	100.0

Following prior research, financial institutions are excluded from the initial sample because of their less comparable balance sheet structure.

From the initial sample of 185 companies, 43 companies are excluded because accounting data or IPO data were not available in *Bloomberg* and could not be obtained elsewhere. Three companies changed their fiscal year prior to IPO and are excluded from the sample. One company is excluded because it is a mutual fund of Internet companies rather than an Internet company itself. After the exclusion of those 47 companies, our sample consists of 138 European Internet IPOs. Table 1 shows that 11 European countries are represented in the sample. Three U.S. firms with a listing on *EASDAQ* are included as well. A large portion of the sample is listed on the German *Neuer Markt*. The appendix contains a list of the Internet companies included in our study.

3.2. Sub-sectors of the Internet industry

We divide the Internet industry in 6 sub-sectors: ISPs, content/portals, e-commerce, IT-infrastructure, Internet software and Internet services (see Figure 1). Table 2 shows the distribution of the companies in the sample over these sub-sectors of the Internet industry.[1] We discuss each sub-sector briefly. *Internet Service Providers*

[1] Some companies operate in more than one sub-sector. For example, one of our sample companies, World Online, offers access to the Internet, Internet services and content/portal services. In those cases, we have chosen the sub-sector which represents the major part of the firm's revenues.

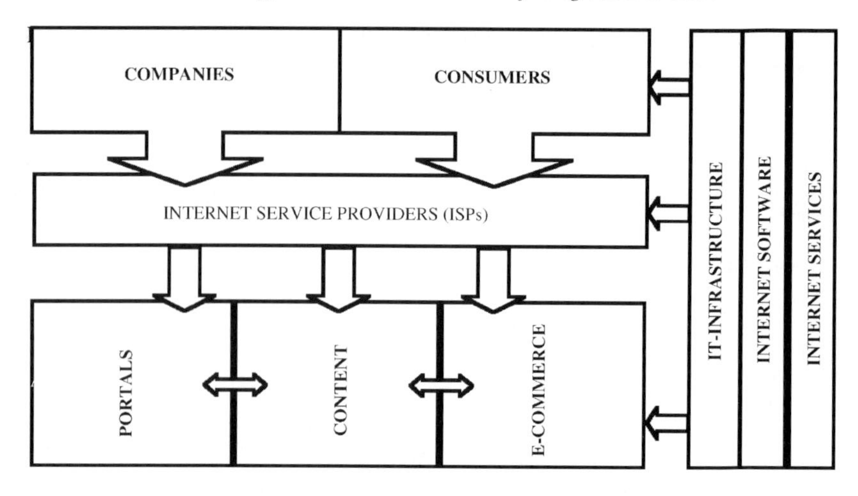

Figure 1. Sub-sectors of the Internet industry.

(ISPs) are network service providers supplying IP-based connections to its network and access to the public Internet to consumers and/or business users. Our sample contains 15 Internet IPOs from the ISP sector. *Content* companies offer free information to users about products and services through their web sites. *Portals* are major starting sites for users when they get connected to the Internet or that users tend to visit as an anchor site. A portal is essentially a content website because it offers free information to navigate the Internet. Because the sources of revenue are the same for these two sub-sectors (primarily advertising revenues), we take content and portals as one sub-sector of Internet industry. There are 19 IPO firms that operate in the content/portals sector. *E-commerce* is the activity where vendors of goods and services and a buyer of such goods and services enter into a commercial

Table 2. Distribution of European Internet IPOs over sub-sectors

Sub-sector	*N*	Percentage
Internet Service Provider	15	10.9
Content/portals	19	13.8
E-commerce	17	12.3
IT-infrastructure	4	2.9
Internet software	39	28.3
Internet services	44	31.9
Total	138	100.0

transaction over a digital infrastructure. Our sample contains 17 companies from the e-commerce sector. Together with ISPs and content/portals firms, e-commerce firms are web-based companies that are expected to earn revenues directly or indirectly by attracting web traffic to their sites.

IT-infrastructure companies provide the physical hardware used to interconnect computer and users. Infrastructure includes the transmission media, including telephone lines, cable television lines, and satellite and antennas, and also the router, aggregator, repeater, and other devices that control transmission paths. The sample consists of only 4 IT-infrastructure companies. *Internet software* companies provide the various kinds of programs used to operate Internet-related computers and devices. Our sample is composed of 39 Internet software companies. Companies in the *Internet services* market can be divided in companies that provide product support and maintenance for software and hardware, and companies that provide professional services. Professional services include consulting, development and integration, education and training, management services, and business management services. Our sample contains 44 Internet services firms.

3.3. Summary statistics

Figure 2 shows the number of IPOs per quarter during the period January 1, 1998 until December 31, 2000. We observe that Internet IPOs are not equally distributed over time. In the first half of the year 2000, the number of IPOs is higher than in other quarters. After the bubble burst in April 2000, the number of Internet IPOs declined rapidly in the second half of 2000.

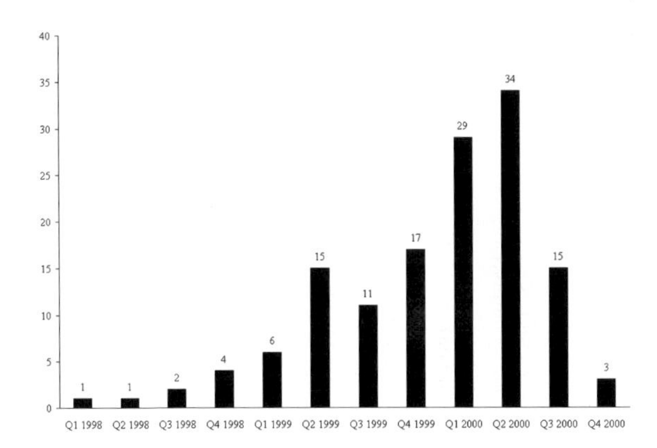

Figure 2. Number of European Internet IPOs during 1998–2000.

Table 3. Summary statistics

	Mean	Median	Std. dev.	Minimum	Maximum	N
Offer value (€ million)	918.437	208.695	3,561.651	7.374	31,960.790	138
Market value (€ million)	1,276.093	238.835	4,872.850	10.852	44,389.99	138
Sales (€ million)	35.835	5.336	130.702	0.000	921.770	138
EBIT (€ million)	−2.514	−0.925	13.258	−108.800	59.020	138
Net income (€ million)	−3.249	−1.235	10.134	−91.200	24.750	138
Total assets (€ million)	41.649	9.017	132.541	0.228	1,101.860	138
Book value of equity (€ million)	15.877	2.265	45.260	−48.484	295.040	138
Number of employees	140.907	60.000	411.012	4.000	4.512	138
Free float (%)	24.877	23.592	7.251	8.204	49.242	133
Ownership of largest shareholder (%)	39.639	33.875	22.407	10.000	92.000	118
First-day return (%)	51.606	14.099	95.689	−84.894	444.444	138

Note: This table shows summary statistics for 138 European Internet IPOs. Offer value is determined as the number of post-IPO shares outstanding times the final offer price. Market value is computed as the number of post-IPO shares outstanding times the market closing price on the first day of trading. Sales, earnings before interest and taxes (EBIT), net income and the number of employees are for the last fiscal year before the IPO. Total assets and the book value of equity are taken from the balance sheet for the fiscal year prior to the IPO. Free float is defined as the number of shares sold in the IPO divided by the number of post-IPO shares outstanding. Ownership of the largest shareholder equals the number of shares retained by the largest owner scaled by the number of shares outstanding after the IPO. The first-day return (underpricing) is calculated as the percentage difference between the first day closing market price and the offer price.

The major part of the accounting data is obtained through *Bloomberg*. Missing data are obtained through *Datastream* or are hand-collected from annual reports and prospectuses. For 10% of the *Bloomberg* data we checked the validity. No errors are found. Table 3 presents descriptive statistics of the variables used in this study. The first row of Table 3 shows that the offer value, measured as the number of post-IPO shares times the final offer price, equals €918.4 million. The median offer value is much lower at €208.7 million. The average (median) market value equals €1,276.1 million (€238.8 million) with a minimum of €10.8 million and a maximum of €44,390 million. The distribution of market values is therefore significantly skewed. The average (median) sales in the last fiscal year before the IPO equal €35.8 (€5.3 million). This indicates that market value and offer values are high when measured against current revenues. The majority of Internet IPO firms (66.7%) report losses in the fiscal year before the IPO. The average (median) earnings before interest and taxes equals −€2.5 million (−€0.9 million). We also look at net income before the IPO. The average (median) net income in the year before the IPO is −€3.2 million (−€1.2 million).

Total assets in the year before the IPO average €41.6 million. Total assets equal €9 million, evaluated at the median. The average (median) book value of equity is €15.9 million (€2.2 million). The average (median) company employs 141 (60) people in the year before the IPO. Table 3 shows that the free float, defined as the number of shares sold at the IPO scaled by the number of post-IPO shares, is 24.9% with a median of 23.6%. The average (median) ownership by the largest shareholder equals 39.6% (33.9%) of post-IPO shares. We also calculate first-day returns. The first-day closing market price is typically higher than the offer price, so that first-day returns are positive. The positive first-day offer-to-close return is commonly known as underpricing. The average (median) first-day return equals 51.6% (14.1%). There are 24 IPOs (17.4%) that double in price on the first trading day (i.e. have a first-day return of 100% or higher). Our average first-day return is lower than the average first-day return reported by Arosio, Giudici and Paleari (2000). Analyzing a sample of 86 Internet IPOs on *Euro-NM* and *EASDAQ*, they find an average first-day return of 76.4%. The difference can be explained by the different sample periods. Arosio, Giudici and Paleari (2000) investigate the period from January 1999 to May 2000, whereas we examine a longer period from January 1998 to December 2000.

4. METHODOLOGY

We build our empirical tests on the well-known residual-income or Ohlson (1995) model. This model states that the stock price is a function of the book value of equity and residual earnings:

$$P_t = \text{BVE}_t + \sum_{i=1}^{\infty} \frac{E(\text{RE}_{t+i})}{(1+r)^t} \tag{1}$$

where P_t is the firm's stock price at the end of the current period t, BVE_t is the book value of equity at that time, and RE_{t+i} is the firm's residual earnings for period $t + i$ (defined as the period's earnings available to shareholders less a charge applied to beginning-of-period book value), r is the firm's required rate of return on equity capital, and $E(\cdot)$ is the expectation operator.

Following Trueman, Wong and Zhang (2000), we take an empirical application of the Ohlson model as the base for empirical tests. In particular, we regress market values on an intercept, book value of equity, current earnings and other explanatory variables. We use offer values as an alternative dependent variable. Table 4 provides the definitions of dependent and independent variables that we use in regression analyses. It is important to note that we do not deflate market values and offer values by book value of equity or sales. Other studies use market-to-book ratios (Core, Guay and Van Buskirk, 2001; Trueman, Wong and Zhang, 2000) or price-to-sales ratios (Demers and Lev, 2000) as a dependent variable. We do not adopt this approach because market-to-book ratios and price-to-sales ratios do not have the same economic interpretation as they have in a cross-section of more established and profitable firms. Book values and sales tend to be small or absent for Internet firms. These ratios therefore tend to "blow up" because of a small denominator

Table 4. Variable definitions

Variable name	Definition
Dependent variables	
MVE	Market value of equity; number of post-IPO shares times the market price on the first day of trading
OVE	Offer value of equity; number of post-IPO shares times the final offer price
Independent variables	
EBIT	Earnings before earnings and taxes; last fiscal year before IPO
NI	Net income; last fiscal year before IPO
ISP	Internet Service Provider; indicator variable that takes on value of one if IPO firm is classified as Internet Service Provider
REV	Sales; last fiscal year before IPO
BE	Book value of equity; last fiscal year before IPO
FLOAT%	Free float; number of shares sold at the IPO scaled by the number of post-IPO shares
TOPSH%	Ownership of largest shareholder; number of shares held by largest owner scaled by number of post-IPO shares

problem. For example, the market-to-book ratio in our sample would range from −16,900 to +15,750.

We use log-linear regression methodology advocated by Hand (2000a, b). There are three important advantages to this methodology. First, Hand (2000a) shows that the log-linear regressions yield lower pricing errors for Internet stocks than do regressions using per-share or non-logged data. Second, log-linear regressions moderate the impact of anomalous or outlier observations in accounting data. Table 3 shows that the accounting data is highly skewed. Using log-transformed accounting data may thus help to resolve this problem. Third, Hand (2000b) argues that log-linear regressions generally achieve greater homoscedasticity in regression residuals. We log transform each dependent and independent accounting variable Z (where Z is defined in € millions) in the following way:

$$\mathrm{LZ} = \log[\mathrm{Z}+1] \quad \text{if } \mathrm{Z} \geq 0, \quad \text{and} \quad -\log[-\mathrm{Z}+1] \quad \text{if } \mathrm{Z}<0 \tag{2}$$

We add €1 million to Z so that LZ is defined when Z is close to at or close to zero. The log-linear regression model may incorporate concavity, linearity or convexity between the dependent and independent variables. Consider the relation between the log-transformed non-negative values of X and Y:

$$\log(Y+1) = \alpha + \beta\log(X+1) \quad \Leftrightarrow \quad \mathrm{LY} = \alpha + \beta\,\mathrm{LX} \tag{3}$$

This implies that the non-logged and unscaled relation between X and Y is given by:

$$Y = \mathrm{e}^{\alpha}(X+1)^{\beta} - 1 \tag{4}$$

The coefficient β measures the degree and type of non-linearity between X and Y. We refer to Hand (2000a, b) for a more detailed discussion of the non-linear regression model. For non-negative values of X, the relation between X and Y in Eq. (4) is concave if $0 < \beta < 1$, linear if $\beta = 1$ and convex if $\beta > 1$. When X is negative and log-transformed using Eq. (2), the relation between X and Y is concave if $-1 < \beta < 0$, linear if $\beta = -1$, and convex if $\beta < -1$. If $\beta = 0$ then X and Y are unrelated regardless of the sign of X. If $\log(Y+1)$ is a function of several independent variables, suppose X and W, then β captures the marginal concavity, linearity or convexity of X (i.e. holding constant W). The next section describes our log-linear regression results.

5. EMPIRICAL RESULTS

5.1. Log-linear regressions for market values

Table 5 presents the log-linear regression results using logged market value as the dependent variable. Market value is defined as the number of post-IPO shares

Table 5. Log-linear regression results for market values

	(1)	(2)	(3)	(4)	(5)	(6) Bubble	(7) Post-bubble
LEBIT	−0.356				−0.332		
	(−4.576)***				(−3.697)***		
LEBIT_POS		−0.078				0.441	−0.359
		(−0.343)				(1.133)	(−1.492)
LEBIT_NEG		−0.484				−0.641	−0.338
		(−4.721)***				(−3.808)***	(−1.590)
LNI			−0.397				
			(−5.441)***				
LNI_POS				−0.120			
				(−0.489)			
LNI_NEG				−0.488			
				(−4.804)***			
FLOAT%					−2.327	−3.579	1.025
					(−1.211)	(−1.690)*	(0.360)
TOPSH%					0.515	0.156	0.800
					(1.080)	(0.227)	(1.310)
ISP	1.694	1.751	1.621	1.663	1.650	1.211	2.162
	(4.308)***	(4.436)***	(4.261)***	(4.336)***	(4.447)***	(2.385)**	(4.442)***
LREV	0.254	0.173	0.193	0.148	0.179	−0.115	0.321
	(2.499)**	(1.350)	(1.974)*	(1.255)	(1.398)	(−0.570)	(1.980)*
LBVE	0.054	0.022	0.057	0.038	0.097	0.107	0.085
	(0.934)	(0.362)	(1.002)	(0.622)	(1.231)	(1.130)	(0.709)
CONSTANT	4.673	4.664	4.725	4.692	5.169	5.914	3.543
	(19.192)***	(18.977)***	(19.700)***	(20.170)***	(7.736)***	(7.120)***	(3.743)***
R^2 adjusted	36.506%	36.993%	36.749%	36.971%	40.832%	38.966%	61.635%
F-test	20.692***	17.087***	20.899***	17.072***	14.000***	7.566***	10.181***
N	138	138	138	138	114	73	41

Note: The table shows the log-linear regression results using logged market values (*LMVE*) as the dependent variable. Independent variables are defined as in Table 4. We put the letter '*L*' in front of the variable name to indicate that the variable has been log-transformed using Eq. (2). *LEBIT_POS* equals *LEBIT* if the earnings before interest and taxes are positive, 0 otherwise. *LEBIT_NEG* equals *LEBIT* if the earnings before interest and taxes are negative, 0 otherwise. *LNI_POS* and *LNI_NEG* are defined similarly but then using net income. We define the bubble period as the period from January 1999 to March 2000. The post-bubble period is from April 2000 to December 2000. White (1980) heteroskedastic-consistent t-statistics are within parentheses. The asterisks (*) denotes significance at the 10% level; (**) at the 5% level and (***) at the 1% level.

times the market closing price on the first day of trading. We assume that market values capture the value assessment of small investors. We put the letter '*L*' in front of the variable name to indicate that the variable has been log-transformed using Eq. (2). In the first model, we regress logged market values on logged earnings before interest and taxes (*LEBIT*), an indicator variable for Internet Service Providers (ISPs), logged sales and logged book value of equity. We include an indicator variable for ISPs in the regression model because these firms have large market values by comparison.[2] We use sales as a proxy for marketplace acceptance and market share. For young, fast growing firms it could be more important to aim for revenues than profits. Book value of equity directly comes from the residual income model that we discussed in the previous section. Moreover, book value of equity is a standard control variable in valuation regressions (Collins, Pincus and Xie, 1999).

The first column shows the results of the model (1). There is a significant and negative relation between logged earnings before interest and taxes (EBIT) and logged market values. Investors therefore value losses at the IPO. The larger the loss, the higher the market value. The coefficient on the indicator variable for ISPs is positive and significant. This reflects that ISPs have higher market values in comparison to other Internet companies. The indicator variable for ISPs is significantly positive in all subsequent regressions. Logged sales are positively associated with logged market values. We argue that firms with higher sales have more marketplace acceptance and market share, which is valued by investors. However, this result is not robust to alternative specifications of the regression model, as we will discuss next. The log of the book value of equity is not related to logged market value. The log of the book value of equity is insignificant in all subsequent regressions.

Next, we distinguish between profit and loss reporting Internet firms (see model (2)). We include a variable *LEBIT_POS* that equals the logged earnings before interest and taxes if it is positive, and a variable *LEBIT_NEG* that equals the logged earnings before interest and taxes if it is negative. We find no relation between *LEBIT_POS* and logged market values, but a strongly negative and thereby concave relation between *LEBIT_NEG* and logged market values. This reinforces that investors are negatively pricing losses. The larger the loss, the higher the market value. Investors appear to recognize that losses reflect strategic expenditures for customer base creation and market share, not poor performance. The coefficient on logged sales is no longer statistically significant at conventional levels.

We include net income in model (3) as an alternative earnings measure. We find that logged net income is negatively related to logged market values. In model (4), we split between Internet firms that report positive net income (*LNI_POS*) and negative net income (*LNI_NEG*). The coefficient on *LNI_POS* is insignificant, while the coefficient on *LNI_NEG* is negative and highly significant. Again, this shows that investors are negatively pricing losses. More sizable losses in the fiscal year

[2] Our sample includes 15 Internet Service Providers, such as T-Online International AG, World Online International N.V. and Terra Networks, S.A. The average (median) market value for ISPs equals €8,004 million (€1,313 million) versus €456 million (€219 million) for other Internet firms. Because we use unscaled market values, it is important to control for this effect in the regressions.

before the IPO are associated with higher market values. This finding conflicts with Bartov, Mohanram and Seethamraju (2001) for U.S. Internet IPOs. They find no relation between earnings per share or cash flow per share and the closing market price on the first trading day.

Model (5) is an augmented version of model (1). We add two non-accounting variables; free float and the post-IPO percentage ownership of the largest shareholder. Due to data availability, the number of observations is reduced to 114 firms. We expect to find that free float is negatively related to market value. If fewer shares are sold at the IPO, the supply of shares is restricted. This may drive up market prices as investors compete for few Internet IPO shares. The largest owner is the party closest to the Internet firm. If the largest owner decides to retain a large fraction of the post-IPO shares, this may signal positive news to investors about the value of the Internet firm. Contrary to expectations, we do not find a significant association between logged market values and free float or post-IPO percentage ownership of the largest owner. However, the negative and significant association between logged earnings before interest and taxes and logged market values remains.

Have things changed since the Internet bubble burst in April 2000? To answer this question, we estimate model (6) for the 73 firms that went public before April 2000 (the bubble period). We observe that in the bubble period, investors priced losses negatively. The larger the loss, the higher the market value. In addition, free float is negatively associated with logged market values in the bubble period. This shows that stock prices were driven by supply and demand. The fewer shares were sold at the IPO, the higher the market value. Model (7) is identical to model (6), but now we estimate the non-linear regression for the sub-sample of 41 Internet companies that went public after April 2000 (the post-bubble period). We find that investors no longer negatively price losses in the post-bubble period. Internet firms with larger losses are no longer valued more by investors. This finding corresponds to the findings of Demers and Lev (2000) for the United States. They report that investors are no longer willing to capitalize marketing and R&D costs as intangible assets in 2000. In contrast to the bubble period, logged sales are positively related to logged market values in the post-bubble period. Investors seem to attach a higher value to Internet firms that have earned more revenues before going public.

Taken together, our regression results suggest that accounting information is relevant to valuing Internet IPOs. This is consistent with the findings of Hand (2000a, b) for the United States. However, the relationships between specific accounting variables and market value appear to have changed after the Internet bubble burst in April 2000. In the bubble period, investors negatively price losses. The larger the loss, the higher the market value. In the post-bubble period, this relation has disappeared. Instead, investors positively price sales.

5.2. Log-linear regressions for offer values

Table 6 shows the log-linear regression results using logged offer value as the dependent variable. Offer value is defined as the number of post-IPO shares times the final offer price. We assume that offer values capture the value assessment of the underwriter and the issuing firm. This allows us to examine the determinants of the offer price that are considered to be key by the most informed parties in the

Table 6. Log-linear regression results for offer values

	(1)	(2)	(3)	(4)	(5)	(6) Bubble	(7) Post-bubble
LEBIT	−0.363				−0.321		
	(−4.869)***				(−3.563)***		
LEBIT_POS		−0.062				0.352	−0.298
		(−0.264)				(0.801)	(−1.269)
LEBIT_NEG		−0.500				−0.634	−0.361
		(−5.197)***				(−3.692)***	(−1.695)*
LNI			−0.417				
			(−6.051)***				
LNI_POS				−0.112			
				(−0.451)			
LNI_NEG				−0.517			
				(−5.487)***			
FLOAT%					−3.510	−4.360	0.243
					(−1.803)*	(−1.993)*	(0.084)
TOPSH%					0.374	0.270	0.657
					(0.871)	(0.431)	(1.113)
ISP	1.744	1.806	1.657	1.702	1.680	1.248	2.094
	(4.944)***	(5.105)***	(5.100)***	(5.205)***	(5.458)***	(3.167)***	(4.641)***
LREV	0.305	0.217	0.247	0.197	0.216	−0.042	0.301
	(3.113)***	(1.711)*	(2.613)***	(1.717)*	(1.746)*	(−0.197)	(2.010)*
LBVE	0.005	−0.030	0.007	−0.014	0.044	−0.003	0.061
	(0.092)	(−0.500)	(0.125)	(−0.232)	(0.599)	(−0.032)	(0.485)
CONSTANT	4.363	4.352	4.400	4.363	5.251	5.697	3.764
	(19.579)***	(19.473)***	(19.960)***	(20.578)***	(7.834)***	(6.818)***	(3.887)***
R^2 adjusted	40.732%	41.497%	41.905%	42.382%	45.573%	41.256%	60.292%
F-test	24.538***	20.435***	25.705***	21.154***	16.769***	8.223***	9.676***
N	138	138	138	138	114	73	41

Note: The table shows the log-linear regression results using logged offer values (*LOVE*) as the dependent variable. Independent variables are defined as in Table 4. We put the letter '*L*' in front of the variable name to indicate that the variable has been log transformed using Eq. (2). *LEBIT_POS* equals *LEBIT* if the earnings before interest and taxes are positive, 0 otherwise. *LEBIT_NEG* equals *LEBIT* if the earnings before interest and taxes are negative, 0 otherwise. *LNI_POS* and *LNI_NEG* are defined similarly but then using net income. We define the bubble period as the period from January 1999 to March 2000. The post-bubble period is from April 2000 to December 2000. White (1980) heteroskedastic-consistent *t*-statistics are within parentheses. The asterisks (*) denotes significance at the 10% level; (**) at the 5% level and (***) at the 1% level.

IPO market. For reasons of brevity, we focus on the results that are different from the results that we discussed earlier for market values.

The regression results for model (1) are presented in the first column. As with market values, there is a negative relation between logged earnings before interest and taxes (*LEBIT*) and the logged offer value. We infer that underwriters and issuing firms also price losses negatively. The larger the loss, the higher the offer value. The indicator variable for Internet Service Providers (ISPs) loads up with a highly significant coefficient. These types of Internet companies have higher offer values. The indicator variable for ISPs is significant in all subsequent regressions. Logged sales are positively related to logged offer values. Market place acceptance appears to be a value driver of the offer price set by underwriters and issuers. In contrast to the market value regressions, the positive relation between sales and offer value does not lose significance in subsequent regressions. As before, the log of the book value of equity is not related to the logged offer value. The coefficient on the log of the book value of equity is insignificant in all subsequent regressions.

In model (2), we include a variable *LEBIT_POS* that equals the logged earnings before interest and taxes if it is positive, and a variable *LEBIT_NEG* that equals the logged earnings before interest and taxes if it is negative. We find no association between logged offer value and *LEBIT_POS*, but a negative and significant relation between logged offer value and *LEBIT_NEG*. Because the coefficient of *LEBIT_NEG* lies between -1 and 0, there is a concave relation between losses and offer value. This reinforces that underwriters and issuers are negatively pricing losses. The larger the loss, the higher the offer value. A difference with the market value regressions is that logged sales remain positively related to offer values. Underwriters and issuers set higher offer prices if the firm earns more revenues.

Model (3) and model (4) use net income as an alternative accounting variable. Results are similar as for the regressions using logged market values as the dependent variable. Again, logged sales remains significant in these regressions. The inverse relation between logged offer value and net income does not correspond to the findings of Bartov, Mohanram and Seethamraju (2001) for U.S. Internet IPOs. They find a negative relation between the offer price and cash flow per share, but not net income per share.

In model (5), we include non-accounting variables. We find that free float, defined as the number of shares sold at the IPO divided by the number of post-IPO shares, is negatively related to logged offer values. Underwriters and issuers seem to anticipate that investors will bid up prices in the light of limited supply. This leads them to set higher offer prices. The percentage post-IPO ownership by the largest shareholder is not significantly related to logged offer values.

We also investigate whether changes have occurred by splitting the sample into a group of 73 Internet IPOs that went public during the Internet bubble and 41 IPOs that went public after the Internet bubble burst in April 2000. Model (6) is estimated for the bubble sample. We find that underwriters and issuers negatively price losses and that float is negatively related to logged offer values. There is no relation between logged sales and logged offer value. Model (7) shows the regression results for the post-bubble sample. The coefficient on *LEBIT_NEG* is statistically significant and negative. This indicates that underwriters continue to price losses negatively. There is no relation between float and logged offer value. This indicates that in the post-bubble period, underwriters and issuers were no longer expecting that they could set higher offer prices when they restricted supply of shares at the IPO. Interestingly, logged sales are positively related to logged offer value in the post-bubble period.

In short, there are only few differences between the way underwriters value European Internet IPOs and the way small investors value these firms. One important difference is that underwriters and issuers have continued to negatively price losses after the bubble burst in April 2000. Investors, on the other hand, do no longer value losses negatively in that period.

5.3. Sensitivity analyses

We have performed several sensitivity checks. First, we have run non-linear regressions with other measures of an Internet firm's earnings as independent variable, such as its operating income and operational cash flow, respectively. The empirical results are qualitatively similar to those presented in Tables 5 and 6. Second, we have included five industry dummies for the different sub-sectors of the Internet industry in the regressions. We choose the sub-sector of Internet services as our reference category to avoid the dummy trap problem. Only the ISP dummy is statistically significant and positive. Third, in order to control for outliers, we have omitted the five firms with the largest and those with the smallest market values or offer values, respectively. Results are qualitatively similar to those of Tables 5 and 6. Fourth, following Trueman, Wong and Zhang (2000) we also use non-logged market-to-book ratios as a dependent variable and scale accounting variables with the book value of equity without the log transformation. Our results are robust to this alternative approach. Fifth, we run a regression using the non-logged price-to-sales ratio as the dependent variable and scale all accounting variables with sales without log transformation. Again, we find similar results.

6. CONCLUSIONS

This paper examines the relevance of accounting information to valuing European Internet IPOs, both before and after the Internet bubble burst in April 2000. We first use market value as our proxy for firm value. We assume that market values capture the value assessment of small investors. We find that accounting variables such as earnings before interest and taxes (EBIT) and net income are relevant to valuing European Internet IPOs. Arguably, this is at odds with the widely held belief that accounting information is of limited use when valuing the IPO shares of Internet companies. In particular, we document a non-linear relation between earnings and market values. We also find a negative pricing of losses in the period before the Internet bubble burst in April 2000. The larger the loss, the higher the market value. But the negative pricing of losses does not extend beyond the Internet bubble period. The inverse relationship between market value of equity and earnings is consistent with an Internet firm's start-up expenditures being considered by the market as assets, not as costs. We use sales as a proxy for marketplace acceptance and market share. We find that sales are positively associated with market values, but only in the post-bubble period after April 2000. The free float is negatively related to market values in the bubble period. This shows that if fewer shares were sold in the IPO, the limited supply together with heavy demand for IPO shares drove market values higher. In the post-bubble period the relation between free float and market values is absent. Contrary to expectations, we do not find any relation between market value and the book value of equity or between market value and the post-IPO percentage ownership of the largest shareholder.

Next, we perform all analyses from the viewpoint of the underwriter and issuing firm (values based on final offer price). Overall, we find qualitatively similar results. One important difference is that underwriters and issuers have continued to negatively price losses after the bubble burst in April 2000. Investors, on the other hand, do no longer value losses negatively in that period. Our results remain qualitatively similar under a number of sensitivity analyses that refer to other measures for the Internet firm's earnings, the inclusion of industry dummies, and using market-to-book ratios and price-to-sales ratios as alternative dependent variables in the regressions.

REFERENCES

Arosio, R., Giudici, G., & Paleari, S. (2000). Why do (or did?) Internet-stock IPOs leave so much 'money on the table'? Working Paper, Politecnico di Milano and University of Bergamo.

Bartov, E., Mohanram, P., & Seethamraju, C. (2001). Valuation of Internet stocks—An IPO perspective, Working Paper, New York University.

Collins, D. W., Pincus, M., & Xie, H. (1999). Equity valuation and negative earnings: The role of book value of equity, *The Accounting Review*, *74*, 29–61.

Core, J. E., Guay, W. R., & Van Buskirk, A. (2001). Market valuations in the New Economy: An investigation of what has changed, Working Paper, University of Pennsylvania.

Demers, E., & Lev, B. (2000). A rude awakening: Internet shakeout in 2000, Working Paper, University of Rochester and New York University.

Hand, J. R. M. (2000a). Profits, losses and the non-linear pricing of Internet stocks, Working Paper, Kenan-Flagler Business School, UNC Chapel Hill.

Hand, J. R. M. (2000b). The role of economic fundamentals, web traffic, and supply and demand in the pricing of U.S. Internet stocks, Working Paper, Kenan-Flagler Business School, UNC Chapel Hill.

Ofek, E., & Richardson, M. (2001). Dotcom mania: A survey of market efficiency in the Internet sector, Working Paper, New York University.

Ohlson, J. (1995). Earnings, book values and dividends in security valuation, *Contemporary Accounting Research*, *11*, 661–687.

Rajgopal, S., Kotha, S., & Venkatachalam, M. (2000). The relevance of web traffic for Internet stock prices, Working Paper, University of Washington.

Schultz, P. H., & Zaman, M. (2001). Do the individuals closest to Internet firms believe they are overvalued? *Journal of Financial Economics*, *59*, 347–381.

Trueman, B., Wong, M. H. F., & Zhang, X. (2000). The eyeballs have it: Searching for the value in Internet stocks, *Journal of Accounting Research*, *38*, 137–170.

White, H. (1980). A heteroskedasticity-consistent covariance matrix estimator and a direct test for heteroskedasticity, *Econometrica*, *48*, 817–838.

APPENDIX: LIST OF EUROPEAN INTERNET COMPANIES INCLUDED IN THIS STUDY

Company name	IPO date	Company name	IPO date
365 Corporation plc	12/2/99	Inferential S.p.A.	8/1/00
Access Commerce S.A.	2/21/00	Infobank International Holdings plc	11/26/98
Actinic plc	5/26/00	InfoVista S.A.	7/7/00
ad pepper media International N.V.	10/9/00	Integra-Net S.A.	6/3/99
AdLINK Internet Media AG	5/10/00	Interactive Investors International plc	2/17/00
ADORI AG	5/10/00	Internet Business Group plc	2/2/00
Affinity Internet Holdings plc	4/20/99	InternetMediaHouse.com AG	7/30/99
ARBOmedia.net AG	5/9/00	Internolix AG	3/27/00
ARTICON Information Systems AG	10/28/98	INTERSHOP COMMUNICATIONS AG	7/16/98
artnet.com AG	5/17/99	Iomart Group plc	4/19/00
Artprice.com S.A.	2/21/00	ISION Internet AG	3/17/00
AuFeminin.com S.A.	7/20/00	Jippii Group Oyj	4/12/00
Best Of Internet S.A.	4/14/00	Jobs & Adverts AG	4/6/00
Biodata Information Technology AG	2/22/00	KaZiBao S.A.	6/27/00
Blue-C New Economy Consulting & Incubation AG	8/24/00	Keyrus-Progiware S.A.	7/20/00
Boss Media AB	6/24/99	Lastminute.com plc	3/14/00
BOV AG	6/20/00	Letsbuyit.com N.V.	7/21/00
Brain Force Software AG	6/10/99	Liberty Surf Group S.A.	3/16/00
BROKAT Infosystems AG	9/17/98	Lycos Europe N.V.	3/22/00
buch.de internetstores AG	11/8/99	MatchNet plc	6/27/00
ComputerLinks AG	7/7/99	Medcost S.A.	6/12/00
Concept! AG	3/27/00	Mediascape Communications AG	5/22/00
Consors France S.A.	4/9/99	Met@Box AG	7/7/99
Crealogix AG	9/6/00	MultiMania S.A.	3/9/00

APPENDIX (*Continued*)

Company name	IPO date	Company name	IPO date
Cross Systems S.A.	11/11/99	Netbenefit plc	6/4/99
CTS Eventim AG	2/1/00	Netlife AG	6/1/99
Cyber Com Consulting Group Scandinavia AB	12/1/99	NetStore plc	4/19/00
Cyberdeck S.A.	6/27/00	NetValue S.A.	1/31/00
CyberSearch S.A.	7/21/00	OnVista AG	2/28/00
Cyrano S.A.	6/11/98	Orchestream Holdings plc	6/28/00
d+s online AG	5/23/00	PIRONET AG	2/22/00
Dada S.p.A.	6/29/00	Pixelpark AG	10/4/99
Dalet S.A.	6/23/00	PopNet Internet AG	2/2/00
DataDesign AG	11/9/98	Prime Response, Inc.	3/2/00
Datasave AG Informationssysteme	2/14/00	Prodacta AG	6/7/99
Day Interactive Holding AG	4/3/00	QS Communications AG	4/18/00
DCI Database for Commerce and Industry AG	3/13/00	QXL.com plc	10/7/99
digital advertising AG	10/29/99	realTech AG	4/26/99
e.biscom S.p.A.	3/30/00	ricardo.de AG	7/21/99
ebookers.com plc	11/12/99	RMR plc	4/12/00
e-district.net plc	3/7/00	Satama Interactive Oyj	3/15/00
Effnet Group AB	4/6/99	Scandinavia Online AB	6/7/00
EMPRISE Management Consulting AG	7/15/99	Secunet Security Networks AG	11/9/99
Endemann!! Internet AG	3/10/99	SinnerSchrader AG	11/2/99
e.Planet S.p.A.	8/3/00	SQLI S.A.	7/24/00
Europstat S.A.	1/4/99	StepStone ASA	3/14/00
Feedback AG	6/27/00	Stilo International plc	8/30/00
Fi System S.A.	10/8/98	Swissquote Group Holding S.A.	5/29/00
Fidelity Net Marketing S.A.	6/3/99	Systems Union Group plc	12/1/99
Fluxx.com AG	9/28/99	Syzygy AG	10/6/00
Focus Digital AG	7/13/00	TeleCity plc	6/15/00
FortuneCity.com Inc.	3/19/99	Telesens AG	3/21/00
Framfab AB	6/23/99	Terra Networks, S.A.	11/17/99

APPENDIX (*Continued*)

Company name	IPO date	Company name	IPO date
Freedomland Internet Television Network S.p.A.	4/19/00	Tiscali S.p.A.	10/27/99
freenet.de AG	12/3/99	tiscon AG Infosystems	10/14/99
Freeserve plc	7/26/99	TJ Group Oyj	9/3/99
GameLoft.com S.A.	6/13/00	TOMORROW Internet AG	11/30/99
Gauss Interprise AG	10/28/99	T-Online International AG	4/17/00
GFT Technologies AG	6/28/99	Travel24.com AG	3/15/00
Gigabell AG	8/11/99	TXT e-solutions S.p.A.	7/12/00
Groupe Neurones S.A.	5/24/00	UBIZEN N.V.	2/10/99
Himalaya S.A.	3/28/00	United Internet AG	3/23/98
Hi-Média S.A.	6/8/00	Uproar Inc.	7/13/99
HUBWOO.com S.A.	9/28/00	UTIMACO Safeware AG	2/16/99
I.net S.p.A.	4/4/00	Vitaminic S.p.A.	10/12/00
IB Group.Com S.A.	6/27/00	Wanadoo S.A.	7/19/00
IBNet plc	3/9/00	WEB.DE AG	2/17/00
I-D Media AG	6/17/99	World Online International N.V.	3/17/00
i:FAO AG	3/1/99	WWL Internet AG	7/15/99

New Venture Investment: Choices and Consequences
A. Ginsberg and I. Hasan (editors)

Chapter 12

Deliberate Underpricing and Price Support: Venture Backed and Nonventured Backed IPOs

BILL B. FRANCIS[a,**], IFTEKHAR HASAN[b,*] and CHENGRU HU[c]

[a] *College of Business Administration, University of South Florida, Tampa, FL 33620, USA*
[b] *Lally School of Management, Rensselaer Polytechnic Institute, Troy, NY 12180-3590, USA*
[c] *School of Management, Rutgers University, Newark, NJ 07012-1982, USA*

ABSTRACT

In this chapter we examine the premarket underpricing phenomenon within a group of venture-backed and a group of non venture-backed initial public offerings (IPOs) using a stochastic frontier approach. Consistent with previous research, we find that venture-backed IPOs are managed by more reputable underwriters and are generally associated with less underwriter compensation. However, unlike other papers in the literature, we find that the initial day returns of venture-backed IPOs are, on average, higher than the non venture-backed group. We observe a significantly higher degree of pre-market pricing inefficiency in the initial offer price of venture-backed IPOs. Further, our results show that a significant portion of the initial day returns is due to deliberate underpricing in the premarket. We also observe that for both venture and non-venture issuers, there is a positive relationship between deliberate underpricing and the probability that underwriters provide support

* Corresponding author. Tel.: +1-518-276-2525; fax: +1-518-276-8661.
** Tel.: +1-813-974-6330.
E-mail addresses: bfrancis@coba.usf.edu (B.B. Francis), hasan@rpi.edu (I. Hasan)

for the issue. This evidence is consistent with the notion that underwriters deliberately underprice the offering to reduce costs of price stabilization in the after-market.

1. INTRODUCTION

One of the most widely investigated phenomena in the finance literature is the price performance of initial public offerings (IPOs) of common stock (see, e.g. Brav and Gompers, 1997; Ibbotson, 1975; Ibbotson, Sindelar and Ritter, 1988; Ritter, 1991, among others). These studies document the presence of significant average underpricing defined as the percent difference between the closing day bid price and the initial offer price over the first day of trading. Work by Barry et al. (1990), Megginson and Weiss (1991) and Brav and Gompers (1997), to name a few, indicate that the degree of underpricing depends on whether or not an IPO is backed by venture capitalists (VC). Specifically, they show that non VC-backed IPOs are more severely underpriced than VC-backed IPOs. The authors contend that this difference in the degree of underpricing is primarily due to the certification role of venture capitalists. An important implication of this is that VC-backed IPOs are characterized by less deliberate underpricing in the pre-market in comparison to non VC-backed IPOs, rather than other factors in the after-market which may contribute to underpricing such as underwriter price support (see, e.g. Hanley, Kumar and Seguin, 1993; Rudd, 1983).[1]

Our purpose here is twofold. First, we investigate whether the difference in underpricing between VC-backed IPOs and non VC-backed IPOs is due to deliberate underpricing in the pre-market or to other factors such as underwriter price support in the after market. Second, we examine the role deliberate underpricing plays in the IPO process as it relates to initial day returns and price stabilization.

We measure underpricing using information available only in the pre-market period. We follow Hunt-McCool, Koh and Francis (1996) to ascertain possible differences in the underpricing between VC- and non VC-backed issuers. In doing so, we attempt to understand whether the source(s) of the difference in underpricing is due to deliberate under-pricing in the pre-market or to forces in the after-market. The evidence indicates VC-backed IPOs with higher underpricing in than the non VC-backed IPOs immediately after the issuance as for the sample the deliberate pre-market underpricing is also significantly higher for VC-backed IPOs. Thus, the

[1] Here we use the term pre-market to indicate the time before the IPO is actually issued. And the after market to indicate the time once trading commences.

results suggest that the documented difference in underpricing between VC-backed and non VC-backed IPOs cannot be fully attributed to post-market phenomena alone. Rather, a significant portion may be due to deliberate underpricing as a means of compensating investors. Importantly, the results indicate that for both VC- and non VC-backed IPOs deliberate underpricing in the pre-market leads to a higher probability of price stabilization in the after-market. However, there is a higher likelihood of price support for non VC-backed IPOs.

The chapter is organized as follows. In Section 2, we briefly review the existing research on the phenomenon of IPO pricing. Section 3 explains the concept of a full-information IPO price frontier and the associated estimator capturing systematic departures from the pricing frontier (deliberate underpricing). Section 4 contains a description of the data and model specifications and univariate analyses. Section 5.4 reports the relationship between deliberate underpricing and price support. Section 5 provides empirical results of the stochastic frontier estimation and the initial returns regressions. Additional evidence on the robustness of our results is provided in Section 5.5. Our conclusions are presented in Section 6.

2. RELATED LITERATURE

2.1. Underpricing of IPOs

The IPO literature generally ascribes the underpricing of IPOs to the existence of pre-market information asymmetry and views the abnormal initial return as an effort to compensate investors. Baron (1982) explains IPO underpricing as an effort by the issuing firm to compensate underwriters for their holding of superior security-related information. Rock (1986) derives a model in which IPO offering prices are reduced so that investors earn a positive return even if they are less informed about the issue. Grinblatt and Hwang (1989) view the underpricing phenomenon as a signal by the issuing firm to the market concerning the quality of their securities. Welch (1989) and Allen and Faulhaber (1989) claim that IPO underpricing is intended for high proceeds in future sales of seasoned equity issues for high-quality firms. Hunt-McCool et al. (1996) find evidence consistent with deliberate underpricing of IPOs in the pre-market.

In general, the underpricing of IPOs is regarded as deliberate underpricing in the pre-market by the issuing firms. However, Ritter (1991) found that IPOs are overpriced in the after market. In replicating Ritter's work within a venture vs. non-venture backed IPO framework, Brav and Gompers (1997) find that the observed after-market under performance is fully attributed to non-venture backed IPOs. That is, venture-backed IPOs do not under perform a portfolio of comparable sized firms.

Megginson and Weiss (1991) present evidence indicating that VC-backed IPOs are associated with higher underwriter prestige, higher institutional holdings, and lower level of underpricing than non VC-backed IPOs. Lerner (1994) examines the timing of initial public offerings and finds that venture capitalists take firms public at market peaks and rely on private financing when the market valuation of equity is low. He further notes that the more experienced the venture capitalists, the more proficient they are in the timing of taking firms public, thus reducing the level of underpricing.

However, venture funds are generally associated with higher levels of risk. Carleton (1986) shows that capital projects of venture capitalists are investments in new, small, and risky companies especially those based on commercial application of technological innovations. The level of riskiness associated with venture investment is reflected in the relatively long period of time between initial investment and final repayment (usually 10 or more years), and in the 20–30% return to venture capitalists. According to Megginson and Weiss (1991), VC-backed IPOs go public at a significantly earlier stage than non VC-backed IPOs, with little information disclosure before the going public date.

The foregoing suggests that there is less public information available concerning venture investments. Thus pre-market information asymmetry, if any, may be more severe for VC-backed offers compared to non VC-backed IPOs. It is therefore an empirical issue as to whether the certification role of venture capitalists dominates the effect that VC-backed IPOs tend to be younger and riskier than non VC-backed IPOs. If the latter effect dominates, then VC-backed IPOs should be characterized by more deliberate underpricing (i.e. larger systematic deviations from the efficient frontier) than non VC-backed IPOs.

2.2. Price stabilization

Recently, the issue of price stabilization of IPOs in the after market has received much attention (see, e.g. Agarwall, 2000; Prabhala and Puri, 2000; Rudd, 1983). Although stabilization of IPOs is a form of price manipulation, it is exempted from the provisions of the 1934 Securities Act, if it is carried out within certain guidelines. Price stabilization has existed since the 1930s, nevertheless not much is understood about its impact on the going public process and the factors that determine whether a particular issue is stabilized in the after market.

Prabhala and Puri (2000) argue that price stabilization has an effect on the IPO process by changing the ex-ante uncertainty of the offering. Specifically, they contend that underwriters in offering price support in the after-market, are, in effect, providing investors with a put option on the IPO being supported. Thus, it is within an underwriter's best interest to reduce the ex-ante price uncertainty and thereby the value of the put-option to investors.[2]

To reduce the ex-ante price uncertainty, and thus the cost of the put option, underwriters can deliberately under-price the IPO. If underwriters use this method, then we would expect a positive relationship between deliberate underpricing in the pre-market and the likelihood of price support. It should be noted that if the conjecture of Prabhala and Puri (2000) and Cantale (1999) were supported, we would expect a negative relationship between deliberate underpricing and ex-ante risk.

3. METHODOLOGY

3.1. The stochastic frontier

A point on the stochastic frontier represents the maximum price that would prevail for a given IPO if all parties in the trade had full information. The difference between any given offer price and the maximum price under full information would be the result of random error alone. There would be no systematic underpricing, and the frontier price could be computed by using ordinary least squares to estimate the expected price given the observable variables. If, however, there is deliberate underpricing (a systematic negative bias) the offer price will fall below the maximum potential by an amount or fraction represented by the one-sided error. In this case, there is both a stochastic error and a systematic, one-sided error. Under the stochastic frontier maximum likelihood estimation, any systematic error, if it exists, will appear in the form of skewness in the residuals and can be separately computed for each IPO. The average deviation from the full information price can also be computed for each type of IPO.[3]

This methodology provides a direct test of the existence of deliberate underpricing. If OLS estimation is equivalent to the maximum likelihood (ML) estimation, no systematic underpricing exists in the pre-market, and the difference in underpricing would be due to forces in the after-market such as price stabilization. It should be noted that if deliberate underpricing in the pre-market is not found, then it would support the notion that non VC-backed IPOs are provided more after-market price support than VC-backed IPOs.

---

[2] Recently Cantale (1999) shows that the empirical regularity of underpricing could be solved.

[3] For details on the methodology and maximum likelihood estimator, see Hunt-McCool et al. (1996) and Hasan and Francis (2001).

3.2. The maximum likelihood estimator

The stochastic frontier model of ALS is estimated via maximum likelihood (ML) methods. The frontier and distributional assumptions can be expressed as:

$$P_i = f(\boldsymbol{X}_i; \boldsymbol{\beta}) + e_i, \quad i = 1, 2, \ldots, n, \tag{1}$$

where, $e_i = v_i + u_i$, $v_i \sim N(O, \sigma_v^2)$, $u_i \sim N[\sqrt{2}/\sqrt{\pi}\sigma_u, \sigma_u^2]$, $u_i = \min('u_i, 0)$.

Here, P_i is the log of observed initial offer price at the time of offering; $\boldsymbol{X}$ is a vector of firm and offering method characteristics; $\boldsymbol{\beta}$ is a vector of coefficients of the IPO pricing frontier; v denotes the symmetric error component; and u is the asymmetric component and is truncated at zero. The composite error term is e. The non-positive error term u_i is interpreted to mean that the actual price must lie on or below the true but unknown frontier, while the two-sided error v_i suggests that observed prices may lie above or below the estimated frontier due to statistical noise.

Following ALS (1977), the density function is:

$$f(e_i)\left(\frac{2}{\sigma}\right)f\left(\frac{e}{\sigma}\right)[1 - F(e_i\lambda\sigma^{-1})], \tag{2}$$

where, $\sigma^2 = \sigma_u^2 + \sigma_v^2$; $\lambda = \sigma_u/\sigma_v$; $-\infty < e_i < +\infty$; and $f(e_i/\sigma)aF(e_i\lambda\sigma^{-1})$ are the standard normal density and distribution functions respectively.

If the P_i's are independently distributed, and the non-stochastic part of P_i is explained by a set of exogenous variables $\boldsymbol{X}$, the log-likelihood function becomes:

$$\ln L(P, \beta, \lambda, \sigma^2) = N \ln\left(\frac{\sqrt{2}}{\sqrt{\pi}}\right) + N \ln s^{-1} + \sum_i \ln[1 - F(e_i\lambda s^{-1})]$$
$$-\left(\frac{1}{2}\sigma^2\right)\sum_i e_i^2, \tag{3}$$

where, $e_i = P_i - \beta' X_i$, $i = 1, \ldots, N$.

ALS (1997) provide a means for obtaining the optimal values of $\boldsymbol{\sigma}^2$, $\boldsymbol{\beta}$, and λ.[4] λ, the skewness in the error component (u_i) relative to the symmetric random disturbance (v_i), measures gains in statistical efficiency of the frontier estimator over OLS. The estimate of λ has a straightforward interpretation for inferring

[4] Note that $M(u_i/e_i)$ is an observation-specific estimate of the asymmetric error. Given that the dependent variable (log of the offer price) in the first stage regression is expressed in logarithmic terms, $u_i^*/P_i^* = 1 - \exp M(u_i/e_i)$.

deliberate IPO underpricing. If λ goes to zero, deviations in actual prices from the frontier are simply due to statistical noise. On the other hand, if λ is greater than zero, deviations from the frontier are characterized by some form of deliberate discounting in the pre-market pricing of IPOs. Thus, as λ becomes large, deliberate discounting in the pricing of IPOs is indicated.

3.3. Specification of initial returns regression

As in Rudd (1983), we interpret the existence of skewness in the measured returns to suggest excess returns. If after-market returns are randomly distributed around the full information offer price, the mean excess return will be zero, and the distribution will be symmetric around this mean. In our sample, 593 of the 843 IPOs have a positive return measured by comparing the first day closing price with the pre-market offer price. Of course, the measurement of excess returns remains dependent upon the pre-market offer price. If excess returns are an artifact of measurement (resulting from failure of the offer price to correctly indicate the economic value of the firm due to deliberate underpricing), they should disappear once adjustments are made for pre-market underpricing. However, if they exist independent of underpricing, the correlation between the measured returns and measured underpricing may be close to zero. To test for a relationship between after-market returns and pre-market underpricing for both VC-backed and non VC-backed IPOs, we first estimate the P_i^*s, the predicted maximum prices for each IPO. Using the modal formula proposed by Jondrow et al. (1982), the level of systematic underpricing for each IPO is computed. The formula is

$$\begin{aligned} U_i^* = \text{Mode} &= -e_i\left(\frac{\sigma_u^2}{\sigma^2}\right) \quad \text{if } e_i \leq 0, \\ &= 0 \qquad\qquad\quad\; \text{if } e_i > 0. \end{aligned} \tag{4}$$

The second stage involves the approach used by Hunt-McCool et al. (1996), where the deliberate underprice score, u_i^*, is used as an independent variable in a regression in which the dependent variable is the initial day return:[5]

$$\text{RET}_i = \alpha + \beta(u_i^*) + \varepsilon_i, \tag{5}$$

[5] It should be noted that care should be taken in interpreting results from this specification. Pagan (1984) points out that the use of a generated regressor introduces errors in variables problem in the regression results. In this case given that there is only one regressor the estimated coefficient will be biased downwards. Thus our test of a significant relationship will be a conservative one.

where, RET_i is measured from the offer price to the close (or bid) price at the end of first trading day; α and β are parameters to be estimated; the ratio of u_i^* is the issue-specific deliberate underpricing and ε_i is the random error term. Separate OLS regressions are estimated for both VC-backed and non VC-backed IPOs.[6]

4. DATA SELECTION

4.1. Description of variables

Our sample consists of firm commitment IPOs that went public during the 1990–1993 period. IPO prospectuses (where available), Investment Dealers Digest (IDD), various issues of Venture Capital Journal were the main data-sources. Additionally, underwriter rankings were obtained from Carter, Federick and Singh (1998). We kept the penny stock offerings out of our sample by taking a minimum of $5. Stock prices of all companies were obtained from the CRSP tape and the Standard & Poor's daily stock price record. We dropped companies with missing values for stock prices or other variables that are used in our estimations. The final sample contains 843 IPOs of which 415 are VC-backed. To examine the degree of pre-market pricing inefficiency associated with VC—and non-VC-backed IPOs, we use information available prior to the commencement of trading. Following Hunt-McCool et al. (1996), the dependent variable we use is the natural log of the initial offer price. The choice of independent variables is motivated by the underlying theory and by the existing literature.

According to the signaling model (Grinblatt and Hwang, 1989), the percentage of insider ownership (INSFR) signals to the market the amount of private information possessed by insiders. The better information insiders have the larger the fraction of shares they will retain. Thus a positive relation between insider fraction and the offer price is expected. Further, because of the direct participation of venture capitalists in monitoring and managing these firms, the relation should be stronger for VC-backed firms.

Underwriters convey firm-related information to the public during the going-public process of the firm. In fact, underwriters are compensated for reducing the information asymmetry between the issuing firm and the market. The larger the information asymmetry, the more important the underwriter's role, and the greater the level of compensation. Thus we should observe a negative relation between the IPO's offer price and the underwriter compensation (COM). The natural log

[6] For more details on the differences see Francis and Hasan (2000).

of the total proceeds from the issue (LPROC) (measured in millions of dollars) is included to control for the level of asymmetry. Tinic (1988) points out that smaller, riskier firms often make small offerings. And according to Dunbar (1997), firm size conveys additional information to outsiders. Thus we expect a positive relation between the IPO offer price and issue size.

Underwriter ranking (UWRANK) has long being recognized as an important factor in determining the degree of underpricing and therefore the offer price. Megginson and Weiss (1991), Carter and Manaster (1990) and Carter et al. (1998), among others, document a negative relation between underwriter ranking and IPO underpricing. These authors also show that the more reputable the underwriter, the lower the degree of underpricing and thus the higher the offer price relative to the full information offer price.

Following Dunbar (1995), we utilize the reciprocal of the issuing company's size (INV) as a proxy for the riskiness of the offerings. We expect that securities that are riskier and characterized by a higher degree of information asymmetry to be priced below the frontier. Finally, we incorporate a binary variable assuming a value of 1 for IPOs floated in New York Stock Exchange (NYSE) and a value of zero for IPOs introduced in other markets (e.g. NASDAQ and AME) to capture the prestige of the initial listing decision. If IPOs are issued on an exchange that is prestigious, well known to investors and has very strict listing requirements, the expected underpricing would be less, and thus would be priced closer to the frontier.

Tables 1 and 2 provide frequency distribution and summary statistics for the IPOs contained in the sample. Venture capital backed IPOs account for approximately one-half of the IPOs in each year, with the number of IPOs generally increasing over the 4-year sample period. The average initial offer price for VC-backed issues is $11.70, significantly different from the non VC-backed IPO price of $13.16. Similarly, the average initial day return of 13.50% by VC-backed IPOs is significantly different than the 10.06 return posted by the non VC-backed IPOs the average initial day return is 10.06%. This finding is significantly different from those reported by Megginson and Weiss (1990) and by Brav and Gompers

Table 1. Distribution of IPOs by year: VC-backed vs. non VC-backed

Year	VC-backed	Percent of all IPOs in the year	Non VC-backed	Percent of all IPOs in the year	Total
1990	40	48.2	43	51.8	83
1991	113	57.1	85	42.9	198
1992	139	74.3	48	25.7	187
1993	123	32.8	252	67.2	375

Table 2. Comparative descriptive statistics

	Full sample	VC-backed IPOs	Non VC-backed IPOs	t-Statistics for difference in means
Price0 (offer price)	12.44 (4.72)	11.70 (3.73)	13.16 (15.07)	4.56**
SIZE (in 000) of the issuing firm	64,925.75 (78,381)	46,529.81 (26,314.94)	78,695.83 (19,364.18)	6.75**
Offering proceeds (in 000)	48,592.48 (81,515.15)	32,245.81 (24,336.90)	64,554.86 (109,785.48)	6.58**
INSFR (insider fraction)	28.83 (16.10)	29.68 (20.55)	27.99 (18.34)	2.87*
UC (underwriter compensation)	6.944 (1.204)	6.930 (1.342)	6.981 (1.536)	1.80#
UWRANK (underwriter ranking)	7.530 (1.983)	8.072 (1.412)	7.530 (1.983)	4.56**
Return (initial return)	11.37854 (16.65)	13.50 (17.98)	10.06 (15.07)	3.00**
Number of observations	843	415	428	–

Note: **, *, # indicates significant at 1, 5, and 10% significance level.

(1997). However these results support the theoretical work of Welch (1989) and Chemmanur (1993) that predicts that high-quality IPOs are more underpriced so that follow-up seasoned equity offers are more favorably received by investors.

On average, VC-backed issues are smaller than those of non VC-backed firms. This is inconsistent with Megginson and Weiss (1991) and Brav and Gompers (1997) findings but consistent with Carleton's (1986) argument that venture capitalists generally bring firms to the market at a relatively earlier stage of development. Non VC-backed firms also issue significantly more shares to the public than VC-backed firms. But the fraction held by insiders is not significantly different between the two subgroups, which indicates that the market reacts more favorably to non VC-backed IPOs. The results also show that on average VC-backed firms hire more prestigious underwriters and pay significantly less compensation for underwriters' services. This is consistent with the notion that VC-backed firms because of the certification provided by venture capitalists can attract more reputable underwriters, thereby reducing the costs of going public.[7] Interestingly, our results show that VC-backed IPOs use more reputable underwriters, they still display a higher level of underpricing. This suggests that the lower level of underpricing of VC-backed IPOs documented in previous studies may not be due to the certification role of underwriters.

5. EMPIRICAL RESULTS

5.1. Estimates of stochastic frontier

Table 3 presents the results of the estimates of the price frontier conditioned on pre-market information. Almost all of the coefficients are statistically significant and have the expected signs. INV (FIRM SIZE) has a negative impact on the maximum potential offer price, implying that the higher the risks of IPOs, the lower the offer price. INSFR is positive and significant. This is consistent with the notion that the better the information insiders have, the larger the fraction of shares they will retain. A positive association is found between the underwriter ranking (UWRANK) and the offer price. This supports the certification role of underwriters, which contends that more prestigious underwriters provide more certification to the market about the quality of the IPOs and thus lessens the pre-market information asymmetry.

[7] In these additional estimates we eliminated independent variables that were highly correlated, e.g. INV highly correlated with LPROCEED (0.79) and UWRANK is highly correlated with COMP (0.66) and so on.

Table 3. Estimated frontier

Independent variables	Parameters	Standard errors
Constant	1.8244	0.3457**
INV (FIRM SIZE)	−5.0829	1.338**
LPROCEED	0.1357	0.0329**
COMP	−0.09E − 03	0.03E − 05**
UWRANK	0.0285	0.0116**
INSFR	0.82E − 03	0.29E − 03*
NYSE (CAPITAL MARKET DUMMY)	0.1147	0.0355**
λ	1.7353	0.1824**
σ	0.4406	0.0692**
Log likelihood function	−753.068	
Number of observations	843	

Note: Dependent variable is log of initial offer price. Independent variable includes INVSIZE, inverse of total size of the offering firm, logarithm of the proceed of the offering firm LPROCEED, underwriter compensation to offering ratio COMP, underwriter rank UWRANK, insider holding to the total offerings INSFR, and CDUM, a binary variable representing the capital market where the offering was introduced. CDUM taking a value of 1 for NYSE market and a value of 0 for other markets. λ indicates whether inefficiency exists or not. Inefficiency is defined as in Section 2.
**, * significant at 0.01 and 0.05 levels respectively. The corrected R^2 of the OLS model = 78.9%.

Our results show an inverse association between the offer price and total compensation. This is consistent with the hypotheses that additional certification function provided by underwriters is less important for high quality issues thus requiring less compensation for underwriting the securities (or the marginal certification role of underwriters would decrease with the offer price). The capital market binary variable (NYSE) also displays the expected positive relationship with the log of offer price.[8]

The coefficient representing the presence of systematic deviations from the maximum potential offer price, λ, is highly significant. This finding is consistent with that of Hunt-McCool et al. (1996) that the offer prices for IPOs are priced below the maximum potential offer price and that IPOs are characterized by deliberate underpricing. Additional estimates of the offer price frontier using different combinations of independent variables also produced a highly significant λ in all estimates thus confirming the robustness of the frontier estimate reported in this paper.

Several authors (see, e.g. Dunbar, 1997) present evidence showing that the level of underpricing is nonlinear in firm size. To test this hypothesis we re-estimated

[8] All of these additional estimates we available upon request.

Table 4. Deliberate underpricing differences between VC-backed and Non VC-backed IPOs

	Full sample	VC-backed IPOs	Non VC-backed IPOs	*t*-Statistics difference in VC–non VC mean
Inefficiency	0.291 (0.203)	0.353 (0.164)	0.271 (0.134)	7.54**

Note: Inefficiency score from regression using stochastic frontier estimation. Dependent variable is log of initial offer price. Independent variable includes INVSIZE, inverse of total size of the offering firm, logarithm of the proceed of the offering firm LPROCEED, underwriter compensation to offering ratio COMP, underwriter rank UWRANK, insider holding to the total offerings INSFR, and CDUM, a binary variable representing the capital market where the offering was introduced. CDUM taking a value of 1 for NYSE market and a value of 0 for other markets. Inefficiency is defined as in Section 2. We also give the test result for the hypothesis that the inefficiency scores from the two regressions are equal and the *t*-statistics for the difference in mean scores is provided in the last column.

**, * significant at 0.01 and 0.05 levels respectively.

the model with a quadratic term (INV^2 or $Size^2$) in the regression. Results not reported indicate that deliberate underpricing (significant parameter λ) still exists. However, the inclusion of the quadratic-term in the regression significantly lowered the overall model goodness of fit statistics (asymptotic corrected R^2) from 0.680 to 0.562. Consequently it is not included in any further analysis.[9]

Next, we investigate whether the level of deliberate underpricing is different for VC- and non VC-backed IPOs. Average statistics for deliberate underpricing is provided in Table 4. On separating our sample in to VC-backed and non VC-backed IPOs, the results indicate that the deliberate underpricing embodied in the offer price of VC-backed IPOs is significantly higher (0.353) than that of non VC-backed IPOs (0.271). This finding, contrary to other studies, is consistent with our earlier findings that VC-backed IPOs are characterized by a greater level of underpricing, and thus supports the notion that, at least for the IPOs in our sample, VC-backed IPOs are more deliberately underpriced than the non VC-backed group.

As noted earlier the estimated stochastic frontier represents the full information price. Given that the level of deliberate underpricing is obtained relative to this frontier, it is plausible that our measure of underpricing (the mode of the distribution) is a statistical noise.[10] We check the robustness of the reported underpricing estimates—obtained using the half-normal distribution—by re-estimating the

[9] We thank the referee for pointing this out to us.

[10] There is no public available data indicating which of the IPOs are price-supported. In keeping with the literature (see, e.g. Hanley et al., 1993; Prabhala and Puri, 2000; Rudd, 1993), we assume that price-supported IPOs have an initial day return of zero. Evidence provided by Asquith, Jones and Kieschnick (1998) and Prabhala and Puri (2000) is consistent with this assumption.

model assuming an exponential distribution of u_i. Under this new assumption, the average score of deliberate underpricing is reported to be 0.266 for the combined sample as oppose to a score of 0.291 obtained using previously reported half-normal distribution assumption. The difference of these scores under two different distribution assumptions however is not statistically significant at 10% significance level. Partitioning the sample into VC-backed and non VC-backed sub-samples, we find the underpricing score to be 0.328 for the VC-backed sample and 0.245 for the non VC-backed sample. These scores and differences between the two groups are qualitatively similar to the reported results (0.353 for the VC-backed sample and 0.271 for the non VC-backed sub-sample) under the half-normal distribution assumption for u_i. Thus, it confirms that the reported deliberate underpricing estimates are robust to different distributional assumptions on u_i.

5.2. Relation between underpricing and initial returns

The above analysis establishes the empirical fact that IPOs are deliberately underpriced in the pre market and that the level of underpricing is significantly higher for VC-backed than for non VC-backed IPOs. Earlier, we suggested that the initial day return might be due to deliberate underpricing in the pre-market. We test this hypothesis estimating OLS regressions using the initial day returns as the dependent variable and the level of deliberate underpricing as the independent variable. Regression results are presented in Table 5. Evidence indicates a positive and significant relationship between initial day return and our pre-market measure of underpricing for the sample. The results indicate that a 1% increase in deliberate underpricing will result in about a 1.98% increase in initial day returns. On separating the sample into VC-backed and non VC-backed groups, we find results consistent with those of the full sample.

Table 5. Initial returns and deliberate underpricing

	Full sample	Venture backed IPOs	Non-venture backed IPOs
Constant	10.266 (4.093)*	8.943 (4.082)*	9.461 (3.647)*
U'	1.975 (0.406)**	3.061 (0.654)**	1.730 (0.698)*
R^2	0.1458	0.1172	0.1039

Note: OLS regressions of initial return on the measure of pre-market deliberate underpricing to initial offer price. Initial return is defined as difference between first trading day closing price and initial offer price divided by initial offer price. Inefficiency is defined as in Section 2. Heteroscedasticity corrected standard errors are reported in the parentheses.

** and * means significant at 0.01 and 0.05 level respectively.

Table 6. Deliberate pre-market underpricing and size

	Deliberate underpricing VC-backed IPOs (quartiles by size)				Deliberate underpricing of non VC-backed IPOs (quartiles by size)			
	1	2	3	4	1	2	3	4
Mean	0.387*	0.346*	0.301*	0.264	0.322	0.306	0.260	0.258
Median	0.357*	0.334*	0.285*	0.260	0.313	0.272	0.258	0.251

Note: Dependent variable is log of initial offer price. Dependent variable is log of initial offer price. Independent variable includes INVSIZE, inverse of total size of the offering firm, logarithm of the proceed of the offering firm LPROCEED, underwriter compensation to offering ratio COMP, underwriter rank UWRANK, insider holding to the total offerings INSFR, and CDUM, a binary variable representing the capital market where the offering was introduced. CDUM taking a value of 1 for NYSE market and a value of 0 for other markets. Inefficiency is defined as in Section 2.
* means significantly different from the non VC-Backed IPOs in the same quartile category at least at the 5% significance level.

5.3. Deliberate underpricing and size

Brav and Gompers (1997) show that based on size, there is a difference between the long-run performance of VC-backed and non VC-backed IPOs. To investigate whether there is a relationship between pre-market underpricing and size, we separate our sample first into VC- and non VC-backed IPOs and then, using each group, form four portfolios. Table 6 shows mean and median efficiency scores by quartile size (proceeds) using the measure of underpricing obtained from our frontier estimates. The data shows an interesting trend. Based on both mean and median estimates of VC- and non-VC backed IPOs, we observe a decrease in pre-market underpricing as size increases. For the VC-backed sub-sample, the mean level of deliberate underpricing in the first (smallest) quartile is 0.387 and 0.264 in the fourth (largest) size group. The median estimates also reveal a similar trend where deliberate underpricing declines from a first quarter high of 0.357 to a fourth quarter low of 0.260. Similar results are observed for the non-VC sub-sample. Consistent with our earlier findings, the VC-backed sub-sample still records a higher degree of pre-market underpricing in both the mean and median estimates for all quartile groups. However, the differences between the two sub-samples are insignificant when the upper-most quartile is compared, which indicates that pre-market underpricing differences are more prominent among the smaller issuers. To the extent that there is more ex-ante pricing uncertainty associated with smaller firms, this finding supports our earlier results that deliberate underpricing is undertaken to reduce ex-ante pricing uncertainty. We also re-estimate Equation 5 for each size quartiles. Evidence in Table 7 shows consistent evidence where the impact of pre-market underpricing

Table 7. Initial return and deliberate underpricing by quartiles

	Quartiles based on size			
	Quartile 1	Quartile 2	Quartile 3	Quartile 4
Constant	8.045 (2.783)*	7.083 (3.801)+	7.645 (3.022)*	8.145 (3.052)*
U'	1.752 (0.476)**	1.408 (0.594)*	2.087 (1.151)+	3.249 (2.450)
R^2	0.0679	0.0656	0.0504	0.0393

Note: The relationship between initial return and inefficiency score investigated for each quartiles of sample based on offering size. Dependent variable is initial return and independent variable is the deliberate underpricing or price inefficiency in the pre-market. Heteroscedasticity corrected standard errors are reported in the parentheses.

**, *, + means significant at 1, 5, and 10% level respectively.

is positively and significantly related to initial returns for all groups except the largest quartiles.

5.4. Deliberate underpricing and price support

In this section, we examine whether deliberate underpricing plays a role in determining the underwriters' decision to provide price support for IPOs. We earlier suggested that by deliberately underpricing the IPOs in the pre-market, investment bankers are reducing the price risk of IPOs brought to the market. That is, by deliberately underpricing the IPOs, underwriters are reducing the value of the put option given to investors. To test this conjecture, we estimate a logit model with the dependent variable taking a value of 1 for IPOs that are supported and 0 otherwise.[11] This is regressed against our estimated measure of deliberate underpricing. Deliberate underpricing in the pre-market is expected to have a positive effect on the probability that the IPO will be price-supported.

The logit estimates are reported in Table 8. Consistent with our conjecture, the results indicate that IPOs that are deliberately underpriced in the pre-market are price-stabilized. This finding applies to the full sample and also to the sample when it is separated into VC- and non VC-backed IPOs. The results indicate that although deliberate underpricing is higher for VC-backed IPOs, it has a higher explanatory power for non VC-backed IPOs. A possible explanation is that the certification

[11] In examining financial institutions, Mester (1993) use a similar estimating approach that focused on the different production technologies among a group of financial institutions based on their organizational forms.

Table 8. Likelihood of receiving price support from the underwriters (logistic regression analysis of price support)

Variables	Parameters (Chi-square statistics)		
	Combined sample	VC-backed IPOs	Non VC-backed IPOs
Constant	−0.973 (4.86)**	−1.038 (4.02)**	−1.544 (5.07)**
Deliberate underpricing	0.016 (3.05)*	0.011 (2.68)*	0.016 (3.34)*
Model statistics			
−2 log likelihood	785.92	363.82	274.01
Model Chi-square	138.06*	119.99*	105.26*
Number of observations	843	415	428

Note: The dependent variable is the price support dummy variable which takes a value of 1 if the initial day return is zero and 0 otherwise. Independent variable is the pre-market pricing inefficiency (deliberate underpricing). Inefficiency scores are based on estimates that assumes all IPOs (VC and non-VC backed samples) have a similar price technology. Chi-square statistics are reported in the parentheses.

**, *, # indicates significant at 1, 5, and 10% significance level.

role of venture capitalists helps to reduce the pre-market pricing uncertainty. In summary, to the extent that deliberate underpricing reduces price uncertainty in the pre-market, our results are consistent with the hypothesis that underwriters provide price support for IPOs that are characterized by less ex-ante pricing uncertainty.

5.5. Additional evidence

So far our results indicate that VC-backed IPOs are different from non VC-backed IPOs. As a result it is possible that VC- and non VC-backed IPOs are characterized by different pricing frontiers.[12] To examine the robustness of our results we estimate additional stochastic frontier regressions allowing the pricing technology and error structure to differ between the VC-backed and non VC-backed IPOs. The λ—which represents systematic deviations from the maximum potential offer price—is significant in both cases, with its significance being stronger for the VC-backed IPOs. Table 9 presents results of additional estimates of the price frontier for VC- and non VC-backed IPO sub-samples based on the assumption that the two groups are characterized by different pricing technologies. The magnitude, strength, and significance of the coefficients are similar to the findings reported in

[12] All of these estimates are available upon request.

Table 9. Estimated frontier

Independent variables	Parameters	
	VC-backed	Non VC-backed
Constant	2.065*	2.661*
INVSIZE	3.382**	4.906**
LPROCEED	0.114**	0.073**
COMP	−0.10E − 04*	−0.05E − 04*
UWRANK	0.024*	0.019**
INSFR	0.72E − 04**	−0.75E − 04*
Capital market dummy	0.082**	0.102**
λ	2.365**	3.072**
σ	0.401*	0.255*
Log likelihood function	−674.920	−588.03
Number of observations	415	428

Note: Dependent variable is log of initial offer price. Independent variable includes INVSIZE, inverse of total size of the offering firm, logarithm of the proceed of the offering firm LPROCEED, underwriter compensation to offering ratio COMP, underwriter rank UWRANK, insider holding to the total offerings INSFR, and CDUM, a binary variable representing the capital market where the offering was introduced. CDUM taking a value of 1 for NYSE market and a value of 0 for other markets. λ indicates whether inefficiency exists or not. Inefficiency is defined as in Section 2. Regression results are given for VC- and non VC-backed sub-samples where Inefficiency estimates here are based on separate estimates of VC and non-VC backed samples, i.e. assuming a different pricing frontier.

**, * significant at 0.01 and 0.05 levels respectively. The corrected R^2 of the OLS model for VC Backed sample was 48.06% and for the non VC Backed sample it was 52.39%.

Table 3. However, in some cases the statistical significance is found to be even more robust in VC-backed sub-samples. The output λ—which represents systematic deviations from the maximum potential offer price—is significant in both cases, with its significance being stronger for the VC-backed IPOs.

Using the inefficiency scores from the separate estimations for the two groups, we also attempt to confirm previous findings that deliberate underpricing affects the initial day returns. The results are documented in Table 10. Irrespective of the sub-sample considered, we find a positive and significant impact of pre-market deliberate underpricing on initial returns in all regressions. Moreover, we find (Table 11) that the significance of pre-market coefficient for the VC-backed sub-sample is stronger for the estimates obtained based on the assumption of different pricing technologies. Thus, there is additional support for the hypothesis that IPO underpricing can be partly explained by deliberate underpricing in the pre-market. Apparently, VC-backed IPOs play a stronger role in deliberate underpricing then their non VC-backed counterparts.

Table 10. Initial returns and deliberate underpricing: Assuming different pricing technology between VC- and non-VC backed IPOs

	Venture backed IPOs	Non-venture-backed IPOs
Constant	10.452 (4.012)*	9.446 (4.072)*
U^*/P_i^*	3.146 (0.295)**	1.907 (0.526)**
R^2	0.1042	0.0927

Note: OLS regressions of initial return on the measure of pre-market deliberate underpricing to initial offer price. Initial return is defined as difference between first trading day closing price and initial offer price divided by initial offer price. Inefficiency is defined as in Section 2. Heteroscedasticity corrected standard errors are reported in the parentheses. Here the inefficiency scores are based on separate estimates of VC and non-VC backed samples, i.e. assuming a different pricing technology for each group. Heteroscedasticity corrected standard errors are reported in the parentheses.

$^{**}, ^{*}, ^{+}$ means significant at 1, 5, and 10% level respectively.

Table 11. Likelihood of receiving price support from the underwriters (logistic regression analysis of price support; different pricing technology for VC and non-VC Sub-samples)

Variables	Parameters (Chi-square statistic)	
	VC-backed IPOs	Non VC-backed IPOs
Constant	−1.176 (4.27)**	−1.480 (6.93)**
Deliberate underpricing (Pre-market inefficiency)	0.019 (4.46)**	0.016 (3.08)*
Model statistics		
−2 log likelihood	370.09	364.56
Model Chi-square	119.56**	104.32**
Number of observations	415	428

Note: The dependent variable is the price support dummy variable which takes a value of 1 if the initial day return is zero and 0 otherwise. Independent variable is the pre-market pricing inefficiency (deliberate underpricing). Inefficiency scores are based on estimates that assumes all IPOs (VC and non-VC backed samples) have a similar price technology. Chi-square statistics are reported in the parentheses.

$^{**}, ^{*}, ^{\#}$ indicates significant at 1, 5, and 10% significance level.

6. CONCLUSION

In this chapter we reexamine the underpricing phenomenon of VC-backed and non VC-backed IPOs. Our finding suggests that the underpricing of IPOs is determined not only by factors such as third party certification and public information about the new offerings, but is also influenced by factors that lead to pre-market deliberate underpricing. The regression results indicate that our measure of pre-market underpricing is higher for VC-backed than for non VC-backed IPOs. In addition, initial day returns are higher for VC-backed IPOs than for non VC-backed IPOs. We attribute such phenomena to the greater pre-market deliberate underpricing that characterizes IPOs brought to the market by venture capitalists.

We also find that there is a positive relationship between deliberate underpricing and the probability that underwriters provide support for the issue. This evidence is consistent with the notion that underwriters deliberately under-price the offering to reduce the costs of price stabilization in the after-market. Although we have presented evidence that price support plays an important role in the deliberate underpricing of IPOs, it is clear that there are other factors. The existing literature suggests that factors such as investor sentiment and hot issue periods, among others, may be important in explaining the deliberate underpricing phenomenon. Nevertheless, we believe that in addition to the observed factors, pre-market inefficiency provides an additional explanation for the underpricing of IPOs.

ACKNOWLEDGMENTS

Some part of this paper has some common elements associated with the paper "Underpricing of Venture and Non Venture Capital IPOs: An Empirical Investigation" published in Journal of Financial Services and Research, 2001. We thank Yakov Amihud, Sris Chatterjee, Ari Ginsberg, Eli Ofek, Christos Pantzalis, N. Prabhala, Gabriel Ramirez, Ajay Singh, Yusif Simaan, and Kluwer Academic Publishers. Usual caveats apply.

REFERENCES

Agarwall, R. (2000). Stabilization activities by underwriters after IPOs, *Journal of Finance*, *55*, 1075–1103.

Aigner, D., Lovell, K., & Schmidt, P. (1977). Formulation and the estimation of stochastic frontier production models, *Journal of Econometrics*, *6*, 21–37.

Allen, F., & Faulhaber, G. (1989). Signaling by underpricing in the IPO market, *Journal of Financial Economics*, *23*, 303–323.

Baron, D. P. (1982). A model of the demand for investment bank advising and distribution services for new issues, *Journal of Finance*, *37*, 955–976.

Barry, C., Muscarella, C., Peavy, H., & Vetsuypens, M. (1990). The role of venture capital in the creation of public companies: Evidence from the going-public process, *Journal of Financial Economics*, *27*, 447–476.

Brav, A., & Gompers, P. A. (1997). Myth or reality? The long-run underperformance of initial public offerings: Evidence from venture and nonventure capital-backed companies, *The Journal of Finance*, *LII*(5), 1791–1821.

Cantale, S. (1999). Putable common stocks, Working Paper, University of Tulane.

Carleton, W. (1986). Issues and questions involving venture capital. *Advances in the Study of Entrepreneurship, Innovation and Economic Growth* (Vol. 1, pp. 59–70). Greenwich, CT.

Carter, R., Federick, D., & Singh, A. (1998). Underwriter reputation, initial returns, and the long-run performance of IPO stocks, *Journal of Finance*, *53*, 285–312.

Carter, R., & Manaster, S. (1990). Initial public offerings and underwriter reputation, *Journal of Finance*, *45*, 1045–1067.

Chemmanur, T. (1993). The pricing of initial public offerings: A dynamic model with information production, *Journal of Finance*, *48*, 678–690.

Dunbar, C. (1995). The use of warrants as underwriter compensation in initial public offerings, *Journal of Financial Economics*, *38*, 59–78.

Dunbar, C. (1997). Overallotment option restrictions and contract choice in initial public offerings, *Journal of Corporate Finance*, *3*, 251–275.

Grinblatt, M., & Hwang, C. (1989). Signaling and pricing of new issues, *Journal of Finance*, *44*, 393–420.

Hunt-McCool, J., Koh, S., & Francis, B. (1996). Testing for deliberate underpricing in the IPO premarket: A stochastic frontier approach, *The Review of Financial Studies*, *9*, 1251–1269.

Jondrow, J., Lovell, K., Materov, I., & Schmidt, P. (1982). On the estimation of technical inefficiency in the stochastic frontier production function model, *Journal of Econometrics*, *19*, 233–238.

Lerner, J. (1994). Venture capitalists and the decision to go public, *Journal of Financial Economics*, *35*, 293–316.

Megginson, W., & Weiss, K. (1991). Venture capitalist certification in initial public offerings, *Journal of Finance*, *6*, 879–903.

Prabhala, N., & Puri, M. (2000). What type of IPOs do underwriters support and why? The role of price support in the IPO process? Working Paper, Stanford University, CA.

Ritter, J. (1991). The long-run performance of initial public offerings, *Journal of Finance*, 3–27.

Rock, K. (1986). Why new issues are underpriced, *Journal of Financial Economics*, *15*, 187–212.

Rudd, J. (1983). Underwriter price support and the IPO underpricing puzzle, *Journal of Financial Economics*, *34*, 135–151.

Welch, I. (1989). Seasoned offerings, imitation costs, and the underpricing of initial public offerings, *Journal of Finance*, *44*, 421–449.